CORROSION ENGINEERING HANDBOOK
SECOND EDITION

CORROSION of LININGS and COATINGS
Cathodic and Inhibitor Protection and Corrosion Monitoring

CORROSION ENGINEERING HANDBOOK
SECOND EDITION

Fundamentals of Metallic Corrosion:
Atmospheric and Media Corrosion of Metals
–ISBN 978-0-8493-8243-7

Corrosion of Polymers and Elastomers
–ISBN 978-0-8493-8245-1

Corrosion of Linings and Coatings:
Cathodic and Inhibitor Protection and Corrosion Monitoring
–ISBN 978-0-8493-8247-5

CORROSION ENGINEERING HANDBOOK
SECOND EDITION

CORROSION of LININGS and COATINGS

Cathodic and Inhibitor Protection and Corrosion Monitoring

Philip A. Schweitzer

CRC Press
Taylor & Francis Group
Boca Raton London New York

CRC Press is an imprint of the
Taylor & Francis Group, an **informa** business

CRC Press
Taylor & Francis Group
6000 Broken Sound Parkway NW, Suite 300
Boca Raton, FL 33487-2742

First issued in paperback 2019

© 2007 by Taylor & Francis Group, LLC
CRC Press is an imprint of Taylor & Francis Group, an Informa business

No claim to original U.S. Government works

ISBN-13: 978-0-8493-8247-5 (hbk)
ISBN-13: 978-0-367-38962-8 (pbk)
Library of Congress Card Number 2006014923

Library of Congress Cataloging-in-Publication Data

Schweitzer, Philip A.
    Corrosion of linings and coatings : cathodic and inhibitor protection and corrosion monitoring / Philip A. Schweitzer.
        p. cm.
    Includes bibliographical references and index.
    ISBN-13: 978-0-8493-8247-5 (alk. paper)
    ISBN-10: 0-8493-8247-5 (alk. paper)
    1. Corrosion and anti-corrosives. 2. Cathodic protection. I. Title.

TA418.74.S3787 2006
620.1'1223--dc22
                                       2006014923

Visit the Taylor & Francis Web site at
http://www.taylorandfrancis.com

and the CRC Press Web site at
http://www.crcpress.com

Preface

Corrosion is both costly and dangerous. Billions of dollars are annually spent for the replacement of corroded structures, machinery, and components, including metal roofing, condenser tubes, pipelines, and many other items. In addition to replacement costs are those associated with maintenance to prevent corrosion, inspections, and the upkeep of cathodically protected structures and pipelines. Indirect costs of corrosion result from shutdown, loss of efficiency, and product contamination or loss.

Although the actual replacement cost of an item may not be high, the loss of production resulting from the need to shut down an operation to permit the replacement may amount to hundreds of dollars per hour. When a tank or pipeline develops a leak, product is lost. If the leak goes undetected for a period of time, the value of the lost product could be considerable. In addition, contamination can result from the leaking material, requiring cleanup, and this can be quite expensive. When corrosion takes place, corrosion products build up, resulting in reduced flow in pipelines and reduced efficiency of heat transfer in heat exchangers. Both conditions increase operating costs. Corrosion products may also be detrimental to the quality of the product being handled, making it necessary to discard valuable materials.

Premature failure of bridges or structures because of corrosion can also result in human injury or even loss of life. Failures of operating equipment resulting from corrosion can have the same disastrous results.

When all of these factors are considered, it becomes obvious why the potential problem of corrosion should be considered during the early design stages of any project and why it is necessary to constantly monitor the integrity of structures, bridges, machinery, and equipment to prevent premature failures.

To cope with the potential problems of corrosion, it is necessary to understand

1. Mechanisms of corrosion
2. Corrosion resistant properties of various materials
3. Proper fabrication and installation techniques
4. Methods to prevent or control corrosion
5. Corrosion testing techniques
6. Corrosion monitoring techniques

Corrosion is not only limited to metalic materials but also to all materials of construction. Consequently, this handbook covers not only metallic materials but also all materials of construction.

Chapter 1 provides information on mortars, grouts, and monolithic surfacings.

Chapter 2 and Chapter 3 explain means of controlling/preventing corrosion through the use of inhibitors or cathodic protection.

In many instances, it is more economical to construct a tank or processing vessel of a less expensive metal, such as carbon steel, rather than an expensive alloy and install a lining to provide protection from corrosion. Chapter 4 through Chapter 6 provides details of various lining materials of both sheet and liquid form, while Chapter 7 through Chapter 10 provides details of coating materials, both organic and metallic. Compatibility charts are provided in all cases.

Because processing conditions can change or upsets take place, it is necessary that any corrosion that may occur be monitored. Chapter 11 discusses various techniques that may be employed to keep check on any corrosion that may be taking place internally in a vessel or pipeline.

It is the intention of this book that regardless of what is being built, whether it is a bridge, tower, pipeline, storage tank, or processing vessel, information for the designer/engineer/maintenance personnel/or whoever is responsible for the selection of material of construction will be found in this book to enable them to avoid unnecessary loss of material through corrosion.

Philip A. Schweitzer

Author

Philip A. Schweitzer is a consultant in corrosion prevention, materials of construction, and chemical engineering based in York, Pennsylvania. A former contract manager and material specialist for Chem-Pro Corporation, Fairfield, New Jersey, he is the editor of the *Corrosion Engineering Handbook* and the *Corrosion and Corrosion Protection Handbook, Second Edition;* and the author of *Corrosion Resistance Tables, Fifth Edition; Encyclopedia of Corrosion Technology, Second Edition; Metallic Materials; Corrosion Resistant Linings and Coatings; Atmospheric Degradation and Corrosion Control; What Every Engineer Should Know About Corrosion; Corrosion Resistance of Elastomers; Corrosion Resistant Piping Systems; Mechanical and Corrosion Resistant Properties of Plastics and Elastomers* (all titles Marcel Dekker, Inc.), and *Paint and Coatings, Applications and Corrosion Resistance* (Taylor & Francis). Schweitzer received the BChE degree (1950) from Polytechnic University (formerly Polytechnic Institute of Brooklyn), Brooklyn, New York.

Contents

1

Chemical Resistant Mortars, Grouts, and Monolithic Surfacings

1.1 Introduction

The industrialization of America following the turn of the century and the subsequent expansion of our agricultural industry created the need for chemical-resistant construction materials. The steel and metalworking, chemical (including explosives), dyestuffs, and fertilizer industries were the initial industries with severe corrosion problems. The pulp and paper, petroleum, petrochemical, and automotive industries followed with similar corrosion problems.

Specifically, the pickling, plating, and galvanizing of metals requires sulfuric, hydrochloric, hydrofluoric, chromic, nitric, and phosphoric acids. The manufacture of explosives, dyestuffs, fertilizer, and other agricultural products requires the same acids as well as other corrosive chemicals.

The pulp and paper industry, from inception of the wood chip into the digester and the subsequent bleaching of pulp, required similar acids. Sodium hydroxide, sodium hypochlorite, and chlorine were additional mandated chemicals for this and other industries.

Rail, automotive, petroleum, and petrochemical operations, and food and beverage sanitation mandates ultimately contributed to additional industrial corrosion problems.

Early reaction vessels in the process industries utilized lead linings with or without further protection from various types of brick, tile, porcelain, and ceramic sheathings. The early jointing materials for installing these ceramic-type linings utilized siliceous fillers mixed with inorganic binders based on various silicates, as well as mortar based on litharge and glycerine. The limitations of these mortars and grouts stimulated research for better setting and jointing materials for installing chemical-resistant brick, tile, and ceramics. The best brick or tile in the world for installing floors or tank linings is only as good as the mortars and grouts used to install it.

In the early 1930s, the first plasticized hot-pour, acid-resistant sulfur mortar was introduced in the United States. Sulfur mortars were immediately accepted by industry; however, they have thermal limitations and lack resistance to alkalies and solvents.

In the late 1930s, the first American acid-, salt-, and solvent-resistant phenolic mortar was introduced. In the early 1940s, the first furan mortar was developed and introduced in the United States and soon became the standard of the industry. It provided outstanding resistance to acids, alkalies, salts, and solvents.

Additional resin-binder systems have been developed for use as chemical-resistant mortars, grouts, and monolithic surfacings.

It is important to understand the vernacular of an industry. "Mortar" and "grout" are terms generally associated with the brick, tile, and masonry trades. Mortars and grouts are used for "setting" and joining various types and sizes of brick and tile.

A mortar can be described as a material of heavy consistency that can support the weight of the brick or tile without being squeezed from the joints while the joint is curing. Chemical-resistant mortar joints are customarily approximately 1/8 in. (3 mm) wide. A mortar is applied by buttering each unit and is generally associated with bricklayer trade.

A grout can be described as a thin or "soupy" mortar used for filling joints between previously laid brick or tile. Grout joints are customarily approximately 1/4 in. (6 mm) wide. A grout is applied by "squeegeeing" it into the open joints with a flat, rectangular, rubber-faced trowel and is generally associated with tilesetting.

Chemical-resistant machinery grouts are also available whose formulations are similar to those of the tile grouts. Machinery grouts generally utilize larger aggregates than tile grouts. Resin viscosities can also vary from those of the tile grouts.

Chemical-resistant monolithic surfacings or toppings are a mixture of a liquid synthetic resin binder, selected fillers, and a setting agent for application to concrete in thicknesses ranging from approximately 1/16 in. (1.5 mm) to 1/2 in. (13 mm). Materials applied in thicknesses greater than 1/2 in. (13 mm) are usually described as polymer concretes. Polymer concretes are defined as a composition of low-viscosity binders and properly graded inert aggregates that when combined and thoroughly mixed, yield a chemical-resistant synthetic concrete that can be precast or poured in place. Polymer concretes can also be used as a concrete surfacing, with the exception of sulfur cement polymer concrete. All polymer concretes can be used for total poured-in-place reinforced or unreinforced slabs. They can also be used for precasting of slabs, pump pads, column bases, trenches, tanks, and sumps, to mention a few. By definition, monolithic surfacings are also polymer concretes.

Chemical-resistant mortars, grouts, and monolithic surfacings are based on organic and inorganic chemistry. The more popular materials of each group will be discussed.

1.2 Materials Selection

The success or failure of a chemical-resistant mortar, grout, or monolithic surfacing is based on proper selection of materials and their application.

To select the proper material, the problem must first be defined:

1. Identify all chemicals that will be present and their concentration. It is not enough to say that pH will be 4, 7, or 11. This determines if it is acid, neutral, or alkaline; it does not identify whether the environment is oxidizing or nonoxidizing, organic or inorganic, alternately acid or alkaline, etc.

2. Is the application fumes and splash or total immersion? Floors can have integral trenches and sumps, curbs, pump pads.

3. What are the minimum or maximum temperatures to which the installation will be subjected?

4. Is the installation indoors or outdoors? Thermal shock and ultraviolet exposure can be deleterious to many resin systems.

5. What are the physical impositions? Foot traffic vs. vehicular traffic, impact from dropping steel plates vs. paper boxes, etc., must be defined.

6. Longevity—how long must it last? Is process obsolescence imminent? This could have a profound effect on cost.

7. Must it satisfy standards organizations such as USDA or FDA? Some systems do not comply.

8. Are the resin systems odoriferous? This could preclude their use in many processing plants such as food, beverage, and pharmaceutical. Many systems are odoriferous.

Answers to these questions will provide the necessary information to make a proper selection from the available resin system.

1.3 Chemical-Resistant Mortars and Grouts

Chemical-resistant mortars and grouts are composed of a liquid resin or an inorganic binder, fillers such as carbon, silica, and combinations thereof, and a hardener or catalyst system that can be incorporated in the filler or added as a separate component.

The workability of a mortar or grout is predicated on properly selected fillers or combinations of fillers, particle size and gradation of these fillers, resin viscosity, and reactivity of catalysts and hardeners. Improper filler gradation and high-viscosity resins produce mortars with poor working

TABLE 1.1

Guide to Chemical Resistance of Fillers

Medium, 20%	Filler		
	Carbon	Silica	Combination Carbon–Silica
Hydrochloric acid	R	R	R
Hydrofluoric acid	R	N	N
Sulfuric acid	R	R	R
Potassium hydroxide	R	N	N
Sodium hydroxide	R	N	N
Neutral salts	R	R	R
Solvents, conc.	R	R	R

R, recommended; N, not recommended.

properties. Hardener systems must be properly balanced for application at thermal ranges of approximately 60–90°F (15–32°C). The higher the temperature, the faster the set; the lower the temperature, the slower the set. Improper ambient and material temperature can also have a deleterious effect on the quality of the final installation, i.e., adhesion to brick, tile or substrate, high or low, rough or porous, or improperly cured joints.

The most popular fillers used are carbon, silica, and combinations thereof. Baraytes has been used in conjunction with carbon and silica for specific end-use application. Irrefutably, carbon is the most inert of the fillers; consequently, in many of the resin and sulfur systems, it is the filler of choice. It provides resistance to most chemicals, including strong alkalies, hydrofluoric acid, and other fluorine chemicals.

The general chemical resistance of various fillers to acids, alkalies, salts, fluorine chemicals, and solvents is enumerated in Table 1.1. The most popular resin systems from which chemical-resistant mortars and grouts are formulated follow.

1.3.1 Organic

1.3.1.1 Epoxy

The most popular epoxy resins used in the formulation of corrosion-resistant mortars, grout, and monolithic surfacings are low-viscosity liquid resins based on

1. Bisphenol A
2. Bisphenol F (epoxy Novolac)
3. Epoxy phenol Novolac

These base components are reacted with epichlorohydrin to form resins of varying viscosity and molecular weight. The subsequent molecular orientation is predicated on the hardener systems employed to effect the cure or solidification of the resin. The hardening systems selected will dictate the following properties of the cured system:

1. Chemical and thermal resistance
2. Physical properties
3. Moisture tolerance
4. Workability
5. Safety during use

Of the three systems enumerated, the bisphenol A epoxy has been the most popular followed by the bisphenol F, sometimes referred to as an epoxy Novolac resin. The epoxy phenol Novolac is a higher viscosity resin that requires various types of diluents or resin blends for formulating mortars, grouts, and some monolithic surfacings. The bisphenol A resin uses the following types of hardeners:

1. Aliphatic amines
2. Modified aliphatic amines
3. Aromatic amines
4. Others

Table 1.2 shows effects of the hardener on the chemical resistance of the finished mortar or grout of bisphenol A systems when exposed to organic, inorganic, and oxidizing acids as well as aromatic solvents.

Table 1.3 provides, summary chemical resistance of optimum chemical-resistant bisphenol A, and aromatic amine cured with bisphenol F resin systems.

TABLE 1.2

Types of Epoxy Hardeners and Their Effects on Chemical Resistance

	Hardeners		
Medium	Aliphatic Amines	Modified Aliphatic Amines	Aromatic Amines
Acetic acid, 5–10%	C	N	R
Benzene	N	N	R
Chromic acid, <5%	C	N	R
Sulfuric acid, 25%	R	C	R
Sulfuric acid, 50%	C	N	R
Sulfuric acid, 75%	N	N	R

R, recommended; N, not recommended; C, conditional.

TABLE 1.3

General Corrosion Resistance of Epoxy Mortars

Corrodent at Room Temperature	Hardeners[a]			
	Aliphatic Amines	Modified Aliphatic Amines	Aromatic Amines Bisphenol	
			A	F
Acetic acid 5–10%	C	U	R	
Acetone	U	U	U	U
Benzene	U	U	R	R
Butyl acetate	U	U	U	R
Butyl alcohol	R	R	R	R
Chromic acid 5%	U	U	R	R
Chromic acid 10%	U	U	U	R
Formaldehyde 35%	R	R	R	R
Gasoline	R	R	R	R
Hydrochloric acid to 36%	U	U	R	R
Nitric acid 30%	U	U	U	U
Phosphoric acid 50%	U	U	R	R
Sulfuric acid 25%	R	U	R	R
Sulfuric acid 50%	U	U	R	R
Sulfuric acid 75%	U	U	U	U
Trichloroethylene	U	U	U	R

[a] R, recommended; U, unsatisfactory.

Amine hardening systems, being alkaline, provide a high degree of compatibility of these systems for application to a multitude of substrates such as poured-in-place and precast concrete, steel, wood, fiberglass-reinforced plastics (FRP), brick, tile, ceramics, etc.

The most popular filler used for epoxy mortars and grouts is silica. Unfortunately, this precludes their use in hydrofluoric acid, other fluorine chemicals, and hot, strong alkalies. Carbon-filled mortars and grouts are available, however, with some sacrifice to working properties. Fortunately, their most popular applications have been industrial and institutional whereby optimum physical properties are required and exposure to elevated temperatures and corrosives are moderate.

Epoxy systems have outstanding physical properties. They are the premier products where optimum adhesion is a service requirement.

Amine hardening systems are the most popular for ambient-temperature-curing epoxy mortars, grouts, and monolithic surfacings. These systems are hygroscopic, and they can present allergenic responses to sensitive skin. These responses can be minimized, or virtually eliminated, by attention to personal hygiene and the use of protective creams on exposed areas of skin, i.e., face, neck, arms, and hands. Protective garments, including gloves, are recommended when using epoxy materials.

The bisphenol F or epoxy Novolac are similar systems to the bisphenol A epoxy systems in that they use alkaline hardeners and the same fillers. The major advantage for the use of the bisphenol F is improved resistance to

1. Aliphatic and aromatic solvents
2. Higher concentrations of oxidizing and nonoxidizing acids

Disadvantages of these systems are that they involve

1. Less plastic with slightly more shrinkage
2. Slightly less resistance to alkaline mediums

The thermal resistance and physical properties are otherwise very similar to the bisphenol A systems.

1.3.1.2 Furans

The polyfurfuryl alcohol or furan resins are the most versatile of all the resins used to formulate corrosion-resistant mortars and grouts. They are used for monolithic surfacings; however, they are not a popular choice because of their brittleness and their propensity to shrink. They provide a broad range of chemical resistance to most nonoxidizing organic and inorganic acids, alkalies, salts, oils, greases, and solvents to temperatures of 360°F (182°C). Table 1.4 provides comparative chemical resistance for furan resin mortars and grouts with 100% carbon and part carbon/ silica fillers.

Of all the room temperature curing resins, furans are one of the highest in thermal resistance with excellent physical properties. Furan resins are unique in that they are agriculturally, not petrochemically, based as are most synthetic resins. Furfuryl alcohol is produced from such agricultural by-products as corn cob, bagasse, rice, and oat hull.

The furan resin mortars and grouts are convenient-to-use, two-component systems consisting of the resin and a filler. The catalyst or hardener system is an acid that is contained in the filler. The most popular fillers are carbon, silica, and a combination of carbon and silica. The 100% carbon-filled furan resin mortars and grouts provide the broadest range of chemical resistance because of the inherent chemical resistance of the resin and the carbon filler to all concentrations of all alkalies, as well as to hydrofluoric acid and other fluorine chemicals. The advantages of mortars with part carbon and part silica fillers are slightly improved workability, physical properties, and cost. Grouts generally utilize 100% carbon filler because of the superior chemical resistance and flow properties.

The acidic catalysts employed in furan systems preclude their use directly on concrete, steel, and other substrates that could react with the acid.

TABLE 1.4

Chemical Resistance Furan Resins Mortars and Grouts: 100% Carbon vs. Part Carbon/Silica Fillers

Medium, RT	100% Carbon	Part Carbon/Silica
Acetic acid, glacial	R	R
Benzene	R	R
Cadmium salts	R	R
Chlorine dioxide	N	N
Chromic acid	N	N
Copper salts	R	R
Ethyl acetate	R	R
Ethyl alcohol	R	R
Formaldehyde	R	R
Fatty acids	R	R
Gasoline	R	R
Hydrochloric acid	R	R
Hydrofluoric acid	R	N
Iron salts	R	R
Lactic acid	R	R
Methyl ethyl ketone	R	R
Nitric acid	N	N
Phosphoric acid	R	R
Sodium chloride	R	R
Sodium hydroxide, to 20%	R	C
Sodium hydroxide, 40%	R	N
Sulfuric acid, 50%	R	R
Sulfuric acid, 80%	C	C
Trichloroethylene	R	R
Trisodium phosphate	R	C
Xylene	R	R

RT, room temperature; R, recommended; N, not recommended; C, conditional.

This limitation is easily circumvented by using various membranes, primers, or mortar bedding systems that are compatible with the substrate.

The process industries use lining systems (membranes) on most substrates onto which brick and tile are installed to ensure total resistance from aggressive environments encountered in such applications as

1. Pickling, plating, and galvanizing tanks in the steel and metal-working industries

2. Absorber towers in sulfuric acid plants

3. Scrubber in flu gas desulfurization applications

4. Floors in wet acid battery and chemical plants

5. Above-grade applications in dairies, food and beverage, and other processing plants

The versatility of furans is further exemplified by these available variations:

1. High-bond-strength materials for optimum physical mandates
2. Normal-bond-strength materials for economy and less demanding physical impositions
3. Hundred percent carbon filled for resistance to all concentrations of alkalies and most fluorine chemicals
4. Different ratios of carbon and silica for applications requiring varying degrees of electrical resistance or conductivity

1.3.1.3 *Phenolics*

The origin of phenolic resins was in Europe dating back to the late 1800s. At the turn of the century, the only chemical-resistant mortar available was based on the inorganic silicates. These materials possess outstanding acid resistance but little or no resistance to many other chemicals. The silicates also exhibited significant physical limitations.

After World War I, the limitation of the silicates prompted further investigation of the phenolics. These resins ceased to be laboratory curiosities and ultimately made their way into a multitude of applications because of their excellent physical properties. Early application for phenolic resins was for molding of telephones and associated electrical applications.

By the 1930s, the chemical process and the steel and metalworking industries mandated more functional chemical-resistant mortars for installing chemical-resistant brick. Besides chemical resistance, they had to have excellent physical properties.

By the mid-1930s, the first chemical-resistant phenolic resin mortar was introduced in the United States. It met the two most important mandates of the chemical, steel, and metalworking industries, i.e.,

1. Provide resistance to high concentrations of acids and in particular to sulfuric acid at elevated temperatures
2. Provide low absorption with good bond strength to various types of brick, tile, and ceramics while possessing excellent tensile, flexural, and compressive properties

To this day, phenolic resin mortars fulfill many of the requirements in the manufacture and use of the many grades and concentrations of sulfuric acid.

The steel and metalworking industries continue to use phenolic resin-based, chemical-resistant mortars for brick-lined pickling, plating, and galvanizing applications.

Phenolic resins are sufficiently functional to permit use of 100% carbon, 100% silica, or part silica and part carbon as fillers in phenolic mortars. Silica fillers are the most dominant for use in high concentrations of sulfuric acid

TABLE 1.5

Comparative Chemical Resistance: Phenolic Mortars vs. Furan Mortars

Medium, RT	Furan		Phenolic	
	Carbon	Silica	Carbon	Silica
Amyl alcohol	R	R	R	R
Chromic acid, 10%	N	N	N	N
Gasoline	R	R	R	R
Hydrofluoric acid, to 50%	R	N	R	N
Hydrofluoric acid, 93%	N	N	R	N
Methyl ethyl ketone	R	R	R	R
Nitric acid, 10%	N	N	N	N
Sodium hydroxide, to 5%	R	R	N	N
Sodium hydroxide, 30%	R	N	N	N
Sodium hypochlorite, 5%	N	N	N	N
Sulfuric acid, to 50%	R	R	R	R
Sulfuric acid, 93%	N	N	R	R
Xylene	R	R	R	R

RT, room temperature; R, recommended; N, not recommended.

and where electrical resistance is required. Carbon fillers are used where resistance to high concentrations of hydrofluoric acid are required. They are also used as adhesive and potting compounds for corrosive electrical conductance applications. Phenolic mortars are similar to the furans in that they are two-component, easy-to-use mortars, with the acid catalyst or curing agent incorporated in the powder. Phenolic resins are seldom used to formulate grouts or monolithic surfacings.

Phenolic resins have a limited shelf life and must be stored at 45°F (7°C).

Phenolic resin mortars, like epoxies, can be allergenic to sensitive skin. This can be minimized or prevented by exercising good personal hygiene and using protective creams. Table 1.5 provides comparative chemical resistance for phenolic mortars compared to furan mortars, carbon vs. silica-filled.

1.3.1.4 Polyesters

Chemical-resistant polyester mortar was developed and introduced in the early 1950s at the specific request of the pulp and paper industry. The request was for a mortar with resistance to a new bleach process utilizing chlorine dioxide. Polyester mortars ultimately became the premier mortar for use where resistance to oxidizing mediums is required.

Unsaturated polyester resins are also used for formulating tile and machinery grouts as well as monolithic surfacing. These applications are formidable challenges for the formulator because of their propensity to cause shrinkage.

Polyester mortars can be formulated to incorporate carbon and silica fillers depending on the end use intended. For applications requiring resistance to hydrofluoric acid, fluorine chemicals, and strong alkalies, such as sodium and potassium hydroxide, 100% carbon fillers are required.

Polyester resins are available in a number of types, the most popular of which are the following:

1. Isophthalic

2. Chlorendic acid

3. Bisphenol A fumarate

The earliest mortars and grouts were based on the isophthalic polyester resin. This resin performed well in many oxidizing mediums. It did, however, present certain physical, thermal, and chemical resistance limitations.

Formulations utilizing the chlorendic and bisphenol A fumarate resins offered improved chemical resistance, higher thermal capabilities, and improved ductility with less shrinkage. The bisphenol A fumarate resins offered significantly improved resistance to alkalies. They provided essentially equivalent resistance to oxidizing mediums.

All polyester resin systems have provided outstanding chemical resistance to a multitude of applications in the pulp and paper, textile, steel and metalworking, pharmaceutical, and chemical process industries. Typical applications have been brick and tile floors, brick and tile lining in bleach towers, scrubbers, pickling and plating, and waste-holding and treating tanks. All the resins provide formulation flexibility to accommodate carbon and silica as fillers. Carbon- and silica-filled mortars and grouts are easily mixed and handled for the various types of installations. They are easily pigmented for aesthetic considerations. The essentially neutral curing systems provide compatibility for application to a multitude of substrates, i.e., concrete, steel, FRPs, etc. Properly formulated polyester resin systems provide installation flexibility to a wide range of temperatures, humidities, and contaminants encountered on most construction sites. They are one of the most forgiving of all of the resin systems.

Polyester mortars and grouts have certain limitations that are inherent in all polyester formulations. They are as follows:

1. Strong aromatic odor that can be offensive for certain indoor and confined space applications

2. Shelf life limitations that can be controlled by low-temperature storage (below 60°F [15°C]) of the resin component

Table 1.6 provides comparative chemical resistance for the previously enumerated polyester resins. The physical properties of the respective systems are of a magnitude that for most mortar and grout, tile, and masonry applications, they can be considered essentially equal.

TABLE 1.6

Comparison of Corrosion Resistance of Polyester Mortars

Corrodent at Room Temperature	Polyester[a]	
	Chlorendic	Bisphenol A Fumarate
Acetic acid, glacial	U	U
Benzene	U	U
Chlorine dioxide	R	R
Ethyl alcohol	R	R
Hydrochloric acid, 36%	R	R
Hydrogen peroxide	R	U
Methanol	R	R
Methyl ethyl ketone	U	U
Motor oil and gasoline	R	R
Nitric acid, 40%	R	U
Phenol, 5%	R	R
Sodium hydroxide, 50%	U	R
Sulfuric acid, 75%	R	U
Toluene	U	U
Triethanolamine	U	R
Vinyl toluene	U	U

[a] R, recommended; U, unsatisfactory.

1.3.1.5 Vinyl Ester and Vinyl Ester Novolac

Chemically, these resins are addition reactions of methacrylic acid and epoxy resin. The chemistry of these resins has prompted their being referred to as acrylated epoxies. They possess many of the properties of epoxy, acrylic, and bisphenol A fumarate polyester resins. Their similarity to these resins means that they exhibit the same outstanding chemical resistance and physical properties of mortars and grouts formulated from these resins.

The vinyl esters are generally less rigid with lower shrinkage than many polyester systems. They favorably compare with the optimum chemical-resistant bisphenol A fumurate polyester mortars and grouts. The major advantage of the various vinyl ester systems are

1. Resistance to most oxidizing mediums
2. Resistance to high concentrations of sulfuric acid, sodium hydroxide, and many solvents

The vinyl ester resin mortars have supplanted the polyester and have become the mortar of choice for brick-lined bleach towers in the pulp and paper industry. Like polyesters, vinyl ester formulations have similar inherent disadvantages:

1. Strong aromatic odor for indoor or confined space applications. Isolation of area where installations are being made may be

necessary to prevent in-plant ventilating systems from carrying the aromatic odor through the facility.

2. Shelf life limitations of the resin require refrigerated storage below 60°F (15°C) to extend useful life.

Table 1.7 provides comparative chemical resistance for two types of polyester and vinyl ester resin mortar and grouts.

1.3.2 Inorganic

Inorganic materials predate their organic counterparts, offer fewer choices, and consequently are somewhat easier to understand. The most popular materials are

1. Hot-pour sulfur mortars
2. Ambiently mixed and applied silicate mortars

For years, all categories were referred to as "acid-proof mortars" because their capabilities are limited to a maximum pH of 7. They are not intended for alkaline or alternately acid and alkaline service.

TABLE 1.7

Comparative Chemical and Thermal Resistance of Polyester vs. Vinyl Ester Mortars and Grouts

Medium, RT	Polyester		Vinyl Ester	
	Chlorendic	Bisphenol A Fumarate	Vinyl Ester	Novolac
Acetic acid, glacial	C	N	N	R
Benzene	C	N	R	R
Chlorine dioxide	R	R	R	R
Ethyl alcohol	R	R	R	R
Hydrochloric acid, 36%	R	R	R	R
Hydrogen peroxide	R	N	R	R
Methanol	R	R	N	R
Methyl ethyl ketone	N	N	N	N
Motor oil and Gasoline	R	R	R	R
Nitric acid, 40%	R	N	N	R
Phenol, 5%	R	R	R	R
Sodium hydroxide, 50%	N	R	R	R
Sulfuric acid, 75%	R	C	R	R
Toluene	C	N	N	R
Triethanolamine	N	R	R	R
Vinyl toluene	C	N	C	R
Maximum temperature, °F (°C)	260 (127)	250 (121)	220 (104)	230 (110)

RT, room temperature; R, recommended; N, not recommended; C, conditional.

1.3.2.1 *Sulfur*

The initial application for sulfur mortars was to replace lead for joining bell and spigot cast iron water lines. Since then, sulfur mortars have been successfully used for installing brick floors, brick-lined tanks, and joining bell and spigot vitrified clay pipe for corrosive waste sewer lines in the chemical, steel, and metalworking industries. Sulfur mortars are plasticized to impart thermal shock resistance. They utilize 100% carbon or 100% silica fillers.

Because of their outstanding resistance to oxidizing acids, the 100% carbon-filled sulfur mortar is the mortar of choice for installing carbon brick in the nitric/hydrofluoric acid picklers in the specialty steel industry.

Both the 100% carbon and the 100% silica mortars have found widespread use in the plating, galvanizing specialty, and carbon steel industries. Advantages to using sulfur mortars when compared to some resin mortars are as follows:

1. Resistance to oxidizing, nonoxidizing, and mixed acids

2. Ease of use

3. High early strength; "cool it–use it"

4. Resistance to thermal shock

5. Unlimited shelf life

6. Prefabrication and in-place construction

7. Economy

The many advantages of sulfur mortars make them ideal for applications such as setting of anchor bolts and posts, capping of concrete test cylinders, "proving" of molds for castings and hubs for grinding wheels.

A disadvantage of sulfur mortars is flammability. When sulfur mortar masonry sheathings are being installed, melting equipment is placed outdoors and molten materials moved to the point of use, therefore minimizing potential for flammability problems.

Installation of brick linings on vertical surfaces with sulfur mortars utilizes the concept of dry-laying brick by placing the brick on sulfur mortar spacer chips, papering the face of the brick to dam the joint, and pouring behind the brick to fill the joints. Horizontal surfaces are poured until the mortar comes up into the joints approximately 1/4 in. from the top of the brick. A final "flood pour" is made over the entire surface, therefore ensuring full flush joints. As soon as the joints cool, the installation is ready for service. Approximately 95% of the compressive value of sulfur mortars is attained 5 min after the mortar has solidified. Table 1.8 provides comparative chemical resistance to environments most commonly experienced in industries where sulfur mortars are used.

TABLE 1.8

Comparative Chemical Resistance: 100% Carbon vs. 100% Silica-
Filled Sulfur Mortars

Medium, RT	100% Carbon	100% Silica
Acetic acid, to 10%	R	R
Aqua regia	N	N
Cadmium salts	R	R
Chromic acid, to 20%	R	R
Gold cyanide	R	R
Hydrochloric acid	R	R
Hydrofluoric acid	R	N
Iron salts	R	R
Magnesium salts	R	R
Methyl ethyl ketone	N	N
Mineral spirits	N	N
Nickel salts	R	R
Nitric acid, to 40%	R	R
Nitric/hydrofluoric acid	R	R
Phosphoric acid	R	R
Silver nitrate	R	R
Sodium hydroxide	N	N
Sodium salts	R	R
Sulfuric acid, 80%	R	R
Toluene	N	N
Trichloroethylene	N	N
Zinc, salts	R	R

RT, room temperature; R, recommended; N, not recommended.

1.3.2.2 Silicates

These mortars are most notable for their resistance to concentrated acids,
except hydrofluoric acid and similar fluorinated chemicals, at elevated
temperatures. They are also resistant to many aliphatic and aromatic
solvents. They are not intended for use in alkaline or alternately acid and
alkaline environments. This category of mortars includes

1. Sodium silicate
2. Potassium silicate
3. Silica (silica sol)

The major applications for these mortars have been in the construction of
sulfuric acid plants and the brick lining of stacks subjected to varying
concentrations of sulfur and nitrogen oxides at elevated temperatures. Their
upper thermal capabilities approach those of refractory mortars.

The sodium and potassium silicate mortars are available as two-
component systems, filler, and binder, with the setting agent in the filler.

Sodium and potassium silicates are referred to as soluble silicates because of their solubility in water. They are not affected by strong acids; however, this phenomenon precludes the use of many formulations in dilute acid service. However, this disadvantage becomes an advantage for formulating single-component powder systems. All that is required is the addition of water at the time of use. Obviously, as the name of these materials implies, the fillers are pure silica.

The sodium silicates can be produced over a broad range of compositions of the liquid binder. These properties and new hardening systems have significantly improved the water resistance of some sodium silicate mortars. These formulations are capable of resisting dilute as well as concentrated acids without compromising physical properties.

The potassium silicate mortars are less versatile in terms of formulation flexibility. They are, however, less susceptible to crystallization in high concentrations of sulfuric acid so long as metal ion contamination is minimal.

Potassium silicate mortars are available with halogen-free hardening systems, thereby eliminating the remote potential for catalyst poisoning in certain chemical processes.

The silica or silica sol type of mortars are the newest of this class of mortars. They consist of a colloidal silica binder with quartz fillers. The principal difference compared to the other mortars is total freedom from metal ion that could contribute to sulfation hydration within the mortar joints in high-concentration sulfuric acid service.

The workability and storage stability is comparable in the sodium and potassium silicates. The silica materials are harder to use, less forgiving as to mix ratio, and highly susceptible to irreversible damage because of freezing in storage.

The chemical resistances of the various silicate mortars are very similar. Silicate mortars will fail when exposed to mild alkaline mediums such as bicarbonate of soda. Dilute acid solution, such as nitric acid, will have a deleterious effect on sodium silicates unless the water-resistant type is used.

Table 1.9 shows the compatibility of various mortars with selected corrodents.

1.4 Chemical-Resistant Monolithic Surfacings and Polymer Concretes

The chemistry of monolithic surfacings is an exploitation of the resin systems used for mortars and grouts. Additional systems will be included.

To reiterate, monolithic surfacings are installed at thicknesses of 1/16 in. (1.5 mm) to 1/2 in. (13 mm). Polymer concretes are installed at thicknesses greater than 1/2 in. (13 mm). These materials are formidable corrosion

TABLE 1.9

Compatibility of Various Mortars with Selected Corrodents

Mortars	Acetic acid, 10%
Silicate	U
Sodium silicate	R———————————————————————————
Potassium silicate	R———————————————————————————
Silica	R———————————————————————————
Sulfur	R————————————
Furan resin	R—————————————————————
Polyester	R————————————
Epoxy	R————————

	°F	60	80	100	120	140	160	180	200	220	240	260	280	300	320	340	360	380	400	420	440	460
	°C	15	26	38	49	60	71	82	93	104	116	127	138	149	160	171	182	193	204	216	227	238

Mortars	Acetic acid, 50%
Silicate	U
Sodium silicate	R———————————————————————————
Potassium silicate	R———————————————————————————
Silica	R———————————————————————————
Sulfur	U
Furan resin	R—————————————————————
Polyester	U
Epoxy	U

	°F	60	80	100	120	140	160	180	200	220	240	260	280	300	320	340	360	380	400	420	440	460
	°C	15	26	38	49	60	71	82	93	104	116	127	138	149	160	171	182	193	204	216	227	238

Mortars	Acetic acid, 80%
Silicate	U
Sodium silicate	R———————————————————————————
Potassium silicate	R———————————————————————————
Silica	R———————————————————————————
Sulfur	U
Furan resin	R—————————————————————
Polyester	U
Epoxy	U

	°F	60	80	100	120	140	160	180	200	220	240	260	280	300	320	340	360	380	400	420	440	460
	°C	15	26	38	49	60	71	82	93	104	116	127	138	149	160	171	182	193	204	216	227	238

(continued)

TABLE 1.9 *Continued*

Mortars	Acetic acid, glacial
Silicate	U
Sodium silicate	R———————————————————
Potassium silicate	R———————————————————
Silica	R———————————————————
Sulfur	U
Furan resin	R—————————————————
Polyester	U
Epoxy	U

°F	60	80	100	120	140	160	180	200	220	240	260	280	300	320	340	360	380	400	420	440	460
°C	15	26	38	49	60	71	82	93	104	116	127	138	149	160	171	182	193	204	216	227	238

Mortars	Acetic anhydride
Silicate	
Sodium silicate	R———————————————————
Potassium silicate	R———————————————————
Silica	R———————————————————
Sulfur	U
Furan resin	U
Polyester	U
Epoxy	U

°F	60	80	100	120	140	160	180	200	220	240	260	280	300	320	340	360	380	400	420	440	460
°C	15	26	38	49	60	71	82	93	104	116	127	138	149	160	171	182	193	204	216	227	238

Mortars	Aluminum chloride, aqueous
Silicate	R———————————————————
Sodium silicate	R———————————————————
Potassium silicate	R———————————————————
Silica	R———————————————————
Sulfur	R———————
Furan resin	R———————————————
Polyester	U———————————
Epoxy	R———

°F	60	80	100	120	140	160	180	200	220	240	260	280	300	320	340	360	380	400	420	440	460
°C	15	26	38	49	60	71	82	93	104	116	127	138	149	160	171	182	193	204	216	227	238

(continued)

TABLE 1.9 *Continued*

Mortars	Aluminum fluoride
Silicate	U
Sodium silicate	U
Potassium silicate	U
Silica	U
Sulfur	R————————
Furan resin	R—————————————————————
Polyester	R———————————
Epoxy	R———————————

°F	60	80	100	120	140	160	180	200	220	240	260	280	300	320	340	360	380	400	420	440	460
°C	15	26	38	49	60	71	82	93	104	116	127	138	149	160	171	182	193	204	216	227	238

Mortars	Ammonium chloride, 10%
Silicate	R————————————————————————————
Sodium silicate	R————
Potassium silicate	R————————————————————————————
Silica	R———————————————
Sulfur	R——————————
Furan resin	R————————————————————————————
Polyester	R———————————
Epoxy	R———————————

°F	60	80	100	120	140	160	180	200	220	240	260	280	300	320	340	360	380	400	420	440	460
°C	15	26	38	49	60	71	82	93	104	116	127	138	149	160	171	182	193	204	216	227	238

Mortars	Ammonium chloride, 50%
Silicate	R————————————————————————————
Sodium silicate	R————
Potassium silicate	R————————————————————————————
Silica	R———————————————————————
Sulfur	R——————————
Furan resin	R————————————————————————————
Polyester	R———————————
Epoxy	R———————————

°F	60	80	100	120	140	160	180	200	220	240	260	280	300	320	340	360	380	400	420	440	460
°C	15	26	38	49	60	71	82	93	104	116	127	138	149	160	171	182	193	204	216	227	238

(continued)

TABLE 1.9 *Continued*

Mortars	Ammonium chloride, sat.
Silicate	R————————————————————————————
Sodium silicate	R——
Potassium silicate	R————————————————————————————
Silica	R————————————————————————————
Sulfur	R—————————————
Furan resin	R————————————————————————
Polyester	R—————————————————
Epoxy	R—————————————

°F	60	80	100	120	140	160	180	200	220	240	260	280	300	320	340	360	380	400	420	440	460
°C	15	26	38	49	60	71	82	93	104	116	127	138	149	160	171	182	193	204	216	227	238

Mortars	Aluminum fluoride, 10%
Silicate	U
Sodium silicate	U
Potassium silicate	U
Silica	U
Sulfur	U
Furan resin	R————————————————————————
Polyester	R—————————
Epoxy	R———————

°F	60	80	100	120	140	160	180	200	220	240	260	280	300	320	340	360	380	400	420	440	460
°C	15	26	38	49	60	71	82	93	104	116	127	138	149	160	171	182	193	204	216	227	238

Mortars	Aluminum fluoride, 25%
Silicate	U
Sodium silicate	U
Potassium silicate	U
Silica	U
Sulfur	U
Furan resin	R————————————————————————
Polyester	R—————————
Epoxy	R———————

°F	60	80	100	120	140	160	180	200	220	240	260	280	300	320	340	360	380	400	420	440	460
°C	15	26	38	49	60	71	82	93	104	116	127	138	149	160	171	182	193	204	216	227	238

(continued)

TABLE 1.9 *Continued*

Mortars	Ammonium hydroxide, 25%
Silicate	U
Sodium silicate	U
Potassium silicate	U
Silica	U
Sulfur	U
Furan resin	R————————————————————————————————
Polyester	R—————————————————
Epoxy	R—————————————

°F	60	80	100	120	140	160	180	200	220	240	260	280	300	320	340	360	380	400	420	440	460
°C	15	26	38	49	60	71	82	93	104	116	127	138	149	160	171	182	193	204	216	227	238

Mortars	Ammonium hydroxide, sat.
Silicate	U
Sodium silicate	
Potassium silicate	U
Silica	U
Sulfur	U
Furan resin	R————————————————————————————
Polyester	R———
Epoxy	R———

°F	60	80	100	120	140	160	180	200	220	240	260	280	300	320	340	360	380	400	420	440	460
°C	15	26	38	49	60	71	82	93	104	116	127	138	149	160	171	182	193	204	216	227	238

Mortars	Aqua regia 3:1
Silicate	R————————————————————————————————————
Sodium silicate	R————————————————————————————————————
Potassium silicate	R————————————————————————————————————
Silica	R————————————————————————————————————
Sulfur	U
Furan resin	U
Polyester	U
Epoxy	U

°F	60	80	100	120	140	160	180	200	220	240	260	280	300	320	340	360	380	400	420	440	460
°C	15	26	38	49	60	71	82	93	104	116	127	138	149	160	171	182	193	204	216	227	238

(continued)

TABLE 1.9 *Continued*

Mortars	Bromine gas, dry
Silicate	
Sodium silicate	R————————————————————————————————
Potassium silicate	
Silica	
Sulfur	U
Furan resin	U
Polyester	U
Epoxy	U

°F	60	80	100	120	140	160	180	200	220	240	260	280	300	320	340	360	380	400	420	440	460
°C	15	26	38	49	60	71	82	93	104	116	127	138	149	160	171	182	193	204	216	227	238

Mortars	Bromine gas, moist
Silicate	
Sodium silicate	R————————————————————————————————
Potassium silicate	
Silica	
Sulfur	U
Furan resin	U
Polyester	U
Epoxy	U

°F	60	80	100	120	140	160	180	200	220	240	260	280	300	320	340	360	380	400	420	440	460
°C	15	26	38	49	60	71	82	93	104	116	127	138	149	160	171	182	193	204	216	227	238

Mortars	Bromine liquid
Silicate	R————————————————————————————————
Sodium silicate	
Potassium silicate	R————————————————————————————————
Silica	R————————————————————————————————
Sulfur	R————————————
Furan resin	U
Polyester	U
Epoxy	U

°F	60	80	100	120	140	160	180	200	220	240	260	280	300	320	340	360	380	400	420	440	460
°C	15	26	38	49	60	71	82	93	104	116	127	138	149	160	171	182	193	204	216	227	238

(continued)

TABLE 1.9 *Continued*

Mortars	Calcium hypochlorite
Silicate	
Sodium silicate	U
Potassium silicate	R————————————————————————————
Silica	R————————————————————————————
Sulfur	U
Furan resin	U
Polyester	
Epoxy	U

°F	60	80	100	120	140	160	180	200	220	240	260	280	300	320	340	360	380	400	420	440	460
°C	15	26	38	49	60	71	82	93	104	116	127	138	149	160	171	182	193	204	216	227	238

Mortars	Carbon tetrachloride
Silicate	R————————————————————————————
Sodium silicate	R————————————————————————————
Potassium silicate	R————————————————————————————
Silica	R————————————————————————————
Sulfur	U
Furan resin	R—————————————————————————
Polyester	R————
Epoxy	R————

°F	60	80	100	120	140	160	180	200	220	240	260	280	300	320	340	360	380	400	420	440	460
°C	15	26	38	49	60	71	82	93	104	116	127	138	149	160	171	182	193	204	216	227	238

Mortars	Chlorine gas, dry
Silicate	R————————————————————————————
Sodium silicate	R————————————————————————————
Potassium silicate	R————————————————————————————
Silica	R————————————————————————————
Sulfur	U
Furan resin	U
Polyester	U
Epoxy	U

°F	60	80	100	120	140	160	180	200	220	240	260	280	300	320	340	360	380	400	420	440	460
°C	15	26	38	49	60	71	82	93	104	116	127	138	149	160	171	182	193	204	216	227	238

(continued)

TABLE 1.9 *Continued*

Mortars	Chlorine gas, wet
Silicate	R—————————————————————————
Sodium silicate	R—————————————————————————
Potassium silicate	R—————————————————————————
Silica	R—————————————————————————
Sulfur	U
Furan resin	U
Polyester	U
Epoxy	U

°F	60	80	100	120	140	160	180	200	220	240	260	280	300	320	340	360	380	400	420	440	460
°C	15	26	38	49	60	71	82	93	104	116	127	138	149	160	171	182	193	204	216	227	238

Mortars	Chlorine liquid
Silicate	R—————————————————————————
Sodium silicate	
Potassium silicate	R—————————————————————————
Silica	R—————————————————————————
Sulfur	U
Furan resin	U
Polyester	U
Epoxy	U

°F	60	80	100	120	140	160	180	200	220	240	260	280	300	320	340	360	380	400	420	440	460
°C	15	26	38	49	60	71	82	93	104	116	127	138	149	160	171	182	193	204	216	227	238

Mortars	Chromic acid, 10%
Silicate	U
Sodium silicate	R—————————————————————————
Potassium silicate	R—————————————————————————
Silica	R—————————————————————————
Sulfur	U
Furan resin	U
Polyester	R——
Epoxy	U

°F	60	80	100	120	140	160	180	200	220	240	260	280	300	320	340	360	380	400	420	440	460
°C	15	26	38	49	60	71	82	93	104	116	127	138	149	160	171	182	193	204	216	227	238

(continued)

TABLE 1.9 *Continued*

Mortars		Chromic acid, 50%
Silicate		R———————————————————————
Sodium silicate		
Potassium silicate		
Silica		R———————————————————————
Sulfur		U
Furan resin		U
Polyester	30%	R———
Epoxy		U

°F	60	80	100	120	140	160	180	200	220	240	260	280	300	320	340	360	380	400	420	440	460
°C	15	26	38	49	60	71	82	93	104	116	127	138	149	160	171	182	193	204	216	227	238

Mortars	Ferric chloride
Silicate	R———————————————————————
Sodium silicate	U
Potassium silicate	R———————————————————————
Silica	R———————————————————————
Sulfur	R———————————
Furan resin	R—————————————————
Polyester	R—————————————
Epoxy	R———————————————

°F	60	80	100	120	140	160	180	200	220	240	260	280	300	320	340	360	380	400	420	440	460
°C	15	26	38	49	60	71	82	93	104	116	127	138	149	160	171	182	193	204	216	227	238

Mortars	Ferric chloride, 50% in water
Silicate	R———————————————————————
Sodium silicate	U
Potassium silicate	R———————————————————————
Silica	R———————————————————————
Sulfur	R———————————
Furan resin	R—————————————————
Polyester	R—————————————
Epoxy	R———————————————

°F	60	80	100	120	140	160	180	200	220	240	260	280	300	320	340	360	380	400	420	440	460
°C	15	26	38	49	60	71	82	93	104	116	127	138	149	160	171	182	193	204	216	227	238

(continued)

TABLE 1.9 *Continued*

Mortars	Hydrobromic acid, 20%
Silicate	R———————————————————————
Sodium silicate	R———————————————————————
Potassium silicate	R———————————————————————
Silica	R———————————————————————
Sulfur	R————————————
Furan resin	R——
Polyester	R——
Epoxy	U

	°F	60	80	100	120	140	160	180	200	220	240	260	280	300	320	340	360	380	400	420	440	460
	°C	15	26	38	49	60	71	82	93	104	116	127	138	149	160	171	182	193	204	216	227	238

Mortars	Hydrobromic acid, 50%
Silicate	R———————————————————————
Sodium silicate	R———————————————————————
Potassium silicate	R———————————————————————
Silica	R———————————————————————
Sulfur	R————————————
Furan resin	R——
Polyester	R————
Epoxy	U

	°F	60	80	100	120	140	160	180	200	220	240	260	280	300	320	340	360	380	400	420	440	460
	°C	15	26	38	49	60	71	82	93	104	116	127	138	149	160	171	182	193	204	216	227	238

Mortars	Hydrochloric acid, 20%
Silicate	R———————————————————————
Sodium silicate	R———————————————————————
Potassium silicate	R———————————————————————
Silica	R———————————————————————
Sulfur	R————————————
Furan resin	R———————————————————
Polyester	R————————
Epoxy	U

	°F	60	80	100	120	140	160	180	200	220	240	260	280	300	320	340	360	380	400	420	440	460
	°C	15	26	38	49	60	71	82	93	104	116	127	138	149	160	171	182	193	204	216	227	238

(continued)

TABLE 1.9 *Continued*

Mortars	Hydrochloric acid, 38%
Silicate	R————————————————————————————
Sodium silicate	R————————————————————————————
Potassium silicate	R————————————————————————————
Silica	R————————————————————————————
Sulfur	R———————————
Furan resin	R————————————————————————
Polyester	R————
Epoxy	U

°F	60	80	100	120	140	160	180	200	220	240	260	280	300	320	340	360	380	400	420	440	460
°C	15	26	38	49	60	71	82	93	104	116	127	138	149	160	171	182	193	204	216	227	238

Mortars	Hydrofluoric acid, 30%
Silicate	U
Sodium silicate	U
Potassium silicate	U
Silica	U
Sulfur	R———————
Furan resin	R————————————————————————
Polyester	R————
Epoxy	U

°F	60	80	100	120	140	160	180	200	220	240	260	280	300	320	340	360	380	400	420	440	460
°C	15	26	38	49	60	71	82	93	104	116	127	138	149	160	171	182	193	204	216	227	238

Mortars	Hydrofluoric acid, 70%
Silicate	U
Sodium silicate	U
Potassium silicate	U
Silica	U
Sulfur	U
Furan resin	U
Polyester	
Epoxy	U

°F	60	80	100	120	140	160	180	200	220	240	260	280	300	320	340	360	380	400	420	440	460
°C	15	26	38	49	60	71	82	93	104	116	127	138	149	160	171	182	193	204	216	227	238

(continued)

TABLE 1.9 *Continued*

Mortars	Hydrofluoric acid, 100%
Silicate	U
Sodium silicate	U
Potassium silicate	U
Silica	U
Sulfur	U
Furan resin	U
Polyester	
Epoxy	U

°F	60	80	100	120	140	160	180	200	220	240	260	280	300	320	340	360	380	400	420	440	460
°C	15	26	38	49	60	71	82	93	104	116	127	138	149	160	171	182	193	204	216	227	238

Mortars	Magnesium chloride
Silicate	R——————————————————— (to 460)
Sodium silicate	
Potassium silicate	R——————————————————— (to 460)
Silica	R——————————————————— (to 460)
Sulfur	R————————— (to ~220)
Furan resin	R——————————————————— (to ~440)
Polyester	R————————— (to ~200)
Epoxy	R——————————————————— (to 460)

°F	60	80	100	120	140	160	180	200	220	240	260	280	300	320	340	360	380	400	420	440	460
°C	15	26	38	49	60	71	82	93	104	116	127	138	149	160	171	182	193	204	216	227	238

Mortars	Nitric acid, 5%
Silicate	R——————————————————— (to 460)
Sodium silicate	R——————————————————— (to 460)
Potassium silicate	R——————————————————— (to 460)
Silica	R——————————————————— (to 460)
Sulfur	R————————— (to ~140)
Furan resin	U
Polyester	R————————— (to ~140)
Epoxy	U

°F	60	80	100	120	140	160	180	200	220	240	260	280	300	320	340	360	380	400	420	440	460
°C	15	26	38	49	60	71	82	93	104	116	127	138	149	160	171	182	193	204	216	227	238

(continued)

TABLE 1.9 *Continued*

Mortars	Nitric acid, 20%
Silicate	R————————————————————
Sodium silicate	R————————————————————
Potassium silicate	
Silica	
Sulfur	R———
Furan resin	U
Polyester	R———
Epoxy	U

°F	60	80	100	120	140	160	180	200	220	240	260	280	300	320	340	360	380	400	420	440	460
°C	15	26	38	49	60	71	82	93	104	116	127	138	149	160	171	182	193	204	216	227	238

Mortars	Nitric acid, 70%
Silicate	R————————————————————
Sodium silicate	R————————————————————
Potassium silicate	
Silica	
Sulfur	U
Furan resin	U
Polyester	U
Epoxy	U

°F	60	80	100	120	140	160	180	200	220	240	260	280	300	320	340	360	380	400	420	440	460
°C	15	26	38	49	60	71	82	93	104	116	127	138	149	160	171	182	193	204	216	227	238

Mortars	Nitric acid, anhydrous
Silicate	R————————————————————
Sodium silicate	R————————————————————
Potassium silicate	
Silica	
Sulfur	U
Furan resin	U
Polyester	U
Epoxy	U

°F	60	80	100	120	140	160	180	200	220	240	260	280	300	320	340	360	380	400	420	440	460
°C	15	26	38	49	60	71	82	93	104	116	127	138	149	160	171	182	193	204	216	227	238

(continued)

TABLE 1.9 *Continued*

Mortars	Oleum
Silicate	
Sodium silicate	R———
Potassium silicate	
Silica	
Sulfur	U
Furan resin	U
Polyester	U
Epoxy	U

°F	60	80	100	120	140	160	180	200	220	240	260	280	300	320	340	360	380	400	420	440	460
°C	15	26	38	49	60	71	82	93	104	116	127	138	149	160	171	182	193	204	216	227	238

Mortars	Phosphoric acid, 50–80%
Silicate	R————————————————————
Sodium silicate	U
Potassium silicate	R————————————————————
Silica	R————————————————————
Sulfur	R———————
Furan resin	R—————————————
Polyester	R—————————
Epoxy	U

°F	60	80	100	120	140	160	180	200	220	240	260	280	300	320	340	360	380	400	420	440	460
°C	15	26	38	49	60	71	82	93	104	116	127	138	149	160	171	182	193	204	216	227	238

Mortars	Sodium chloride
Silicate	U
Sodium silicate	R————————————————————
Potassium silicate	R————————————————————
Silica	R————————————————————
Sulfur	R———————
Furan resin	R—————————————
Polyester	R—————————
Epoxy	R———————

°F	60	80	100	120	140	160	180	200	220	240	260	280	300	320	340	360	380	400	420	440	460
°C	15	26	38	49	60	71	82	93	104	116	127	138	149	160	171	182	193	204	216	227	238

(continued)

TABLE 1.9 *Continued*

Mortars	Sodium hydroxide, 10%
Silicate	U
Sodium silicate	U
Potassium silicate	U
Silica	U
Sulfur	U
Furan resin	R—————————————————————
Polyester	R———
Epoxy	R—————————————

°F	60	80	100	120	140	160	180	200	220	240	260	280	300	320	340	360	380	400	420	440	460
°C	15	26	38	49	60	71	82	93	104	116	127	138	149	160	171	182	193	204	216	227	238

Mortars	Sodium hydroxide, 50%
Silicate	U
Sodium silicate	U
Potassium silicate	U
Silica	U
Sulfur	U
Furan resin	R—————————————————————
Polyester	R———
Epoxy	R—————————————

°F	60	80	100	120	140	160	180	200	220	240	260	280	300	320	340	360	380	400	420	440	460
°C	15	26	38	49	60	71	82	93	104	116	127	138	149	160	171	182	193	204	216	227	238

Mortars	Sodium hydroxide, conc.
Silicate	U
Sodium silicate	U
Potassium silicate	U
Silica	U
Sulfur	U
Furan resin	
Polyester	
Epoxy	R—

°F	60	80	100	120	140	160	180	200	220	240	260	280	300	320	340	360	380	400	420	440	460
°C	15	26	38	49	60	71	82	93	104	116	127	138	149	160	171	182	193	204	216	227	238

(continued)

TABLE 1.9 *Continued*

Mortars	Sodium hypochlorite, 20%
Silicate	U
Sodium silicate	U
Potassium silicate	U
Silica	U
Sulfur	U
Furan resin	U
Polyester	U
Epoxy	U

°F	60	80	100	120	140	160	180	200	220	240	260	280	300	320	340	360	380	400	420	440	460
°C	15	26	38	49	60	71	82	93	104	116	127	138	149	160	171	182	193	204	216	227	238

Mortars	Sodium hypochlorite, conc.
Silicate	U
Sodium silicate	U
Potassium silicate	U
Silica	U
Sulfur	U
Furan resin	U
Polyester	U
Epoxy	U

°F	60	80	100	120	140	160	180	200	220	240	260	280	300	320	340	360	380	400	420	440	460
°C	15	26	38	49	60	71	82	93	104	116	127	138	149	160	171	182	193	204	216	227	238

Mortars	Sulfuric acid, 10%
Silicate	U
Sodium silicate	R———————————————————————————————
Potassium silicate	
Silica	
Sulfur	R————————————
Furan resin	R—————————————————
Polyester	R——————————
Epoxy	R—————

°F	60	80	100	120	140	160	180	200	220	240	260	280	300	320	340	360	380	400	420	440	460
°C	15	26	38	49	60	71	82	93	104	116	127	138	149	160	171	182	193	204	216	227	238

(continued)

TABLE 1.9 *Continued*

Mortars	Sulfuric acid, 50%
Silicate	U
Sodium silicate	R———————————————————————————
Potassium silicate	
Silica	
Sulfur	R————————
Furan resin	R———————————————————————————
Polyester	R—————————————————
Epoxy	U

°F	60	80	100	120	140	160	180	200	220	240	260	280	300	320	340	360	380	400	420	440	460
°C	15	26	38	49	60	71	82	93	104	116	127	138	149	160	171	182	193	204	216	227	238

Mortars	Sulfuric acid, 70%
Silicate	U
Sodium silicate	R—————————
Potassium silicate	R—————————————————
Silica	R—————————————————
Sulfur	R————————
Furan resin	R————————
Polyester	R————————
Epoxy	U

°F	60	80	100	120	140	160	180	200	220	240	260	280	300	320	340	360	380	400	420	440	460
°C	15	26	38	49	60	71	82	93	104	116	127	138	149	160	171	182	193	204	216	227	238

Mortars	Sulfuric acid, 90%
Silicate	U
Sodium silicate	R————
Potassium silicate	
Silica	
Sulfur	U
Furan resin	U
Polyester	U
Epoxy	U

°F	60	80	100	120	140	160	180	200	220	240	260	280	300	320	340	360	380	400	420	440	460
°C	15	26	38	49	60	71	82	93	104	116	127	138	149	160	171	182	193	204	216	227	238

(continued)

TABLE 1.9 *Continued*

Mortars	Sulfuric acid, 98%
Silicate	U
Sodium silicate	R———
Potassium silicate	
Silica	
Sulfur	U
Furan resin	U
Polyester	U
Epoxy	U

	°F	60	80	100	120	140	160	180	200	220	240	260	280	300	320	340	360	380	400	420	440	460
	°C	15	26	38	49	60	71	82	93	104	116	127	138	149	160	171	182	193	204	216	227	238

The table is arranged alphabetically according to corrodent. Unless otherwise noted, the corrodent is considered pure, in the case of liquids, and a saturated aqueous solution in the case of solids. All percentages shown are weight percents.

Corrosion is a function of temperature. When using the tables note that the vertical lines refer to temperatures midway between the temperatures cited. An entry of R indicates that the material is resistant to the maximum temperature shown. An entry of U indicates that the material is unsatisfactory. A blank indicates that no data are available.

barriers. Monolithic surfacings are not intended to replace brick floors in heavy-duty chemical or physical applications. However, they are economical corrosion barriers for a broad range of applications.

The most popular monolithic surfacings are formulated from the following resins:

1. Epoxy including epoxy Novolac
2. Polyester
3. Vinyl ester, including vinyl ester Novolac
4. Acrylic
5. Urethane, rigid and flexible

Chemical-resistant polymer concretes are formulated from some of the same generic resins. The more popular resins used are

1. Furan
2. Epoxy, including epoxy Novolac
3. Polyester
4. Vinyl ester, including vinyl ester Novolac
5. Acrylics
6. Sulfur

The major advantages to be derived from the use of chemical-resistant monolithic surfacings and polymer concretes are as follows:

1. These formulations provide flexibility, giving aesthetically attractive materials with a wide range of chemical resistances, physical properties, and methods of application.

2. These formulations provide high early development of physical properties. Compressive values with some systems reach 5,000 psi (35 MPa) in 2 h and 19,000 psi (133 MPa) as ultimate compressive value.

3. Most systems are equally appropriate for applications to new and existing concrete including pour-in-place and precasting.

4. Systems offer ease of installation by in-house maintenance personnel.

5. Systems offer economy when compared to many types of brick and tile installations.

6. Systems are available for horizontal, vertical, and overhead applications.

Furan polymer concrete is inherently brittle and, in large masses, has a propensity to shrink. It is used when resistance to acids, alkalies, and solvents such as aromatic and aliphatic solvents is required. It has been successfully used in small areas in the chemical, electronic, pharmaceutical, steel, and metal working industries.

Polyester, vinyl ester, and acrylic polymer concrete have strong aromatic odors that can be offensive to installation and in-plant personnel. Fire codes, particularly for acrylics, must be scrutinized to ensure compliance.

Sulfur cement polymer concrete is flammable; consequently, the potential for oxides of sulfur would preclude its use for most indoor applications. Sulfur cement polymer concrete is not recommended as a concrete topping.

Polymer concretes are not to be misconstrued with polymer-modified Portland cement concrete. Polymer concretes are totally chemical-resistant, synthetic resin compounds with outstanding physical properties. Polymer concretes pass the total immersion test at varying temperatures for sustained periods. Polymer-modified Portland cement concrete can use some of the same generic resins as used in polymer concretes, but with different results.

The major benefits to be derived from polymer-modified Portland cement concrete are as follows:

1. They permit application of concrete in thinner cross-sections.

2. They provide improved adhesion for pours onto existing concrete.

3. They lower absorption of concrete.

4. They improve impact resistance.

5. They improve resistance to salt but not to aggressive corrosive chemicals such as hydrochloric or sulfuric acid.

The success of monolithic surfacing installations is very much predicated on the qualifications of the design, engineering, and installation personnel, be they in-house or outside contractors.

The following fundamental rules are important to the success of any monolithic surfacing installation:

1. Substrate must be properly engineered to be structurally sound, free of cracks, and properly sloped to drains.

2. New as well as existing slabs must be clean and dry, free of laitance and contaminants, with a coarse surface profile.

3. Ambient slab and materials to be installed should be 65–85°F (18–29°C). Special catalyst and hardening systems are available to accommodate higher or lower temperatures, if required.

4. Thoroughly prime substrate before applying any monolithic surfacing. Follow manufacturer's instructions.

5. Thoroughly mix individual and combined components at a maximum speed of 500 rpm to minimize air entrainment during installation.

6. Uncured materials must be protected from moisture and contamination.

Monolithic surfacings are unique, versatile materials used primarily as flooring systems. They are available in a variety of formulations to accommodate various methods of application. They employ many of the installation methods that have been successfully utilized in the Portland cement concrete industry. The most popular methods are:

1. Hand-troweled

2. Power-troweled

3. Spray

4. Pour-in-place/self-level

5. Broadcast

Hand-troweled applications are approximately 1/4 in. (6 mm) thick and are suggested for small areas or areas with multiple obstructions such as piers, curbs, column foundations, trenches, and sumps. The finished application is tight and dense, and it has a high-friction finish. Topcoat sealers are recommended to provide increased density and imperviousness with a smooth, easy-to-clean finish. High- and low-friction finishes are easily accomplished predicated on end use requirements.

Power trowel installations are the fastest, most economical method for large areas with minimum obstructions. Minimum thickness is 1/4 in. (6 mm). Appropriate sealers are available to improve density of finish.

Spray applications are also ideal for areas where corrosion is aggressive. Spray applications are applied, minimum 1/8 in. (3 mm) thick, in one pass on horizontal surfaces. On vertical and overhead areas including structural components, the material can be spray-applied 1/16 to 3/32 in. thick (1.5–2.4 mm) in one pass and without slump. The mortarlike consistency of the material can be varied to control slump and type of finish. Floors installed in this manner are dense, smooth, safe finishes for people and vehicular traffic.

Pour-in-place and self-level materials are intended for flat areas where the pitch to floor drains and trenches are minimal. They are intended for light-duty areas with minimum process spills. The completed installation is 1/8 to 3/16 in. (3–5 mm) thick with a very smooth, high-gloss, easy-to-clean, aesthetically attractive finish.

Broadcast systems are simple to install, economical, aesthetically attractive floors applied in thicknesses of 3/32 to 1/8 in. (2–3 mm). Resins are squeegee-applied to the concrete slab into which filler and colored quartz aggregates of varying color and size are broadcast or sprinkled into the resin. After the resin has set, the excess filler and quartz aggregate is removed by vacuuming or sweeping for reuse. The process is repeated until the desired floor thickness is achieved. The finished floor is easy to clean and outstanding for light industrial, laboratory, cleanroom, and institutional applications.

1.4.1 Chemical Resistance

Monolithic surfacings and polymer concretes mirror image the chemical resistance of their mortar and grout counterparts. Refer to corrosion tables previously presented and Table 1.10 which follows.

Table 1.10 provides comparative chemical resistance for the most popular resins used as monolithic surfacings and polymer concrete. Those systems without chemical-resistant mortar and grout counterparts are as follows:

1. Acrylic
2. Urethane, rigid and flexible

Acrylic monolithic surfacing and polymer concretes are installed in thicknesses of 1/8 to 1/2 in. (3–13 mm) and 1/2 in. (13 mm) and greater, respectively. These flooring systems are an extension of methyl methacrylate chemistry popularized by Rohm and Haas Co.'s Plexiglass and Dupont's Lucite. They are intended for protection against moderate corrosion environments.

TABLE 1.10

Comparative Chemical Resistance

1-A=bisphenol A epoxy—aliphatic amine hardener
1-B=bisphenol A epoxy—aromatic amine hardener
1-C=bisphenol F epoxy (epoxy novolac)
2-D=polyester resin—chlorendic acid type
2-E=polyester resin—bisphenol A fumarate type
3-F=vinyl ester resin
3-G=vinyl ester novolac resin

	1			2		3	
Medium, RT	**A**	**B**	**C**	**D**	**E**	**F**	**G**
Acetic acid, to 10%	R	R	R	R	R	R	R
Acetic acid, 10–15%	C	R	C	R	R	C	R
Benzene	C	R	R	R	N	R	R
Butyl alcohol	R	C	R	R	R	N	R
Chlorine, wet, dry	C	C	C	R	R	R	R
Ethyl alcohol	R	C	R	R	R	R	R
Fatty acids	C	R	C	R	R	R	R
Formaldehyde, to 37%	R	R	R	R	R	R	R
Hydrochloric acid, to 36%	C	R	R	R	R	R	R
Kerosene	R	R	R	R	R	R	R
Methyl ethyl ketone, 100%	N	N	N	N	N	N	N
Nitric acid, to 20%	N	N	R	R	R	R	R
Nitric acid, 20–40%	N	N	R	R	N	N	C
Phosphoric acid	R	R	R	R	R	R	R
Sodium hydroxide, to 25%	R	R	R	N	R	R	R
Sodium hydroxide, 25–50%	R	C	R	N	R	C	R
Sodium hypochlorite, to 6%	C	R	R	R	R	R	R
Sulfuric acid, to 50%	R	R	R	R	R	R	R
Sulfuric acid, 50–75%	C	R	R	R	C	R	R
Xylene	N	R	R	R	R	N	R

RT, room temperature; R, recommended; N, not recommended; C, conditional.

The principal advantages for their use are as follows:

1. They are the easiest of the resin systems to mix and apply by using pour-in-place and self-leveling techniques.

2. Because of their outstanding weather resistance, they are equally appropriate for indoor and outdoor applications.

3. They are the only system that can be installed at below freezing temperatures, 25°F (−4°C), without having to use special hardening or catalyst systems.

4. They are the fastest set and cure of all resin systems. The monolithics will support foot and light-wheeled traffic in 1 h, whereas, the thicker cross-section polymer concrete will also

support foot and light-wheeled traffic in 1 h while developing 90% of its ultimate strength in 4 h.

5. They are the easiest to pigment and with the addition of various types of aggregate, they can be aesthetically attractive.

6. They are equally appropriate for maintenance and new-construction applications. They bond well to concrete. They are ideal for rehabilitating, manufacturing, warehouse, and loading dock floors to impart wear resistance and ease of cleaning.

As previously indicated, a disadvantage inherent in acrylic systems is the aromatic odor for indoor or confined space applications.

The urethane systems are intended to be monolithic floors with elastomeric properties installed in thicknesses of 1/8 to 1/4 in. (3–6 mm). Similar to the acrylic systems, the urethanes are intended for protection against moderate to light corrosion environments. Standard systems are effective at temperatures of 10–140°F (−24 to 60°C). High-temperature systems are available for exposure to temperatures of 10–180°F (−24 to 82°C). Many of the urethane systems are capable of bridging cracks up to 1/16 in. (1.6 mm).

The monolithic urethane flooring systems offer the following advantages:

1. They are easy to mix and apply using the pour-in-place, self-level application technique.

2. Systems are available for indoor and outdoor applications.

3. The elastomeric quality of the systems provides underfoot comfort for production line flooring applications.

4. Because of their being elastomeric, they have excellent sound-deadening properties.

5. They have outstanding resistance to impact and abrasion.

6. They are excellent waterproof flooring systems for above-grade light- and heavy-duty floors. They are equally appropriate for maintenance and new-construction applications.

7. They are capable of bridging cracks in concrete 1/16 in. (1.5 mm) wide.

Urethane materials are demanding systems during installation. Mix ratio of components, temperature, and humidity controls are mandatory for successful installations.

The comparative chemical resistances of acrylic and urethane systems are presented in Table 1.11.

The physical properties of acrylic systems are substantially different from those of the urethanes. The acrylic flooring systems are extremely hard and should be considered too brittle for applications subjected to excessive physical abuse such as impact from large-diameter steel pipe, steel plate,

TABLE 1.11

Comparative Chemical Resistance: Urethane vs. Acrylic Systems

		Urethane	
Medium, RT	**Acrylic**	**Standard**	**High-Temperature**
Acetic acid, 10%	G	G	C
Animal oils	G	G	N
Boric acid	E	E	E
Butter	G	F	N
Chromic acid, 5–10%	C	C	C
Ethyl alcohol	N	N	N
Fatty acids	F	F	N
Gasoline	E	N	N
Hydrochloric acid, 20–36%	F	C	C
Lactic acid, above 10%	F	C	C
Methyl ethyl ketone, 100%	N	N	N
Nitric acid, 5–10%	G	C	F
Sulfuric acid, 20–50%	G	C	C
Water, fresh	E	E	E
Wine	G	G	F

RT, room temperature; E, excellent; G, good; F, fair; C, conditional, N; not recommended.

and heavy castings. The inherent flexibility and impact resistance of urethanes offer potential for these types of applications.

Precast and poured-in-place polymer concrete has been successfully used in a multitude of indoor and outdoor applications in various industries.

Acrylic polymer concrete has been used for precast trenches and covers, delta bus supports, and insulators. Ease of pigmenting to match corporate colors, good weather resistance, resistance to airborne SO_2, SO_3, and NO_x, and dielectric properties have made acrylics particularly attractive in the electric utility industry.

The chemical, steel, electronic, automotive, and pharmaceutical industries have taken advantage of the ease of mixing and placing, as well as the high early strength and chemical resistance properties, of various polymer concretes.

1.4.2 Applications

Carbon steel and reinforced concrete are outstanding general construction materials. They have an enviable record of success in a multitude of industries and applications. Unfortunately, steel and concrete will corrode when oxygen and water are present. Weather and chemicals are catalysts that accelerate the corrosion process.

The cost attributable to corrosion in the United States is estimated to be in a range of $9–$90 billion. This figure was confirmed from a study initiated several years ago by the National Institute of Standards and Technology (formerly National Bureau of Standards).

The range includes corrosion attributable to chemical processes, to corrosion of highways and bridges from de-icing chemicals, to atmospheric corrosion of a steel fence. The economic losses attributable to corrosion are confirmed by various technical organizations such as NACE International and others.

Chemical-resistant mortars, grouts, and monolithic surfacings are used for protecting steel and concrete in a host of applications. Chemical-resistant mortars and grouts for installing all sizes and shapes of brick, tile, and ceramics provide the premier thermal, physical, and chemical-resistant sheathing for protecting linings and membranes applied to steel and concrete. Typical applications include the following:

1. Below, on, and above grade floors
2. Pickling, plating, storage, and chemical process tanks and towers
3. Waste holding and treatment tanks
4. Dual containment for outdoor process and storage vessels, including leak detection systems
5. Stacks and scrubbers from incineration and treatment of toxic waste fumes
6. Above- and below-grade trenches, pipelines, sumps, and manholes

Polymer concretes can be used as fast-set, totally chemical-resistant, poured-in-place slabs or as a topping for refurbishing of existing concrete slabs. They are highly chemical-resistant materials of exceptional physical properties that can also be used for a multitude of precast applications such as curbs, piers, foundations, pump pads, stair treads, trenches and covers, sumps, and manholes.

Chemical-resistant, nonmetallic construction materials are formidable, economical corrosion barriers for protecting steel and concrete when compared to most alloys.

Nonmetallic construction systems can incorporate in their design dual containment and leak detection to meet all federal and state environmental mandates.

The pursuit of clean air and water will continue unabated. Environmental laws will continue to be stringent. Recycling, waste treatments, and incineration will require close attention to corrosivity of all processes.

The agricultural industry, producers of various fertilizers and agricultural chemicals, relies on brick-lined floors and tanks in the production of sulfuric and phosphoric acids. Chemical storage and waste treatment facilities require protection from aggressive chemicals and waste byproducts.

The pharmaceutical, food, and beverage industries are plagued by corrosion from chemicals and food acids as well as corrosion from acid and alkaline cleaning and sanitizing chemicals. The Food and Drug Administration and U.S. Department of Agriculture will be unrelenting in upholding sanitation mandates to ensure the health and welfare of the population. The healthcare industry is under severe cost containment pressure. The best cost containment in the food, beverage, and pharmaceutical industries is their continued high standards of excellence in maintaining standards of cleanliness. Maintaining these standards is not without cost from corrosion to concrete and steel.

Chemical-resistant mortars, grouts, monolithic surfacings, and polymer concrete are proven solutions to a host of these types of corrosion problems.

2

Cathodic Protection

2.1 Introduction

A tremendous investment exists world-wide in underground metallic structures such as pipelines, storage facilities, well casings, structure supports, communication cables, etc. When these members are in direct contact with the soil and unprotected, they are subject to corrosion.

Attempts have been made to overcome this problem by the use of various types of coatings applied to the surface. Although such coatings can be effective, invariably small holes or holidays are usually present. These holidays provide access to the uncovered base metal and permit corrosion.

Subsequent repairs can be extremely costly. In addition, undetected failure (or leakage) can cause accidents or lead to safety hazards, such as causing roadways to collapse as a result of an underground water leak, or to environmental hazards resulting from oil leaks.

By cathodic protection, bare metals and holidays in coated metals can be protected from corrosion.

2.2 Background

Cathodic protection is a major factor in metals' corrosion control. When an external electric current is applied, the corrosion rate can be reduced to practically zero. Under these conditions, the metal can indefinitely remain in a corrosive environment without deterioration.

In practice, cathodic protection can be utilized with such metals as steel, copper, brass, lead, and aluminum against corrosion in all soils and almost all aqueous media. Although it cannot be used above the waterline (because the impressed electric current cannot reach areas outside of the electrolyte), it can effectively be used to eliminate corrosion fatigue, intergranular corrosion, stress-corrosion cracking, dezincification of brass, or pitting of stainless steels in seawater or steel in soil.

In 1824, Sir Humphry David reported that by coupling iron or zinc to copper, the copper could be protected against corrosion. The British Admiralty had blocks of iron attached to the hulls of copper-sheathed vessels to provide cathodic protection. Unfortunately, cathodically protected copper is subject to fouling by marine life, which reduced the speed of vessels under sail and forced the admiralty to discontinue the practice. However, the corrosion rate of the copper had been appreciably reduced. Unprotected copper supplies a different number of copper ions to poison fouling organisms.

In 1829, Edmund Davy was successful in protecting the iron portions of buoys by using zinc blocks, and in 1840, Robert Mallete produced a zinc alloy that was particularly suited as a sacrificial anode. The fitting of zinc slabs to the hulls of vessels became standard practice as wooden hulls were replaced. This provided localized protection, specifically against the galvanic action of a bronze propeller. Overall protection of seagoing vessels was not investigated until 1950 when the Canadian Navy determined that the proper use of anti-fouling paints in conjunction with corrosion resistant paints made cathodic protection of ships feasable and could reduce maintenance costs.

About 1910–1912 the first application of cathodic protection by means of an impressed current was undertaken in England and the United States. Since that time, the general use of cathodic protection has been widespread. There are thousands of miles of buried pipe and cables that are protected in this manner.

This form of protection is also used for water tanks, submarines, canal gates, marine piling, condensers, and chemical equipment.

2.3 Theory

Cathodic protection is achieved by applying electrochemical principles to metallic components buried in soil or immersed in water. It is accomplished by flowing a cathodic current through a metal–electrolyte interface favoring the reduction reaction over the anodic metal dissolution. This enables the entire structure to work as a cathode.

The flux of electrons can be provided by one of two methods. By use of a rectifier, a direct current may be impressed on an inert anode and the components. These components receive an excess of electrons and are, thereby, cathodically protected.

The alternative method is to couple the components with a more active metal, such as zinc or magnesium, to create a galvanic cell. Under these conditions, the active metal operates as an anode and is destroyed while protecting the components that are cathodic. Such an anode is referred to as a *sacrificial anode*.

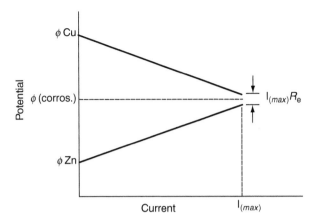

FIGURE 2.1
Polarization for copper–zinc cell.

The basis of cathodic protection is shown in the polarization diagram for a Cu–Zn cell (Figure 2.1). If polarization of the cathode is continued by use of an external current beyond the corrosion potential to the open-circuit potential of the anode, both electrodes reach the same potential and no corrosion of the zinc can take place. Cathodic protection is accomplished by supplying an external current to the corroding metal on the surface of which local action cells operate as shown in Figure 2.2. Current flows from the auxiliary anode and enters the anodic and cathodic areas of the corrosion

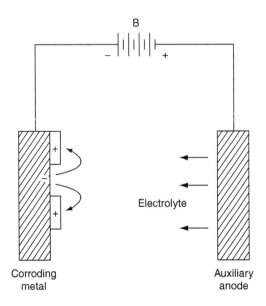

FIGURE 2.2
Cathodic protection using impressed current on local action cell.

cells, returning to the source of the DC current, B. Local action current will cease to flow when all the metal surface is at the same potential as a result of the cathode areas being polarized by an external current to the open-current potential of the anodes. As long as this external current is maintained, the metal cannot corrode.

The corrosion rate will remain at zero if the metal is polarized slightly beyond the open circuit potential of the anode. However, this excess current has no value and may be injurious to amphoteric metals or coatings. Therefore, in actual practice, the impressed current is maintained close to the theoretical minimum.

If the current should fall below that required for complete protection, some protection will still be afforded.

2.4   Method of Application

A source of direct current and an auxiliary electrode are required to provide cathodic protection. The auxiliary electrode (anode) is usually composed of graphite or iron and is located some distance away from the structure being protected. The anode is connected to the positive terminal of the DC source, and the structure being protected is connected to the negative terminal. This permits current to flow from the anode, through the electrolyte, to the structure. Applied voltages are not critical as long as they are sufficient to supply an adequate current density to all parts of the protected structure. The resistivity of the soil will determine the required applied voltage; when the end of a long pipeline is to be protected, the voltage will have to be increased.

Current is usually supplied by a rectifier supplying low-voltage direct current of several amperes.

2.4.1   Impressed Current Systems

For impressed current systems, the source of electricity is external. A rectifier converts high voltage AC current to low voltage DC current. This direct current is impressed between buried anodes and the structure to be protected.

It is preferred to use inert anodes that will last for the longest possible times. Typical materials used for these anodes are graphite, silicon, titanium, and niobium plated with platinum.

For a given applied voltage, the current is limited by electrolyte resistivity and by the anodic and cathodic polarization. With the impressed current system, it is possible to impose whatever potential is necessary to obtain the current density required by means of the rectifier.

Electric current flows in the soil from the buried anode to the underground structure to be protected. Therefore, the anode must be connected to the positive pole of the rectifier and the structure to the negative pole. All cables from the rectifier to the anode and to the structure must be electrically insulated. If not, those from the rectifier to the anode will act as an anode and deteriorate rapidly, whereas, those from the rectifier to the structure may pick up some of the electric current that would then be lost for protection.

2.4.1.1 Current Requirements

The specific metal and environment will determine the current density required for complete protection. The applied current density must always exceed the current density equivalent to the measured corrosion rate under the same conditions. Therefore, as the corrosion rate increases, the impressed current density must be increased to provide protection.

There are several factors that affect the current requirements. These include:

1. The nature of the electrolyte
2. The soil resistivity
3. The degree of aeration

The more acidic the electrolyte, the greater the potential for corrosion and the greater the current requirement.

Soils that exhibit a high resistance require a lower cathode current to provide protection. However, in areas of violent agitation or high aeration, an increase in current will be required. The required current to provide cathodic protection can vary from 0.5 to 20 mA/ft.^{2} of bare surface.

Field testing may be required to determine the necessary current density to provide cathodic protection in a specific area. However, these testing techniques will provide only an approximation. After completion of the installation, it will be necessary to conduct a potential survey and make necessary adjustments to provide the desired degree of protection.

For cathodically controlled corrosion rates, the corrosion potential approaches the open-circuit anodic potential, and the required current density is only slightly greater than the equivalent corrosion current. The required current can be considerably greater than the corrosion current for mixed control, and for anodically controlled corrosion reactions, the required current is even greater.

When a protective current causes precipitation of an inorganic scale on the cathode surface, such as in hard water or seawater, the total current required is gradually reduced. This is the result of an insulating coating being formed. However, the current density at the exposed metal areas does not change; only the total current density per apparent unit area is less.

2.4.2 Sacrificial Anodes

It is possible, by selection of an anode constructed of a metal more active in the galvanic series than the metal to be protected, to eliminate the need for an external DC current. A galvanic cell will be established with the current direction, exactly as described, by using an impressed electric current. These sacrificial anodes are usually composed of magnesium or magnesium-based alloys. Occasionally, zinc and aluminum have been used. Because these anodes are essentially sources of portable electrical energy, they are particularly useful in areas where electric power is not available or where it is uneconomical or impractical to install power lines for this purpose.

Most sacrificial anodes in use in the United States are of magnesium construction. Approximately 10 million pounds of magnesium is annually used for this purpose. The open-circuit potential difference between magnesium and steel is about 1 V. This means that one anode can protect only a limited length of pipeline. However, this low voltage can have an advantage over higher impressed voltages in that the danger of over-protection to some portions of the structure is less, and, because the total current per anode is limited, the danger of stray-current damage to adjoining metal structures is reduced.

Magnesium anode rods have also been placed in steel hot-water tanks to increase the life of these tanks. The greatest degree of protection is afforded in "hard" waters where the conductivity of the water is greater than in "soft" waters.

2.4.2.1 Anode Requirements

To provide cathodic protection, a current density of a few milliamps (mA) is required. Therefore, to determine the anodic requirements, it is necessary to know the energy content of the anode and its efficiency. From this data, the necessary calculations can be made to size the anode, determine its expected life, and determine the number of anodes required. As previously indicated, the three most common metals used as sacrificial anodes are magnesium, zinc, and aluminum. The energy content and efficiency of these metals are as follows:

Metal	Theoretical Energy Content (A h/lb)	Anodic Efficiency (%)	Practical Energy (PE) Constant (A h/lb)
Magnesium	1000	50	500
Zinc	370	90	333
Aluminum	1345	60	810

The number of pounds of metal required to provide a current of 1 A for a year can be obtained from the following equation:

$$\text{lb metal/A} - \text{yr} = \frac{8760 \text{ h/yr}}{\text{PE}}.$$

For magnesium this would be

$$\text{lb Mg/A} - \text{yr} = \frac{8760}{500} = 17.52.$$

The number of years (YN) for which 1 lb of metal can produce a current of 1 mA is determined from the following equation:

$$\text{YN} = \frac{\text{PE}}{10^{-3}\text{A } 8760 \text{ h/yr}}.$$

For magnesium this would be

$$\frac{500}{10^{-3}(8760)} = 60 \text{ years.}$$

The life expectancy (L) of an anode of W lb, delivering a current of i mA is calculated as follows:

$$L = \frac{\text{YN}(W)}{i}.$$

For magnesium this would be

$$L_{\text{Mg}} = \frac{60(W)}{i}$$

that is based on a 50% anodic efficiency. Because actual efficiencies tend to be somewhat less, it is advisable to apply a safety factor and multiply the result by 0.75.

The current required to secure protection of a structure and the available cell voltage between the metal structure and sacrificial anode determine the number of anodes required. This can be illustrated by the following example:

Assume that an underground pipeline has an external area of 200 ft.^{2} and a soil resistivity of 600 Ωcm. Field tests indicate that 6 mA/ft.^{2} is required for protection. To provide protection for the entire pipeline (6 mA/ft.^{2}) (200 ft.^{2}) = 1200 mA. Magnesium anodes used in this particular soil have a voltage of -1.65 V, or a galvanic cell voltage of

$$E_{\text{cell}} = E_C - E_A = -0.85 - (-1.65) = +0.8 \text{ V.}$$

The resistance is therefore

$$R = \frac{V}{I} = \frac{0.8}{1.2} = 0.67 \ \Omega.$$

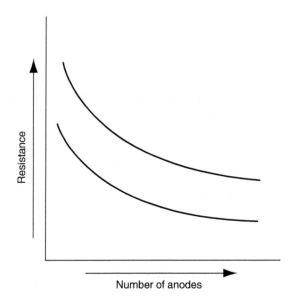

FIGURE 2.3
Plot of Sunde equation.

As the number of anodes are increased, the total resistance of the system decreases. Each anode that is added provides a new path for current flow, parallel to the existing system. The relationship between the resistance of the system and the number of anodes is shown in the Sunde equation.

$$R = \frac{0.00521P}{NL}\left(2.3 \log \frac{8L}{d-1} + \frac{2L}{S} 2.3 \log 0.656N\right),$$

where R, resistance in Ω; P, soil resistivity in Ωcm; N, number of anodes; L, anode length (ft.); d, diameter of anode (ft.); S, distance between anodes (ft.).

Figure 2.3 shows the typical plotting of the results of this equation. Different anodic shapes will have different curves.

2.4.2.2 Anode Materials and Backfill

The use of magnesium as a sacrificial anode has already been discussed. For use with impressed current, auxiliary anodes are usually formed of scrap iron or graphite. Scrap iron is consumed at a considerably faster rate than graphite (15–20 lb/A-yr vs. 2 lb/A-yr); however, graphite costs more both initially and in operating expense. Graphite requires more power than scrap iron. It is also more fragile, and greater care must be taken during installation. Under certain conditions, the advantage of the 8- to 10-times longer life far outweighs the added costs, particularly in areas where replacement poses problems.

Platinum-clad or 2% silver–lead electrodes that use impressed current have been employed for the protection of structures in seawater. The latter anodes are estimated to last ten years, whereas, sacrificial magnesium anodes require replacement every two years. On occasion, aluminum electrodes have been used in freshwaters.

Because the effective resistivity of soil surrounding an anode is limited to the immediate area of the electrode, this local resistance is generally reduced by using backfill. For impressed current systems, the anode is surrounded with a thick bed of coke mixed with three to four parts of gypsum to one part of sodium chloride. The consumption of the anode is reduced somewhat because the coke backfill is a conductor and carries part of the current. If the anode is immersed in a river bed, lake, or ocean, backfill is not required.

Auxiliary anodes need not be consumed to fulfill their purpose. Conversely, sacrificial anodes are consumed no less to supply an equivalent current than is required by Faraday's law.

For magnesium anodes, backfill has the advantage of reducing the resistance of insulating corrosion-product films as well as increasing the conductivity of the immediate area. A typical backfill consists of a mixture of approximately 20% bentonite (for retention of moisture), 75% gypsum, and 5% sodium sulfate.

2.5 Use with Coatings

Insulating coatings are advantageous to use with either impressed current or sacrificial anodes when supplying cathodic protection. These coatings need not be pore free because the protective current flows preferentially to the exposed metal areas that require the protection. Such coatings are useful in distributing the protective current, in reducing total current requirements, and in extending the life of the anode. For example, in a coated pipeline, the current distribution is greatly improved over that of a bare pipeline, the number of anodes and total current required is less, and one anode can protect a much longer section of pipeline. Because the earth is a good electrical conductor and the resistivity of the soil is localized only within the region of the pipeline or electrodes, the limiting length of pipe protected per anode is imposed by the metallic resistance of the pipe and not the resistance of the soil.

One magnesium anode is capable of protecting approximately 100 ft. (30 m) of a bare pipeline, whereas, it can provide protection for approximately 5 miles (8 km) of a coated pipeline.

In a hot-water tank coated with glass or an organic coating, the life of the magnesium anode is extended and more uniform protection is supplied to the tank. Without the coating, the tendency is for excess current to flow to the side and insufficient current flows to the top and bottom. Because of these factors cathodic protection is usually provided with coated surfaces.

2.6 Testing for Completeness of Protection

There are several ways that the effectiveness of protection can be checked. The first two methods are qualitative and do not provide data about whether enough or more than enough current is being supplied. Potential measurements, the third method, is of prime importance in practice.

2.6.1 Coupon Tests

A metal coupon is shaped to conform to the contour of the pipe, weighed, and attached by a brazed-connected cable to the pipe. Both the cable and the surface between the coupon and pipe are coated with coal tar. The coupon is allowed to remain buried for weeks or months, recovered, cleaned, and weighed. The weight loss, if any, is an indication of whether or not the cathodic protection of the pipeline is complete.

2.6.2 Colormetric Tests

A piece of absorbent paper soaked in potassium ferricyanide solution is placed in contact with a cleaned section of the buried pipeline and the soil replaced. After a relatively short time, the paper is retrieved. A blue ferrous ferricyanide reaction indicates incomplete cathodic protection, whereas, an absence of blue on the paper indicates that the cathodic protection is complete.

2.6.3 Potential Measurements

By measuring the potential of the protected structure, the degree of protection, including overprotection, can be quantitatively determined. This measurement is the generally accepted criterion and is used by corrosion engineers. The basis for this determination is the fundamental concept that cathodic protection is complete when the protected structure is polarized to the open-circuit anodic potential of local action cells.

The reference electrode for making this measurement should be placed as closely as possible to the protected structure to avoid and to minimize an error caused by internal resistance (IR) drop through the soil. Such IR drops through corrosion product films or insulating coatings will still be present regardless of precautions taken, tending to make the measured potential more active than the actual potential at the metal surface. For buried pipelines, a compromise location is taken directly over the buried pipe at the soil surface because cathodic protection currents flow mostly to

TABLE 2.1

Calculated Minimum Potential, ϕ, for Cathodic Protection

Metal	$E°$ (V)	Solubility Product, M(OH)$_2$	ϕH_2 Scale (V)	ϕ vs. Cu–CuSO$_4$ Reference Electrode (V)
Iron	0.440	1.8×10^{-15}	-0.59	-0.91
Copper	-0.337	1.6×10^{-19}	0.16	-0.16
Zinc	0.763	4.5×10^{-17}	-0.93	-1.25
Lead	0.126	4.2×10^{-15}	-0.27	-0.59

Source: From H.H. Uhlig. 1971. *Corrosion and Corrosion Control*, 2nd ed., New York: Wiley.

the lower surface and are minimum at the upper surface of the pipe buried a few feet below the soil surface.

The potential for steel is equal to -0.85 V vs. the copper-saturated copper sulfate half-cell, or 0.53 V on the standard hydrogen scale. The theoretical open-circuit anodic potential for other metals may be calculated using the Nernst equation. Several typical calculated values are shown in Table 2.1.

2.7 Overprotection

Overprotection of steel structures, to a moderate degree, usually does not cause any problems. The primary disadvantages are waste of electric power and increased consumption of auxiliary anodes. When the overprotection is excessive, hydrogen can be generated at the protected structure in sufficient quantities to cause blistering of organic coatings, hydrogen embrittlement of the steel, or hydrogen cracking. Damage to steel by hydrogen absorption is more prevalent in environments where sulfides are present.

Overprotection of systems with amphoteric metals (e.g., aluminum, zinc, lead, and tin) will damage the metal by causing increased attack instead of reduction of corrosion. This emphasizes the need for making potential measurements of protected structures.

2.8 Economics

The cost of cathodic protection is more than recovered by reduced maintenance costs or by reduced installation costs, or both. For buried pipelines, the guarantee that there will be no corrosion on the soil side of the pipe has made it economically feasible to transport oil and high-pressure natural gas across the American continent. It has also permitted the use of thinner-walled pipe. Wall thicknesses need only be sufficient to withstand the internal pressures. No extra allowance has to be added for external corrosion. This saving alone has sometimes more than paid for the installation of the cathodic protection equipment.

Similarly, other cathodic protection systems have more than paid for their installation costs by reduced maintenance costs or by longer operating periods between routine inspections and maintenance periods.

3

Corrosion Inhibitors

3.1 Introduction

Corrosion of metallic surfaces can be reduced or controlled by the addition of chemical compounds to the corrodent. This form of corrosion control is called *inhibition* and the compounds added are known as *corrosion inhibitors*. These inhibitors will reduce the rate of either anodic oxidation or cathodic reduction, or both. The inhibitors themselves form a protective film on the surface of the metal. It has been postulated that the inhibitors are adsorbed into the metal surface either by physical (electrostatic) adsorption or chemosorption.

Physical adsorption is the result of electrostatic attractive forces between the organic ions and the electrically charged metal surface. Chemosorption is the transfer, or sharing of the inhibitor molecule's charge to the metal surface, forming a coordinate-type bond. The adsorbed inhibitor reduces the corrosion rate of the metal surface either by retarding the anodic dissolution reaction of the metal by the cathodic evolution of hydrogen, or both.

Inhibitors can be used at pH values of acid from near neutral to alkaline. They can be classified in many different ways according to

1. Their chemical nature (organic or inorganic substances)
2. Their characteristics (oxidizing or nonoxidizing compounds)
3. Their technical field of application (pickling, descaling, acid cleaning, cooling water systems, and the like)

The most common and widely known use of inhibitors is their application in automobile cooling systems and boiler feedwaters.

By considering the electrochemical nature of corrosion processes, constituted by at least two electrochemical partial reactions, inhibition may be defined on an electrochemical basis. Inhibitors will reduce the rates of either or both of these partial reactions (anodic oxidation and/or cathodic reduction). As a consequence, there can be anodic, cathodic, and mixed inhibitors. Other tentative classification of inhibitors have been made by taking into consideration the chemical nature (organic or inorganic substances), their characteristics (oxidizing or nonoxidizing compounds), or their technological field of application (pickling, descaling, acid cleaning, cooling water systems, etc.).

Inhibitors can be used in electrolytes at different pH values, from acid to near-neutral or alkaline solutions. Because of the very different situations created by changing various factors such as medium and inhibitor in the system metal/aggressive medium/inhibitor, various inhibition mechanisms must be considered.[1-6]

An accurate analysis of the different modes of inhibiting electrode reactions including corrosion was carried out by Fischer.[7] He distinguished among various mechanisms of action such as:

Interface inhibition
Electrolyte layer inhibition
Membrane inhibition
Passivation

Subsequently, Lorenz and Mansfield[8] proposed a clear distinction between interface and interphase inhibition, representing two different types of retardation mechanisms of electrode reactions including corrosion. Interface inhibition presumes a strong interaction between the inhibitor and the corroding surface of the metal.[1,7,9] In this case, the inhibitor adsorbs as a potential-dependent, two-dimensional layer. This layer can affect the basic corrosion reactions in different ways:

By a geometric blocking effect of the electrode surface because of the adsorption of a stable inhibitor at a relatively high degree of coverage of the metal surface

By a blocking effect of the surface sites because of the adsorption of a stable inhibitor at a relatively low degree of coverage

By a reactive coverage of the metal surface. In this case, the adsorption process is followed by electrochemical or chemical reactions of the inhibitor at the interface

According to Lorenze and Mansfield,[8] interface inhibition occurs in corroding systems exhibiting a bare metal surface in contact with the corrosive medium. This condition is often realized for active metal dissolution in acid solutions.

Interphase inhibition presumes a three-dimensional layer between the corroding substrate and the electrolyte.[7,10,11] Such layers generally consist of weakly soluble corrosion products and/or inhibitors. Interphase inhibition is mainly observed in neutral media with the formation of porous or nonporous layers. Clearly, the inhibition efficiency strongly depends on the properties of the formed three-dimensional layer.

3.2 Inhibitor Evaluation

Because there may be more than one inhibitor suitable for a specific application, it is necessary to have a means of comparing the performance of each. This can be done by determining the inhibitor efficiency according to the following correlation:

$$I_{eff} = \frac{R_o - R_i}{R_o} \times 100$$

where

I_{eff} = efficiency of inhibitor, %;

R_o = corrosion rate of metal without inhibitor present;

R_i = corrosion rate of metal with inhibitor present.

R_o and R_i can be determined by any of the standard corrosion testing techniques. The corrosion rate can be measured in any unit, such as weight loss (mpy), as long as units are consistent across both tests.

3.3 Classification of Inhibitors

Inhibitors can be classified in several ways as previously indicated. Inhibitors will be classified and discussed under the following headings:

1. Passivation inhibitors
2. Organic inhibitors
3. Precipitation inhibitors

3.3.1 Passivation Inhibitors

Passivation inhibitors are chemical oxidizing materials such as chromate ($Cr_2O_4^{2-}$) and nitrite (NO_2^-) or substances such as Na_3PO_4 or $NaBrO_7$. These materials favor adsorption on the metal surface of dissolved oxygen.

This is the most effective and, consequently, the most widely used type of inhibitor. Chromatics are the least expensive inhibitors for use in water systems and are widely used in the recirculation-cooling systems of internal combustion engines, rectifiers, and cooling towers. Sodium chromate, in concentrations of 0.04–0.1%, is used for this purpose. At higher temperatures or in freshwater that has chloride concentrations above 10 ppm, higher concentrations are required. If necessary, sodium hydroxide is added to adjust the pH to a range of 7.5–9.5. If the concentration of chromate falls below a concentration of 0.016%, corrosion will be accelerated. Therefore, it is essential that periodic colorimetric analysis be conducted to prevent this from occurring.

Recent environmental regulations have been imposed on the use of chromates. They are toxic and, on prolonged contact with the skin, can cause a rash. It is usually required that the Cr^{6+} ion be converted to Cr^{3+} before discharge. The Cr^{3+} ion is insoluble and can be removed as a sludge, whereas, the Cr^{6+} ion is water-soluble and toxic. Even so, the Cr^{3+} sludge is classified as a hazardous waste and must be constantly monitored. Because of the chromate ions' cost of conversion, the constant monitoring required, and the disposal of the hazardous wastes, the economics of the use of these inhibitors are not as attractive as they formerly were.

Because most antifreeze solutions contain methanol or ethylene glycol, the chromates cannot be used in this application because the chromates have a tendency to react with organic compounds. In these applications, borax $(Na_2B_4O_7 \cdot 10H_2O)$ that has added sulfonated oils to produce an oily coating, and mercaptobenzothiazole are used. The latter material is a specific inhibitor for the corrosion of copper.

Nitrites are also used in antifreeze-type cooling water systems because they have little tendency to react with alcohols or ethylene glycol. Because they are gradually decomposed by bacteria, they are not recommended for use in cooling tower waters. Another application for nitrites is as a corrosion inhibitor of the internal surfaces of pipelines used to transport petroleum products or gasoline. Such inhibition is accomplished by continuously injecting a 5–30% sodium nitrite solution into the line.

At lower temperatures, such as in underground storage tanks, gasoline can be corrosive to steel as dissolved water is released. This water, in contact with the large quantities of oxygen dissolved in the gasoline, corrodes the steel and forms large quantities of rust. The sodium nitrite enters the water phase and effectively inhibits corrosion.

The nitrites are also used to inhibit corrosion by cutting oil-water emulsions used in the machining of metals.

Passivation inhibitors can actually cause pitting and accelerate corrosion when concentrations fall below minimum limits. For this reason, it is essential that constant monitoring of the inhibitor concentration be performed.

3.3.2 Organic Inhibitors

These materials build up a protective film of adsorbed molecules on the metal surface that provide a barrier to the dissolution of the metal in the electrolyte. Since the metal surface covered is proportional to the inhibitor concentrates, the concentration of the inhibitor in the medium is critical. For any specific inhibitor in any given medium there is an optimal concentration. For example, a concentration of 0.05% sodium benzoate, or 0.2% sodium cinnamate, is effective in water that has a pH of 7.5 and contains 17 ppm sodium chloride or 0.5% by weight of ethyl octanol.

The corrosion because of ethylene glycol cooling water systems can be controlled by the use of ethanolamine as an inhibitor.

3.3.3 Precipitation Inhibitors

Precipitation inhibitors are compounds that cause the formation of precipitates on the surface of the metal, thereby, providing a protective film. Hard water that is high in calcium and magnesium is less corrosive than soft water because of the tendency of the salts in the hard water to precipitate on the surface of the metal and form a protective film.

If the water pH is adjusted in the range of 5–6, a concentration of 10–100 ppm of sodium pyrophosphate will cause a precipitate of calcium or magnesium orthophosphate to form on the metal surface providing a protective film. The inhibition can be improved by the addition of zinc salts.

3.4 Inhibition of Acid Solution

The inhibition of corrosion in acid solutions can be accomplished by the use of a variety of organic compounds. Among those used for this purpose are triple-bonded hydrocarbons; acetylenic alcohols, sulfoxides, sulfides, and mercaptans; aliphatic, aromatic, or heterocyclic compounds containing nitrogen; and many other families of simple organic compounds and of condensation products formed by the reaction between two different species such as amines and aldehydes.

Incorrect choice or use of organic inhibitors in acid solutions can lead to corrosion stimulation and/or hydrogen penetration into the metal. In general, stimulation of corrosion is not related to the type and structure of the organic molecule. Stimulation of iron's acid corrosion has been found with mercaptans, sulfoxides, azole, and triazole derivatives, nitrites, and quinoline. This adverse action depends on the type of acid. For example, bis(4-dimethylamino-phenyl) antipyrilcarbinol and its derivatives at a 10^{-4} M concentration inhibited attack of steel in hydrochloric acid solutions but stimulated attack in sulfuric solutions. Much work has been done studying the inhibiting and/or stimulating phenomena of organic compounds on ferrous as well as nonferrous metals. Organic inhibitors have a critical concentration value, below which inhibition ceases and stimulation begins. Therefore, it is essential that when organic inhibitors are used, constant monitoring of the solution should take place to ensure that the inhibitor concentration does not fall below the critical value.

Generally, it is assumed that the first stage in the action mechanism of the inhibitors in aggressive acid media is adsorption of the inhibitors onto the metal surface. The process of adsorption of inhibitors is influenced by the nature and surface charge of the metal, by the chemical structure of the inhibitor, and by the type of aggressive electrolyte. Physical (or electrostatic) adsorption and chemisorption are the principal types of interaction between an organic inhibitor and a metal surface.

3.4.1 Physical Adsorption

Physical adsorption is the result of electrostatic attractive forces between inhibiting organic ions or dipoles and the electrically charged surface of the metal. The surface charge of the metal is due to the electric field at the outer Helmholtz plane of the electrical double layer existing at the metal/solution interface. The surface charge can be defined by the potential of the metal (E_{corr}) vs. its zero-charge potential (ZCP) ($E_{q=0}$).[12] When the difference $E_{corr} - E_{q=0} = \phi$ is negative, cation adsorption is favored. Adsorption of anions is favored when ϕ becomes positive. This behavior is related not only to compounds with formal positive or negative charge, but also to dipoles whose orientation is determined by the value of the ϕ potential.

According to Antropov,[12] at equal values of ϕ for different metals, similar behavior of a given inhibiting species should be expected in the same environment. This has been verified for adsorption of organic charged species on mercury and iron electrodes at the same potential for both metals.

In studying the adsorption of ions at the metal/solution interface, it was first assumed that ions maintained their total charge during the adsorption, giving rise in this way to a pure electrostatic bond. Lorenz[13–15] suggested that a partial charge is present in the adsorption of ions; in this case, a certain amount of covalent bond in the adsorption process must be considered. The partial charge concept was studied by Vetter and Schulze,[16–19] who defined as electrosorption valency the coefficient for the potential dependence and charge flow of electrosorption processes. The term *electrosorption valency* was chosen because of its analogy with the electrode reaction valency that enters into Faraday's law as well as the Nernst equation.

Considering the concepts discussed above in relation to corrosion inhibition, when an inhibited solution contains adsorbable anions, such as halide ions, these adsorb on the metal surface by creating oriented dipoles and, consequently, increase the adsorption of the organic cations on the dipoles. In these cases, a positive synergistic effect arises; therefore, the degree of inhibition in the presence of both adsorbable anions and inhibitor cations is higher than the sum of the individual effects. This could explain the higher inhibition efficiency of various organic inhibitors in hydrochloric acid solutions compared to sulfuric acid solutions (Table 3.1).[20]

TABLE 3.1

Inhibition Efficiency of Some Pyridinium Derivatives at the Same Molar Concentration (1×10^{-4} M) on Armco Iron in Hydrochloric and Sulfuric Acid Solutions at 25°C

	Inhibition Efficiency (%)	
Additive	1 N HCl	1 N H$_2$SO$_4$
n-Decylpyridinium bromide	87.6	20.0
n-Decyl-3-hydroxypyridinium bromide	94.8	57.5
n-Decyl-3-carboxypyridinium bromide	92.7	76.5
n-Decyl-3,5-dimethylpyridinium bromide	92.5	30.2

Source: From A. Frignani, G. Trabanelli, F. Zucchi, and M. Zucchini. 1980. *Proceedings of the 5th European Symposium on Corrosion Inhibitors*, Ann. Univ. Ferrara, N.S., Sez. V. Suppl. 7, 1185.

3.4.2 Chemisorption

Another type of metal/inhibitor interaction is chemisorption. This process involves charge sharing or charge transfer from the inhibitor molecules to the metal surface in order to form a coordinate type of bond.

The chemisorption process takes place more slowly than electrostatic adsorption and with a higher activation energy. It depends on the temperature, and higher degrees of inhibition should be expected at higher temperatures. Chemisorption is specific for certain metals and is not completely reversible.[22] The bonding occuring with electron transfer clearly depends on the nature of the metal and the nature of the organic inhibitor. In fact, electron transfer is typical for metals having vacant, low energy electron orbitals. Concerning inhibitors, electron transfer can be expected with compounds having relatively loosely bound electrons. This situation may arise because of the presence in the adsorbed inhibitor of multiple bonds or aromatic rings, whose electrons have a π character. Clearly, even the presence of heteroatoms with lone-pair electrons in the adsorbed molecule will favor electron transfer. Most organic inhibitors are substances with at least one functional group regarded as the reaction center for the chemisorption process. In this case, the strength of the adsorption bond is related to the heteroatom electron density and to the functional group polarizability. For example, the inhibition efficiency of homologous series of organic substances differing only in the heteroatom is usually in the following sequence:

$$P > Se > S > N > O.$$

An interpretation may be found in the easier polaribility and lower electronegativity of the elements on the left of the above sequence. On this basis, a surface bond of a lewis acid-base type, normally with the inhibitor as the electron donor and the metal as the electron acceptor, has been postulated.[21]

3.4.3 Interactions between Adsorbed Inhibitors

When the coverage of the metal surface by the adsorbed inhibitor species increases, lateral interactions between inhibitor molecules may arise, influencing efficiency.

Attractive lateral interactions usually give rise to stronger adsorption and higher inhibition efficiency. This effect has been shown in the case of compounds containing long hydrocarbon chains, because of attractive van der Waals forces.[22,23] In the presence of ions or molecules containing dipoles, repulsive interactions may occur, weakening the adsorption and diminishing the inhibiting efficiency.

3.4.4 Relationships between Inhibitor Reactivity and Efficiency

The nature of the inhibitor initially present in acid solutions may change with time and/or the electrode potential as a consequence of reduction reactions, polymerization reactions, or the formation of surface products. The inhibition because of the reaction products is usually called secondary inhibition, whereas, primary inhibition is attributed to the compound initially added to the solution. Secondary inhibition may be higher or lower than primary inhibition, depending on the effectiveness of the reaction products.[24]

An example of inhibitors undergoing electrochemical reduction is that of sulfoxides, the most important being dibenzyl sulfoxide, whose reduction gives rise to a sulfide that is more effective than the primary compound.[25]

On the contrary, the reduction of thiourea and its alkyl derivatives give rise to HS^- ions, whose accelerating effect is known.[26] In some cases, the reduction reaction may be followed by polymerization reactions at the metal/electrolyte interface. This mechanism of action is generally accepted for acetylenic derivatives.[27,28] Electrochemical measurements on iron electrodes in sulfuric acid solutions inhibited by alkynes[29] showed that acetylenic compounds act as cathodic inhibitors, giving rise to a surface barrier phenomenon. Duwell et al.[27] found hydrogenation and dehydration reaction products in heptane extracts of acid/iron powder/ethynylcyclohexan-1-ol. According to Duwell et al., the efficiency of ethynylcyclohexan-1-ol as a corrosion inhibitor apparently depends on the properties and rates of formation of the reaction products. Putilova et al.[28] used a similar extraction technique to show that acetylene reacted at the iron/hydrochloric acid interface to form thick polymolecular films on the iron. Podobaev et al.,[30] analyzing the results obtained for a large number of acetylene derivatives, concluded that these compounds act according to an adsorption-polymerization mechanism; the adsorption on iron takes place mainly via a triple bond. According to Podobaev et al.,[30] the presence in the molecule of polar groups weakens the stability of the triple bond and increases the probability of adsorption and polymerization. The effect of the polar groups on the triple bond reactivity is maximum when the substituent is in position 3 relative to the triple bond.

3.5 Inhibition of Near Neutral Solutions

Because of the differences in the mechanisms of the corrosion process between acid and near-neutral solutions, the inhibitors used in acid solutions usually have little or no inhibition effect in near-neutral solutions. In acid solutions, the inhibition action is due to adsorption on oxide-free metal surfaces. In these media, the main cathodic process is hydrogen evolution.

In almost neutral solutions, the corrosion process of metals results in the formation of sparingly soluble surface products such as oxides, hydroxides, or salts. The cathodic partial reaction is oxygen reduction.

Inorganic or organic compounds as well as chelating agents are used as inhibitors in near-neutral aqueous solutions. Inorganic inhibitors can be classified according to their mechanisms of action:

1. Formation and maintenance of protective films can be accomplished by the addition of inorganic anions such as polyphosphates, phosphates, silicates, and borates.

2. Oxidizing inhibitors such as chromates and nitrites cause self-passivation of the metallic material. It is essential that the concentration of these inhibitors be maintained above a "safe" level. If not, severe corrosion can occur as a result of pitting or localized attack caused by the oxidizer.

3. Precipitation of carbonates on the metal surfaces forming a protective film. This usually occurs because of the presence of Ca^{2+} and Mg^{2+} ions usually present in industrial waters.

4. Modification of surface film protective properties is accomplished by the addition of Ni^{+2}, Co^{2+}, Zn^{2+}, or Fe^{2+}.

The sodium salts of organic acids such as benzoate, salicylate, cinnamate, tartrate, and azelate can be used as alternatives to the inorganic inhibitors, particularly in ferrous solutions. When using these particular compounds in solutions containing certain anions such as chlorides or sulfates, the inhibitor concentration necessary for effective protection will depend on the concentration of the aggressive anions. Therefore, the critical pH value for inhibition must be considered rather than the critical concentration. Other formulations for organic inhibition of near-neutral solutions are given in Table 3.2.

TABLE 3.2

Organic Inhibitors for Use in Near-Neutral Solutions

Inhibitor	Type of Metal Protected
Organic phosphorus-containing compounds, salts of amino-methylenephosphonic acid, hydroxyethylidenediphosphonic acid, phosphenocarboxylic acid, polyacrylate, poly-methacrylate	Ferrous
Borate or nitrocinnamate anions (dissolved oxygen in solution required)	Zinc, zinc alloys
Acetate or benzoate anions	Aluminum
Heterocylic compounds such as benzotriazole and its derivatives, 2-mercaptobenzothiazole, 2-mercaptobenzimidazole	Copper and copper-based alloys

Chelating agents of the surface-active variety also act as efficient corrosion inhibitors when insoluble surface chelates are formed. Various surface-active chelating agents recommended for corrosion inhibition of different metals are given in Table 3.3.

3.6 Inhibition of Alkaline Solutions

All metals whose hydroxides are amphoteric and metals covered by protective oxides that are broken in the presence of alkalies are subject to caustic attack. Localized attack may also occur as a result of pitting and crevice formation.

Organic substances such as tannions, gelatin, saponin, and agar–agar are often used as inhibitors for the protection of aluminum, zinc, copper, and iron. Other materials that have also been found effective are thiourea,

TABLE 3.3

Chelating Agents Used as Corrosion Inhibitors in Near-Neutral Solutions

Chelating Agent	Type of Metal Protected
Alkyl-catechol derivatives, sarcosine derivatives, carboxymethylated fatty amines, and mercaptocarboxylic acids	Steel in industrial cooling systems
Azo compounds, cupferron, and rubeanic acid	Aluminum alloys
Azole derivatives and alkyl esters of thioglycolic acid	Zinc and galvanized steel
Oximes and quinoline derivatives	Copper
Cresolphthalexon and thymolphthalexon derivatives	Titanium in sulfuric acid solutions

substituted phenols and naphthols, β-diketones, 8-hydroxyquinoline, and quinalizarin.

3.7 Temporary Protection with Inhibitors

Occasions arise when temporary protection of metallic surfaces against atmospheric corrosion is required. Typical instances are in the case of finished metallic materials or of machinery parts during transportation and/or storage prior to use. When ready for use, the surface treatment or protective layer can be easily removed.

It is also possible to provide protection by controlling the aggressive gases or by introducing a vapor phase inhibitor. This latter procedure can only be accomplished in a closed environment such as sealed containers, museum showcases, or similar enclosures.

Organic substances used as contact inhibitors or vapor inhibitors are compounds belonging to the following classes:

1. Aliphatic, cycloaliphatic, aromatic, and heterocyclic amines
2. Amine salts with carbonic, carbamic, acetic, benzoic, nitrous, and chromic acids
3. Organic esters
4. Nitro derivatives
5. Acetylenic alcohols

3.8 Inhibition of Localized Corrosion

Corrosion inhibitors are usually able to prevent general corrosion, but their effect on localized corrosion processes is limited.[31] Generally, a higher inhibitor concentration is required to prevent localized corrosion processes than is necessary to inhibit general corrosion.

Staehle[32] examined the possibility of inhibiting localized attack that does not depend on metallurgical structures. He considered phenomena related to:

> Geometric effects, such as galvanic corrosion and crevice corrosion.
> Simple localization of the attack, such as pitting and dezincification.
> Effects of relative motion, such as erosion and cavitation.

The application of organic inhibitors to control galvanic corrosion presents several problems. It is well known that an inhibitive treatment efficient on a single metal, because of the specific action of the inhibitor, may fail to control corrosion if dissimilar metals are in contact. The behavior of a zinc-steel couple in sodium benzoate solution was studied by Brasher and Mercer.[33] Steel is protected in benzoate solution, but usually it corrodes when coupled to zinc, as in galvanized iron, in that solution. An interpretation of the phenomenon was given,[33] emphasizing that rusting of steel coupled to zinc in benzoate solution can be prevented by bubbling air through the solution for one to two days immediately after immersion. After this time, the system should have become stabilized, and no corrosion of the steel should take place when the air stream is discontinued. Venezel and Wranglen[34] studied the steel-copper combination in a hot-water system. The best inhibitor efficiency was obtained by using a mixture of benzoate and nitrite.

It is well known[35] that pitting depends not only on the concentration of aggressive anions in the solution, but also on the concentration of the non-aggressive anions. For this reason, special attention was paid to the effect of inhibitive anions in the aggressive solution. The presence of nonaggressive anions produces different effects:[35] (1) shifting the pitting potential to more positive values; (2) increasing the induction period for pitting; and (3) reducing the number of pits.

In studying the behavior of various anions in pitting, Strehblow[36] found that pit nucleation on iron electrodes in phthalate buffer containing chloride ions was prevented by picrate ions. Strehblow and Titze[37] emphasized that the pitting potential of iron in borate buffer containing chloride ions was ennobled by the presence of capronate. On the other hand, Vetter and Strahblow[38] showed that sulfate ions can reduce the current density at the pit bottom area, inhibiting the propagation stage of pitting corrosion.

Dezincification is the most common example of selective leaching. It is usually prevented by using less-susceptible alloys. The phenomenon can also be minimized by reducing the aggressiveness of the environment with corrosion inhibitors. Kravaeva et al.[39] demonstrated that the addition of surface-active substances such as saponin, dextrin, or benzotriazole can inhibit dezincification of single-phase and diphasic brass in 0.5 N NaCl and HCl solutions. The highest inhibition efficiency was obtained by benzotriazole.

3.9 Summary

Corrosion inhibitors are usually able to prevent general or uniform corrosion. However, they are very limited in their ability to prevent localized corrosion such as pitting, crevice corrosion, galvanic corrosion, dezincification, or stress corrosion cracking. Additional research work is being undertaken in the use of inhibitors to prevent these types of corrosion. The importance of these studies is realized when it is taken into account that only approximately 30% of all failures because of corrosion in chemical plants result from general corrosion. The remaining 70% are due to stress corrosion cracking, corrosion fatigue, pitting, and erosion-corrosion. Attack on metals by general corrosion can be predicted and life spans of the equipment determined and/or the corrosion rates reduced by use of inhibitors. This is not the case with other types of corrosion.

The use of inhibitors can be advantageous in certain cases. However, before using inhibitors, it is essential that the efficiency of the inhibitor to be used be determined to ensure that inhibition will take place.

References

1. G. Trabanelli and V. Carassiti. 1970. *Advances in Corrosion Science and Technology*, M.G. Fontana and R.W. Staehle, Eds., Vol. 1, New York: Plenum Press, pp. 147–229.
2. Z.A. Foroulis. 1969. *Proceedings of the Symposium on Basic and Applied Corrosion Research*, Houston: NACE International.
3. O.L. Riggs, Jr. 1973. *Corrosion Inhibitors*, C.C. Nathan, Ed., Houston: NACE International, p. 7.
4. J.G. Thomas. 1976. *Corrosion*, J.L. Shreir, Ed., 1976. Vol. 2, London: NewnesButterworths, p. 18.3.
5. E. McCafferty. 1979. *Corrosion Control by Coatings*, H. Leidheiser, Ed., Princeton: Science Press, p. 279.
6. I.L. Rosenfeld. 1981. *Corrosion Inhibitors*, New York: McGraw-Hill.
7. H. Fischer. 1972. *Werkst. Korros.*, **23**: 445–465.
8. W.J. Lorenz and F. Mansfield. 1983. International Conference on Corrosion Inhibition. National Association of Corrosion Engineers, Dallas, paper 2.
9. W.J. Lorenz and F. Mansfield. 1980. 31st ISE Meeting, Venice.
10. P. Lorbeer and W.J. Lorenz. 1980. *Electrochimica Acta*, **25**: 375.
11. R.H. Hausler. 1983. International Conference on Corrosion Inhibition. National Association of Corrosion Engineers, Dallas, paper 19.
12. L.I. Antropov. 1962. *First International Congress of Metallic Corrosion*, London: Butterworths, p. 147.
13. W. Lorenz. 1962. *Zeitschrift für Physikalische Chemie*, **219**: 421.
14. W. Lorenz. 1963. *Zeitschrift für Physikalische Chemie*, **224**: 145.
15. W. Lorenz. 1970. *Zeitschrift für Physikalische Chemie*, **244**: 65.
16. (a) K.L. Vetter and J.W. Schulze. 1972. *Berichte der Bunsengesellschaft fur Physikalische Chemie*, **76**: 920.
    (b) K.L. Vetter and J.W. Schulze. 1972. *Berichte der Bunsengesellschaft fur Physikalische Chemie*, **76**: 927.
17. G.W. Schulze and K.J. Vetter. 1973. *Journal of Electroanalytical Chemistry and Interfacial Electrochemistry*, **44**: 68.
18. K.J. Vetter and J.W. Schulze. 1974. *Journal of Electroanalytical Chemistry and Interfacial Electrochemistry*, **53**: 67.
19. (a) J.W. Schulze and K.D. Koppitz. 1976. *Electrochimica Acta*, **21**: 327.
    (b) J.W. Schulze and K.D. Koppitz. 1976. *Electrochimica Acta*, **21**: 337.

20. A. Frignani, G. Trabanelli, F. Zucchi, and M. Zucchini. 1980. *Proceedings of the 5th European Symposium on Corrosion Inhibitors*, Ann. Univ. Ferrara, N.S., Sez. V, Suppl. 7, 1185.

21. N. Hackerman and R.M. Hurd. 1962. *First International Congress on Metallic Corrosion*, London: Butterworths, p. 166.

22. T.P. Hoar and R.P. Khera. 1961. *Proceedings of the 1st European Symposium on Corrosion Inhibitors*, Ann. Univ. Ferrara, N.S., Sez. V, Suppl. 3, 73.

23. N. Hackerman, D.D. Justice, and E. McCafferty. 1975. *Corrosion*, **31:** 240.

24. W.J. Lorenz and H. Fischer. 1969. *Proceedings of the 3rd International Congress on Metals Corrosion*, Vol. 2, p. 99.

25. G. Trabanelli, F. Zucchi, G.L. Zucchini, and V. Carassiti. 1967. *Electrochemicals and Methods*, **2:** 463.

26. A. Frignani, G. Trabanelli, F. Zucchi, and M. Zucchini. 1975. *Proceeding of the 4th European Symposium on Corrosion Inhibitors*, Ann. Univ. Ferrara, N.S., Sez. V. Suppl. 6, 652.

27. E.J. Duwell, J.W. Todd, and H.C. Butzke. 1964. *Corrosion Science*, **4:** 435.

28. I.N. Putflova, N.V. Rudenko, and A.N. Terentev. 1964. *Russian Journal of Physical Chemistry*, **38:** 263.

29. F. Zucchi, G.L. Zucchini, and G. Trabanelli. 1970. *Proceedings of the 3rd European Symposium on Corrosion Inhibitors*, Ann. Univ. Ferrara, N.S. Sez. V, Suppl. 5, 415.

30. I.I. Podobaev, A.G. Voskresenkil, and G.F. Semikolenkov. 1967. *Protection of Metals*, **3:** 88.

31. G. Trabanelli. 1978. *Proceedings of the 7th International Congress on Metallic Corrosion,*. Abraço, Rio de Janeiro, p. 83.

32. R.W. Staehl. 1975. *Proceedings of the 4th European Symposium on Corrosion Inhibitors*, Ann. Univ. Ferrara, N.S. Sez. V, Suppl. 6, 709.

33. D.M. Brasher and A.D. Mercer. 1969. *Proceedings of the 3rd International Congress on Metallic Corrosion*, Vol. 2, Moscow: MIR, p. 21.

34. J. Venczel and G. Wranglen. 1967. *Corrosion Science*, **7:** 461.

35. S. Szklaraska-Smialowska. 1974. *Localized Corrosion*, Houston: NACE International, p. 329.

36. H.H. Strehblow. 1976. *Werkst. Korros.*, **27:** 792.

37. H.H. Strehblow and B. Titze. 1977. *Corrosion Science*, **17:** 461.

38. K.J. Vetter and H.H. Strehblow. 1970. *Berichte der Bunsengesellschaft fur Physikalische Chemie*, **74:** 1024.

39. A.P. Kravaeva, I.P. Marshakov, and S.M. Mel'nik. 1968. *Protection of Metals*, **4:** 191.

4

Liquid-Applied Linings

Of the various coating applications, the most critical is that of a tank lining. The coating must be resistant to the corrodent and be free of pinholes through which the corrosive could penetrate and reach the substrate. The severe attack that many corrosives have on the bare tank emphasizes the importance of using the correct procedure in lining a tank to obtain a perfect coating. It is also essential that the tank be designed and constructed in the proper manner to permit a perfect lining to be applied.

Liquid-applied linings or coatings may be troweled on or spray applied. In a lined tank there are usually four areas of contact with the stored material. Each area has the potential of developing a different form of corrosive attack. These areas are the bottom of the tank (where moisture and other contaminants of greater density may settle), the liquid phase (the area constantly immersed), the interphase (the area where liquid phase meets the vapor phase), and the vapor phase (the area above the liquid). Each of these areas can be more severely attacked than the rest at one time or another. The type of material contained, the nature of impurities that may be present, and the amount of water and oxygen present are all factors affecting the attack. In view of this, it is necessary to understand the corrosion resistance of the lining material under each condition and not only the immersed condition.

Other factors that have an effect on the performance of the lining material are vessel design, vessel preparation prior to lining, application techniques of the coating, curing of the coating, inspection, operating instructions, and temperature limitations. In general, the criteria for tank linings are given in Table 4.1.

4.1 Design of the Vessel

All vessels to be lined internally should be of welded construction. Riveted tanks will expand or contract, thus damaging the lining and causing leakage. Butt welding is preferred, but lap welding can be used providing a fillet

TABLE 4.1

Criteria for Tank Linings

1.	Design of the vessel
2.	Lining selection
3.	Shell construction
4.	Shell preparation
5.	Lining application
6.	Cure of the lining material
7.	Inspection of the lining
8.	Safety
9.	Causes of failure
10.	Operating instructions

weld is used and all sharp edges are ground smooth (see Figure 4.1). Butt welds need not be ground flush but they must be ground smooth to a rounded contour. A good way to judge a weld is to run your finger over it. Sharp edges can be detected easily. All weld spatter must be removed (see Figure 4.2). Any sharp prominence may result in a spot where the film thickness will be inadequate and noncontinuous, thus causing premature failure.

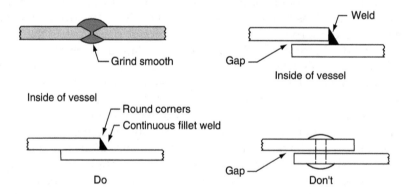

FIGURE 4.1
Butt welding is preferred rather than lap welding or riveted construction.

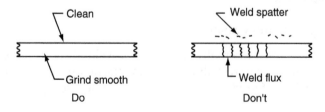

FIGURE 4.2
Remove all weld spatter and grind smooth.

If possible, avoid the use of bolted joints. Should it be necessary to use a bolted joint, it should be of corrosion-resistant materials and sealed shut. The mating surface of steel surfaces should be gasketed. The lining material should be applied prior to bolting.

Do not use construction that will result in the creation of pockets or crevices that will not drain or that cannot be properly sandblasted and lined (see Figure 4.3).

All joints must be continuous and solid welded. All welds must be smooth with no porosity, holes, high spots, or pockets (see Figure 4.4). All sharp angles must be ground to a minimum of 1/8 in. radius (Figure 4.5).

All outlets must be of the flanged or padded type rather than threaded. If pressure requirements permit, use slip-on flanges because the inside diameter

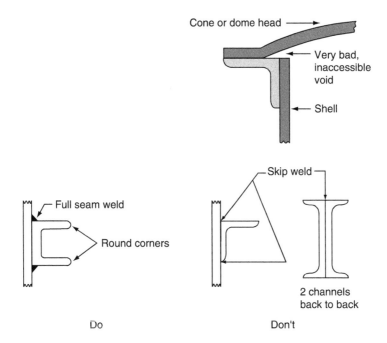

FIGURE 4.3
Avoid all pockets or crevices that cannot be properly sandblasted and lined.

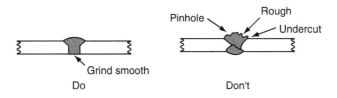

FIGURE 4.4
All joints must be continuous solid welded and ground smooth.

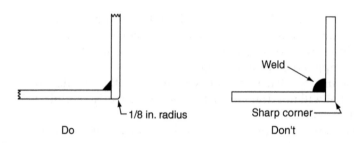

FIGURE 4.5
Grind all sharp edges to a minimum 1/8 in. radius.

of the attaching weld is readily available for radiusing and grinding. If pressure dictates the use of weld neck flanges, the inside diameter of the attach-weld is in the throat of the nozzle. It is therefore more difficult to repair surface irregularities, such as weld undercutting by grinding (see Figure 4.6).

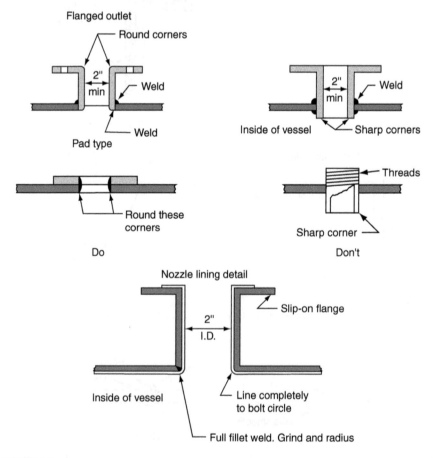

FIGURE 4.6
Typical vessel outlets.

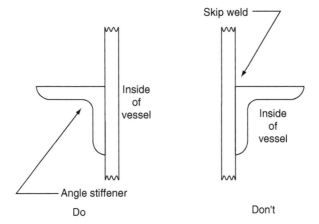

FIGURE 4.7
Stiffening members should be on outside of vessel.

Stiffening members should be placed on the outside of the vessel rather than on the inside (Figure 4.7).

Tanks larger than 25 ft. in diameter may require three man-ways for working entrances. Usually, two are located at the bottom, 180° apart, and one at the top. The minimum opening is 20 in., but 30 in. openings are preferred.

On occasion, an alloy is used to replace the steel bottom of the vessel. Under these conditions, galvanic corrosion will take place. If a lining is applied to the steel and for several inches, usually 6–8 in. onto the alloy, any discontinuity in the lining will become anodic. After corrosion starts, it progresses rapidly because of the bare alloy cathodic area. Without the lining, galvanic corrosion would cause the steel to corrode at the weld area, but at a much lower rate. The recommended practice, therefore, is to line the alloy completely as well as the steel, thereby eliminating the possible occurrence of a large cathode to small anode area (see Figure 4.8).

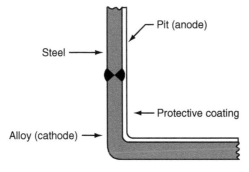

FIGURE 4.8
Potential galvanic action.

It is important that processing liquor is not directed against the side of the vessel, but rather toward the center. Other appurtenances inside the vessel must be located for accessibility of the lining. Heating elements should be placed with a minimum clearance of 6 in. Baffels, agitator base plates, pipes, ladders, and other items can either be lined in place or detached and coated before installation. The use of complex shapes such as angle channels and I-beams should be avoided. Sharp edges should be ground smooth and should be fully welded. Spot welding or intermittent welding should not be allowed. Gouges, hackles, deep scratches, slivered steel, or other surface flaws should be ground smooth.

Concrete tanks should be located above the water table. They require special lining systems. Unless absolutely necessary, expansion joints should be avoided. Small tanks do not normally require expansion joints. Larger tanks can make use of a chemically-resistant joint such as a polyvinyl chloride (PVC). Any concrete curing compound must be compatible with the lining material, or removed before application. Form joints must be as smooth as possible. Adequate steel reinforcement must be used in a strong, dense concrete mix to reduce movement and cracking. The lining manufacturer should be consulted for special instructions. Concrete tanks should be coated by a licensed applicator.

4.2   Vessel Preparation

For the lining material to obtain maximum adhesion to the substrate surface, it is essential that the surface be absolutely clean. All steel surfaces to be coated must be abrasive-blasted to a white metal in accordance with SSPC Specification SP5-63 or NACE Specification 1. A white metal blast is defined as removing all rust, scales, paints, and so on, to a clean white metal that has a uniform gray-white appearance. No streaks or stains of rust or any other contaminants are allowed. At times, a near-white, blast-cleaned surface equal to SSPC SP-10 may be used. If this is permitted by the manufacturer of the lining material it should be used as it is less expensive.

All dust and spent abrasive must be removed from the surface by vacuum cleaning or brushing. After blasting, all workers coming into contact with the clean surface should wear clean, protective gloves and clothing to prevent contamination of the cleaned surface. Any contamination may cause premature failure by osmotic blistering or adhesion loss. The first coat must be applied before the surface starts to rust.

Concrete surfaces must be clean, dry, and properly cured before applying the lining. All protrusions and form joints must be removed. All surfaces must be toughened by sandblasting to remove all loose, weak, or powdery concrete to open all voids and provide the necessary profile for mechanical adhesion of the coating. All dust must be removed by brushing or vacuuming. The coating manufacturer should be contacted for special priming and caulking methods.

4.3 Lining Function

The primary function of a coating system is to protect the substrate. An equally important consideration is product purity protection. The purity of the liquid must not be contaminated with byproducts of corrosion or leachate from the lining system itself. Selection of a lining material for a stationary storage tank dedicated to holding one product at more or less constant durations and temperatures is relatively easy because such tanks present predictable conditions for coating selection. Conversely, tanks that see intermittent storage of a variety of chemicals or solvents, such as product carriers, present a more difficult problem because the parameters of operation vary. Consideration must be given to the effect of cumulative cargoes. In addition, abrasion resistance must be considered if the product in the tank is changed regularly. With complete cleaning between loads, the coating will be abraded.

4.4 Lining Selection

To properly specify a lining material, it is necessary to know specifically what is being handled and under what conditions. The following information must be known about the material being handled:

1. What are the primary chemicals to be handled and at what concentrations?
2. Are there any secondary chemicals, and if so at what concentrations?
3. Are there any trace impurities or chemicals?
4. Are there any solids present, and if so what are the particle sizes and concentrations?
5. Will there be any agitation?
6. What are the fluid purity requirements?
7. What will be present in the vapor phase above the liquid?

The answers to these questions will narrow the selection to those coatings that are compatible. Table 4.2 provides a list of typical lining materials and their general area of application. However, the answers to the next set of questions will narrow the selection down to those materials that are

TABLE 4.2

Typical Lining Materials

Lining Material	Applications
High-bake phenolic	Excellent resistance to acids, solvents, food products, beverages and water. Most widely used lining material, but has poor flexibility compared with other lining materials.
Modified air-dry phenolics (catalyst required)	Nearly equal in resistance to high-bake phenolics. May be formulated for excellent resistance to alkalies, solvents, salt water, deionized water, fresh water, and mild acids. Excellent for dry products.
Modified PVC polyvinyl chloroacetals, air-cured	Excellent resistance to strong mineral acids and water. Most popular lining for water storage tanks; used in water immersion service (potable and marine) and beverage processing.
PVC plastisols	Acid resistant; must be heat cured.
Hypalon	Chemical salts.
Epoxy (amine catalyst)	Good alkali resistance. Fair to good resistance to solvents, mild acids, and dry food products. Finds application in covered hopper-car linings and nuclear containment facilities.
Epoxy polyamide	Good resistance to water and brines. Used in storage tanks and nuclear containment facilities. Poor acid resistance and fair alkali resistance.
Epoxy polyester	Good abrasion resistance. Used for covered hopper-car linings. Poor solvent resistance.
Epoxy coal tar	Excellent resistance to mild acids, mild alkalies, salt water, and fresh water. Poor solvent resistance. Used for crude oil storage tanks, sewage disposal plants, and water works.
Coal tar	Excellent water resistance. Used for water tanks.
Asphalts	Good acid and water resistance.
Neoprene	Good acid and flame resistance. Used for chemical processing equipment.
Polysulfide	Good water and solvent resistance.
Butyl rubber	Good water resistance.
Styrene-butadiene polymers	Finds application in food and beverage processing and in the lining of concrete tanks.
Rubber latex	Excellent alkali resistance. Finds application in caustic tanks (50–73%) at 180°F (82°C) to 250°F (121°C).
Urethanes	Superior abrasion resistance. Excellent resistance to strong mineral acids and alkalies. Fair solvent resistance. Used to line dishwashers and washing machines.
Vinyl ester	Excellent resistance to strong acids and better resistance up to 350°F (193°C) to 400°F (204°C), depending upon thickness.
Vinyl urethanes	Finds application in food processing, hopper cars, and wood tanks.
Fluoropolymers	High chemical resistance and fire resistance. Used in SO_2 scrubber service.
Vinylidene chloride latex	Excellent fuel oil resistance.
Alkyds, epoxy esters, oleoresinous primers	Water immersion applications and as primers for other top coats.

(continued)

TABLE 4.2 *Continued*

Lining Material	Applications
Inorganic zinc, water-based post-cure, and water-based self-cure	Jet fuel storage tanks and petroleum products.
Inorganic zinc, solvent-based self-curing	Excellent resistance to most organic solvents (aromatics, ketones, and hydrocarbons), excellent water resistance. Difficult to clean. May be sensitive to decomposition products of materials stored in tanks.
Furan	Most acid resistant organic polymer. Used for stack linings and chemical treatment tanks.

compatible as well as to those coatings that have the required mechanical and/or physical properties.

1. What is the normal operating temperature and temperature range?
2. What peak temperatures can be reached during shutdown, startup, process upset, etc.?
3. Will any mixing areas exist where exothermic or heat of mixing temperatures may develop?
4. What is the normal operating pressure?
5. What vacuum conditions and range are possible during operation, startup, shutdown, and upset conditions?
6. Is grounding necessary?

Other factors must also be considered before the final decisions can be made as to which coating to use. After the previous questions have been answered, there will still be several potential materials from which to choose.

Service-life expectation must be considered. Different protective coating options afford different degrees of protection for different periods of time at a variety of costs. Factors such as maintenance cycle, operating cycles, and reliability of the coating must all be considered. Can the facility tolerate any downtime for inspection and maintenance? If so, how often and for how long?

When these questions have all been answered, an appropriate decision can be made as to which coating material will be used.

The size of the vessel must also be considered in the coating selection. If the vessel is too large, it may not fit in a particular vender's oven for curing the coating. Also, nozzle diameters 4 in. or less are too small to spray-apply a liquid coating. When a lining is to be used for corrosion protection, it is necessary to review the corrosion rate of the immersed environment on the bare substrate. Assuming that the substrate is carbon steel with a corrosion rate of less than 10 mil per year (mpy) at the operating temperature, pressure, and concentration of corrodent, then a thin film lining of less than 20 mil can be used. For general corrosion, this corrosion rate is not considered severe. However, if a pinhole should be present through the lining, a concentration of the corrosion current density occurs as a result of the large ratio of cathode area to anode area. The pitting corrosion rate will rapidly increase above the 20 mpy rate and through-wall penetration can occur in months.

When the substrate exhibits a corrosion rate in excess of 10–20 mpy, a thick film coating exceeding 20 mil in thickness is used. These thicknesses are less susceptible to pinholes.

Thin linings are used for overall corrosion protection as well as combating localized corrosion such as pitting and stress cracking of the substrate. Thin fluoropolymer coatings are used to protect purity and to provide nonstick surfaces for easy cleaning.

Among the materials available for thin coatings are those based on epoxy and phenolic resins that are 0.15–0.30 mm (0.006–0.12 in.) thick. They are either chemically cured or heat baked. Baked phenolic linings are used to protect railroad tank cars transporting sulfonic acid. Tanks used to store caustic soda (sodium hydroxide) have a polyamide cured epoxy coating.

Thin coatings of sprayed and baked FEP, PFA, and ETFE are also widely used. They are applied to primed surfaces as sprayed water-borne suspension or electrostatically charged powders sprayed on a hot surface. Each coat is baked before the next is applied. Other fluoropolymers can also be susceptible to delamination in applications where temperatures cycle frequently between ambient and steam. Table 4.3 presents details of these linings. Fluoropolymer thin coatings can be applied as thick coatings.

When the corrosion rate of the substrate exceeds 10-mpy-thick coatings exceeding 2.5 mil (0.025 in.) are recommended. One such coating is vinyl ester reinforced with glass cloth or woven roving. Coatings greater than 125 mil (0.125 in.) thick can be sprayed or troweled. Maximum service temperature is 170°F (73°C). These coatings can be applied in the field and are used in service with acids and some organics.

Another thick coating material for service with many acids and bases is plasticized PVC. This has a maximum operating temperature of 150°F (66°C).

Sprayed and baked electrostatic powder coating of fluoropolymers, described under thin coatings, can also be applied as thick coatings. One such coating is PVDF and glass or cotton fabric.

Manufacturers and/or other corrosion engineers should be consulted for case histories of identical applications. Included in the case history should be the name of the applicator who applied the coating, application conditions, type of equipment used, degree of application difficulty, and other required special procedures. A lining with superior chemical resistance will fail rapidly if it cannot be properly applied, so it is advantageous to learn from the experience of others.

To maximize sales, lining manufacturers formulate their products to meet as broad a range as possible of chemical and solvent environments. Consequently, a tank coating may be listed as suitable for more than 100

TABLE 4.3
Fluoropolymer Lining Systems

Lining System	Thickness, in. (mm)	Maximum Size	Design Limits	Installation	Repair Considerations
Sprayed Dispersions					
FEP	0.04 (1.0)				
PFA	0.01–0.04 (0.25–1.0)				
PVDF	0.025–0.03 (0.06–0.76)				
PFA with mesh and carbon	0.08 (2.0)				
PVDF with mesh and carbon	0.04–0.09 (1.0–2.30)	8 ft. (2.4 m) dia. 40 ft. (12.2 m) length	Pressure allowed. Vacuum rating undetermined	Primer and multiple coats with combination 2 psig equipment. Each coat is baked	Hot patching is possible, but testing is recommended
Electrostatic Spray Powders					
ETFE	Up to 0.09 (2.3)	8 ft. (2.4 m) dia. 40 ft. (12.2 m) length	Pressure allowed. Vacuum rating undetermined	Primer and multiple coats applied with electrostatic spraying equipment. Each coat is baked	Hot patching is possible, but testing is recommended
FEP	0.01–0.04 (0.28)				
PFA	0.01 (0.28)				
ECTFE	0.06–0.07 (1.5–1.8)				
PVDF	0.025 (0.64)				

products with varying degrees of compatibility. However, there is potential danger of failure if the list is viewed only from the standpoint of the products approved for service. If more than one of these materials listed as being compatible with the lining is to be used, consideration must be given to the sequence of use in which the chemicals or solvents will be stored or carried in the tank. This is particularly critical when the cargo is water miscible, for example methanol or cellosolve, and is followed by a water ballast. A sequence such as this creates excessive softening of the film and makes recovery of the lining film more difficult and thus prone to early failure.

Certain tank coating systems may have excellent resistance to specific chemicals for a given period after which they must be cleaned and allowed to recover for a designated period of time to return to their original resistance level. Thirty days is a common period of time for this process between chemicals such as acrylonitrile and solvents such as methanol.

When case histories are unavailable, or manufacturers are unable to make a recommendation, it will be necessary to conduct tests. This can occur in the case of a proprietary material being handled or if a solution might contain unknown chemicals. Sample panels of several coating systems should be tested for a minimum of 90 days, with a 6 month test being preferable. Because of time requirements, 90 days is standard.

The test must be conducted at the maximum operating temperature to which the lining will be subjected and should simulate operating conditions, including washing cycles, cold wall, and the effects of insulation.

Other factors to consider in lining selection include service life, maintenance cycles, operating cycles, and the reliability of the lining. Different protective linings provide different degrees of protection for various periods of time, at a variety of costs. Allowable downtime of the facility for inspection and maintenance must also be considered in terms of frequency and length of time.

After the coating system has been selected, recommendations from the manufacturer for a competent applicator should be requested and contact should made with previous customers.

4.5 Lining Application

In lining a vessel, the primary concern is to deposit a void-free film of the specified thickness on the surface. Any area that is considerably less than the specified thickness may have a noncontinuous film. In addition, pinholes in the coating may cause premature failure.

If the film is too thick, there is always the danger of solvents being entrapped, which can lead to bad adhesion, excessive brittleness, improper cure, and subsequent poor performance. Dry spraying of the coating should be avoided because it causes the coating to be porous. Poor film formation may be caused if thinners other than those recommended by the coating manufacturer are used. Do not permit application to take place below the temperature recommended by the manufacturer.

During application, the film thickness should be checked. This can be accomplished by use of an Elcometer or Nordsen wet film thickness gauge. If the wet film thickness meets specifications, in all probability the dry film will also be within specification limits.

All gauges used to measure dry film thickness must be calibrated before use, following the manufacturer's recommended procedure. Readings should be taken at random locations on a frequent basis. Special attention should be given to hard-to-coat areas.

4.6 Inspection

Proper inspection requires that the inspector be involved with the job from the beginning. An understanding of the design criteria of the vessel and the reasons for the specific design configuration is helpful. The inspector should participate in the prework meeting, prejob inspection, surface preparation inspection, and coating application inspection. Daily inspection reports should be prepared along with a final acceptance report.

Before the coating is applied, the inspector should verify that the vessel has been properly prepared for lining. Welds must be ground smooth with a rounded contour. It is not necessary that they be ground flush. Sharp protrusions should be rounded and weld crevices manually opened so that the coating can penetrate. If this is not possible, the projection should be removed by grinding. Back-to-back angles, tape, or stitch welding cannot be properly cleaned and coated. They should be sealed with caulking to prevent crevice corrosion.

After the vessel has been sandblasted, the inspector should work quickly so that the application of the coating to the surface is not delayed.

The inspector should examine coatings during and after application. It is important to check for porosities. The first visual inspection is mandatory to detect pinholing and provide recoat instructions. Visual inspections are performed either with the unaided eye or by the use of a magnifying glass.

After repairs of visible defects, inspection should be carried out using low voltage (75 V or less) holiday detectors that ring, buzz, or light up to show electrical contact through a porosity within the coating to the metal surface. This check should be performed after the primer or second coat, so that these areas can be touched up and made free of porosities before the final top coat. These visual techniques permit the inspector to identify areas that have been missed, damaged areas, or thin areas. In some instances, white primers have been used to spot areas of low film thickness or inadequate coverage of the substrate. The final test, employing instruments, will provide the inspector with an accurate appraisal as to whether or not the proper film thickness has been achieved.

The inspector should also be involved in the selection of the applicator. Because a tank lining requires nearly perfect application, a knowledgeable and conscientious contractor is required. Again, the lowest bidder is not necessarily the best choice. Evaluate the applicator before awarding the contract to assure that the tank lining contractor is experienced in applying the specified lining. Before placing a contractor on the bidder's list, review his qualifications. Inquire as to what jobs he has done using the selected lining material and follow up with those particular applications. If possible, visit his facilities and inspect his workmanship. An experienced inspector is helpful during this evaluation. Taking these steps will help ensure installation of a tank lining that will provide the desired performance.

4.7   Curing

Proper curing is essential if the lining is to provide the corrosion protection for which it was selected. Each coat must be cured using proper air circulation techniques. Fresh air over 50°F (10°C) and having a relative humidity of less than 89% should be supplied to an opening at the top of the vessel with an exhaust at the bottom. The air flow should be fed by forced air fans and should be downward because the solvents used in coatings are usually heavier than air. Because of this, proper exhaustion can only be obtained with downward flow.

To prevent solvent entrapment between coats and to ensure a proper final cure, the curing time and temperature must be in accordance with the manufacturer's instructions for the specific coating material. A warm forced-air cure between coats and as a final cure will provide a dense film and tighter cross-linking, which provides superior resistance to solvents and moisture permeability. Before placing the vessel in service, the lining should be washed down with water to remove any loose overspray.

Linings must be allowed sufficient time to obtain a full cure before being placed in service. This usually requires 3–7 days. Do not skimp on this time.

When the tank is placed in service, operating instructions should be prepared and should include the maximum temperature to be used. The outside of the tank should be labeled; "Do not exceed X°F (X°C). This tank has been lined with_____. It is to be used only for _____ service."

4.8   Safety during Application

Many lining formulations contain solvents, making it necessary to take certain safety precautions. It is necessary that all lining materials and thinners be kept away from any source of open flame. This means that welding in adjacent areas must be discontinued during application and "no smoking" must be the rule during application.

A power air supply and ventilation must be provided during the application of the lining. The vapor concentration inside of the vessel should be checked on a regular basis to ensure that the maximum allowable vapor concentration is not reached. For most solvents, a vapor concentration of between 2% and 12% in the air is sufficient to cause an explosion. As long as the vapor concentration is kept below the lower level, no explosion will take place. All electrical equipment must be grounded.

Because flammable solvents are being exhausted from the tank, precautions must be taken on the exterior of the tank. These flammable vapors will travel a considerable distance along the ground. No flames, sparks, or ungrounded equipment can be nearby.

Those applying the lining should wear fresh airline respirators and protective cream on exposed parts of the body. Water should be available for flushing accidental spills on the skin. Never allow one worker in the tank alone.

OSHA issues a form called the Material Data Sheet. The manufacturers supply this form by listing all toxicants or hazardous materials and provide a list of the solvents used. Also included are the threshold limit values (TLVs) for each substance. Explosive hazards, flash points, and temperature limits are established for safe application of each lining material. These Material Safety Data Sheets should be kept on file in the job superintendents office and at the first aid station.

4.9 Causes of Lining Failure

Most lining failures are the result of the misuse of the tank lining, which results in blistering, cracking, hardening or softening, peeling, staining, burning, and undercutting. A frequent cause of failure is overheating during operation. When a heavily pigmented surface or thick film begins to shrink, stresses are formed on the surface that result in cracks. These cracks do not always expose the substrate and may not penetrate. Under these conditions, the best practice is to remove these areas and recoat according to standard repair procedures.

Aging or poor resistance to the corrosive can result in hardening or softening. As the coating ages, particularly epoxy and phenolic amines, it becomes brittle and may chip from the surface. Peeling can result from improperly cured surfaces, poor surface preparation, or a wet or dry surface. Staining results when there is a reaction of the corrosive on the surface of the lining or slight staining from impurities in the corrosive. The true cause must be determined by scraping or detergent washing the lining. If the stain is removed and softening of the film is not apparent, failure has not occurred.

Any of the above defects can result in undercutting. After the corrosive penetrates to the substrate, corrosion will proceed to extend under the film areas that have not been penetrated or have not failed. Some linings are more resistant than others to undercutting or underfilm corrosion. Usually, if the lining exhibits good adhesive properties, and if the primer coat is chemically-resistant to the corrosive environment, underfilm corrosion will be greatly retarded.

In addition, a tank lining must not impart any impurities to the material contained within it. The application is a failure if any taste, smell, color, or other contamination is imparted to the product, even if the lining is intact. Such contamination can be caused by the extraction of impurities from the lining, leading to blistering between coats to the metal.

If the lining is unsuited to the service, complete failure may occur by softening, dissolution, and finally complete disintegration of the lining. This type of problem is prevalent between the interphase and bottom of the tank. At the bottom of the tank and throughout the liquid phase, penetration is of great concern.

The vapor phase of the tank is subject to corrosion from concentrated vapors mixed with any oxygen present and can cause extensive corrosion.

4.10 Specific Liquid Linings

Most liquid-applied lining materials are capable of being formulated to meet requirements for specific applications. Corrosion data referring to the suitability of a lining material for a specific corrodent indicates that a formulation is available to meet these conditions. Because all formulations may not be suitable, the manufacturer must be checked as to the suitability of his formulation. The most common linings are discussed.

4.10.1 Phenolics

Synthetic phenolic resins were developed and commercialized in the early 1900s by Leo Bakeland.[2] The reduction of phenol and formaldehyde produces a product that forms a highly cross-linked, three-dimensional product when cured. The resins have found use in various applications in the lining industry because of their excellent heat resistance, chemical resistance, and electrical properties. They also offer good adhesion to many substrates and have good compatibility with other polymers.

Phenolic resins have two basic classifications: resoles and novalaks. Resoles, or heat reactive resins, are made using an excess of formaldehyde and a base catalyst. The polymer that is produced has reactive methylol groups that form a thermoset structure when heat is applied.

Novalaks are made using an excess of phenol and an acid catalyst. Reaction occurs by the protonation of formaldehyde,[3] and the intermediate is characterized by methylene linkages rather than methylol groups. These products are not heat reactive and require additional cross-linking agents such as hexamethylenetetramine to become thermosetting.

Both polymerization reactions evolve water during curing. This condensation reaction serves to limit film thicknesses to approximately 3 mil because the volatiles will cause blistering while curing takes place.

The outstanding phenolic systems are those that are baked at approximately 450°F (230°C) to provide a 3–5 mil (75–130 μm) lining of high chemical resistance.

Phenolic coatings have a wet temperature resistance to 200°F (93°C). They are odorless, tasteless, nontoxic, and suitable for food use. They must be baked at a metal temperature ranging from 350 to 450°F (175–230°C). The coating must be applied in a thin layer (approximately 1 mil (0.03 mm)) and partially baked between coats. Multiple thin coats are necessary to allow removal of water from the condensation reaction. The cured coating is difficult to patch due to extreme solvent resistance.

A brown color results upon baking that can be used to indicate the degree of cross-linking. It can be modified with epoxies and other resins to enhance water, chemical, and heat resistance.

Phenolic resins exhibit excellent resistance to most organic solvents, especially aromatic and chlorinated solvents. Organic polar solvents capable of bonding, such as alcohols and ketones, can attack phenolics. Although the phenolics have an aromatic character, the phenolic hydroxyls provide sites for hydrogen bonding and attack by caustics.

Phenolics are not suitable for use in strong alkaline environments. Strong mineral acids also attack the phenolics, and acids such as nitric, chromic, and hydrochloric can cause severe degradation. Sulfuric and phosphoric acids may be suitable under some conditions. There is some loss of properties when phenolics are in contact with organic acids such as acetic, formic, or oxalic.

Although attacked by oxidants and even by dilute alkalies, the phenolics provide both corrosion and contamination protection in a wide variety of chemical and petroleum services. Refer to Table 4.4 for the compatibility of phenolics with selected corrodents.

Modified air-dried phenolics are nearly equivalent to high-baked phenolics with a dry heat resistance of 150°F (65°C). They can be formulated for excellent resistance to alkalies, solvents, fresh water, salt water, deionized water, and mild acid resistance.

4.10.2 Epoxy

Epoxy resins can be formulated with a wide range of properties. These medium to high-priced resins are noted for their adhesion. Epoxy linings provide excellent chemical and corrosion resistance. They exhibit good resistance to alkalies, non-oxidizing acids, and many solvents. Typically epoxies are compatible with the following materials at 200°F (93°C) unless otherwise noted:

Acids

Acetic, 10% to 150°F (66°C)	Benzoic
Butyric	Fatty acids
Hydrochloric, 10%	Oxalic
Rayon spin bath	Sulfuric, 20% to 180°F (82°C)

Bases

Sodium hydroxide, 50% to 180°F (82°C)	Sodium sulfide, 10%
Trisodium phosphate	Magnesium hydroxide

Salts, Metallic salts

Aluminum	Calcium
Iron	Magnesium
Potassium	Sodium
Most ammonium salts	

Alcohols, Solvents

Methyl	Ethyl acetate, to 150°F (66°C)
Ethyl	Naphtha
Isopropyl, to 150°F (66°C)	Toluene
Benzene, to 150°F (66°C)	Xylene

Miscellaneous

Distilled water	Sea water
Jet fuel	Gasoline
White liquor	Diesel fuel
Sour crude oil	Black liquor

Epoxies are not satisfactory for use with:

Bromine water	Chromic acid
Bleaches	Fluorine
Methylene chloride	Hydrogen peroxide
Sulfuric acid above 70%	Wet chlorine gas
Wet sulfur dioxide	

Refer to Table 4.5 for the compatibility of epoxy with selected corrodents, and Reference [1] for a more comprehensive listing.

Epoxy resins must be cured with cross-linking agents (hardeners) or catalysts to develop the desired properties. Cross-linking takes place at the epoxy and hydroxyl groups that are the reaction sites. The primary types of curing agents used for linings are the aliphatic amines and catalytic curing agents. Epoxies (amine catalyst) are widely used because curing of the epoxies takes place at room temperature. High exothermic reactions develop during the curing reaction, which limits the mass of material that can be cured. Amine-cured coatings exhibit good resistance to alkalies, and fair to good resistance to mild acids, solvents, and dry food products. They are widely used for hopper car linings and nuclear containment facilities. The maximum allowable temperature is 275°F (135°C).

Catalytic curing agents require a temperature of 200°F (93°C) or higher to react. These baked epoxies exhibit excellent to acids, alkalies, solvents,

TABLE 4.4

Compatibility of Phenolics with Selected Corrodents

Chemical	Maximum Temperature °F	°C
Acetic acid, 10%	212	100
Acetic acid, glacial	70	21
Acetic anhydride	70	21
Acetone	X	X
Aluminum chloride, aqueous	90	32
Aluminum sulfate	300	149
Ammonia gas	90	32
Ammonium carbonate	90	32
Ammonium chloride, 10%	80	27
Ammonium chloride, 50%	80	27
Ammonium chloride, sat.	80	27
Ammonium hydroxide, 25%	X	X
Ammonium hydroxide, sat.	X	X
Ammonium nitrate	160	71
Ammonium sulfate, 10–40%	300	149
Aniline	X	X
Benzaldehyde	70	21
Benzene	160	71
Benzenesulfonic acid, 10%	70	21
Benzoic acid		
Benzyl alcohol	70	21
Butadiene	X	X
Butyl phthalate	160	71
Calcium chlorate	300	149
Calcium hypochlorite, 10%	X	X
Carbon dioxide, dry	300	149
Carbon dioxide, wet	300	149
Carbon tetrachloride	200	93
Carbonic acid	200	93
Chlorine gas, wet	X	X
Chlorine, liquid	X	X
Chlorobenzene	260	127
Chloroform	160	71
Chromic acid, 50%	X	X
Chromyl chloride	X	X
Citric acid, conc.	160	71
Copper acetate	160	71
Cresol	300	149
Ethylene glycol	70	21
Ferric chloride, 50% in water	300	149
Hydrobromic acid, dil.	200	93
Hydrobromic acid, 20%	200	93
Hydrobromic acid, 50%	200	93
Hydrochloric acid, 20%	300	149
Hydrochloric acid, 38%	300	149
Hydrofluoric acid, 30%	X	X

(continued)

TABLE 4.4 *Continued*

Chemical	Maximum Temperature	
	°F	°C
Hydrofluoric acid, 60%	X	X
Hydrofluoric acid, 100%	X	X
Lactic acid, 25%	160	71
Methyl ethyl ketone	160	71
Methyl isobutyl ketone	X	X
Muriatic acid	160	71
Nitric acid, 5%	300	149
Nitric acid, 20%	X	X
Nitric acid, 70%	X	X
Nitric acid, anhydrous	X	X
Nitrous acid, conc.	X	X
Phosphoric acid, 50–80%	X	X
Picric acid	212	100
Sodium hydroxide, 10%	300	149
Sodium hydroxide, 50%	X	X
Sodium hydroxide, conc.	X	X
Sodium hypochlorite, 15%	X	X
Sodium hypochlorite, conc.	X	X
Sulfuric acid, 10%	250	121
Sulfuric acid, 50%	250	121
Sulfuric acid, 70%	200	93
Sulfuric acid, 90%	70	21
Sulfuric acid, 98%	X	X
Sulfuric acid, 100%	X	X
Sulfurous acid	80	27
Thionyl chloride	200	93
Zinc chloride	300	149

The chemicals listed are in the pure state or in a saturated solution unless otherwise indicated. Compatibility is shown to the maximum allowable temperature for which data is available. Incompatibility is shown by an X. A blank space indicates that data is unavailable.

Source: From P.A. Schweitzer. 2004. *Corrosion Resistance Tables,* Vols. 1–4, 5th ed., New York: Marcel Dekker.

inorganic salts, and water. The maximum operating temperature is 325°F (163°C), somewhat higher than that of the amine-cured epoxies.

4.10.3 Furans

Furan polymers are derivatives of furfuryl alcohol and furfural.[4] Using an acid catalyst, polymerization occurs by the condensation route that generates heat and water by-product.

All furan linings must be post-cured to drive out the reaction "condensate" to achieve optimum properties.

TABLE 4.5

Compatibility of Epoxy with Selected Corrodents

Chemical	Maximum Temperature	
	°F	°C
Acetaldehyde	150	66
Acetamide	90	32
Acetic acid, 10%	190	88
Acetic acid, 50%	110	43
Acetic acid, 80%	110	43
Acetic anhydride	X	X
Acetone	110	43
Acetyl chloride	X	X
Acrylic acid	X	X
Acrylonitrile	90	32
Adipic acid	250	121
Allyl alcohol	X	X
Allyl chloride	140	60
Alum	300	149
Aluminum chloride, aq. 1%	300	149
Aluminum chloride, dry	90	32
Aluminum fluoride	180	82
Aluminum hydroxide	180	82
Aluminum nitrate	250	121
Aluminum sulfate	300	149
Ammonia gas, dry	210	99
Ammonium bifluoride	90	32
Ammonium carbonate	140	60
Ammonium chloride, sat.	180	82
Ammonium fluoride, 25%	150	66
Ammonium hydroxide, 25%	140	60
Ammonium hydroxide, sat.	150	66
Ammonium nitrate, 25%	250	121
Ammonium persulfate	250	121
Ammonium phosphate	140	60
Ammonium sulfate, 10–40%	300	149
Ammonium sulfite	100	38
Amyl acetate	80	27
Amyl alcohol	140	60
Amyl chloride	80	27
Aniline	150	66
Antimony trichloride	180	82
Aqua regia, 3:1	X	X
Barium carbonate	240	116
Barium sulfide	300	149
Barium chloride	250	121
Barium hydroxide, 10%	200	93
Barium sulfate	250	121
Benzaldehyde	X	X
Benzene	160	71
Benzenesulfonic acid, 10%	160	71
Benzoic acid	200	93
Benzyl alcohol	X	X

(continued)

TABLE 4.5 *Continued*

Chemical	Maximum Temperature	
	°F	°C
Benzyl chloride	60	16
Borax	250	121
Boric acid, 4%	200	93
Bromine gas, dry	X	X
Bromine gas, moist	X	X
Bromine, liquid	X	X
Butadiene	100	38
Butyl acetate	170	77
Butyl alcohol	140	60
n-Butylamine	X	X
Butyric acid	210	99
Calcium bisulfide		
Calcium bisulfite	200	93
Calcium carbonate	300	149
Calcium chlorate	200	93
Calcium chloride, 37.5%	190	88
Calcium hydroxide, sat.	180	82
Calcium hypochlorite, 70%	150	66
Calcium nitrate	250	121
Calcium sulfate	250	121
Caprylic acid	X	X
Carbon bisulfide	100	38
Carbon dioxide, dry	200	93
Carbon disulfide	100	38
Carbon monoxide	80	27
Carbon tetrachloride	170	77
Carbonic acid	200	93
Cellosolve	140	60
Chloroacetic acid, 92% water	150	66
Chloroacetic acid	X	X
Chlorine gas, dry	150	66
Chlorine gas, wet	X	X
Chlorobenzene	150	66
Chromic acid, 50%	X	X
Chloroform	110	43
Chlorosulfonic acid	X	X
Chromic acid, 10%	110	43
Citric acid, 15%	190	88
Citric acid, 32%	190	88
Copper acetate	200	93
Copper carbonate	150	66
Copper chloride	250	121
Copper cyanide	150	66
Copper sulfate, 17%	210	99
Cresol	100	38
Cupric chloride, 5%	80	27
Cupric chloride, 50%	80	27
Cyclohexane	90	32
Cyclohexanol	80	27
Dichloroacetic acid	X	X

(continued)

TABLE 4.5 *Continued*

Chemical	Maximum Temperature	
	°F	°C
Dichloroethane (ethylene dichloride)	X	X
Ethylene glycol	300	149
Ferric chloride	300	149
Ferric chloride, 50% in water	250	121
Ferric nitrate, 10–50%	250	121
Ferrous chloride	250	121
Ferrous nitrate		
Fluorine gas, dry	90	32
Hydrobromic acid, dil.	180	82
Hydrobromic acid, 20%	180	82
Hydrobromic acid, 50%	110	43
Hydrochloric acid, 20%	200	93
Hydrochloric acid, 38%	140	60
Hydrocyanic acid, 10%	160	71
Hydrofluoric acid, 30%	X	X
Hydrofluoric acid, 70%	X	X
Hydrofluoric acid, 100%	X	X
Hypochlorous acid	200	93
Ketones, general	X	X
Lactic acid, 25%	220	104
Lactic acid, conc.	200	93
Magnesium chloride	190	88
Methyl chloride	X	X
Methyl ethyl ketone	90	32
Methyl isobutyl ketone	140	60
Muriatic acid	140	60
Nitric acid, 5%	160	71
Nitric acid, 20%	100	38
Nitric acid, 70%	X	X
Nitric acid, anhydrous	X	X
Nitrous acid, conc.	X	X
Oleum	X	X
Perchloric acid, 10%	90	32
Perchloric acid, 70%	80	27
Phenol	X	X
Phosphoric acid, 50–80%	110	43
Picric acid	80	27
Potassium bromide, 30%	200	93
Salicylic acid	140	60
Sodium carbonate	300	149
Sodium chloride	210	99
Sodium hydroxide, 10%	190	88
Sodium hydroxide, 50%	200	93
Sodium hypochlorite, 20%	X	X
Sodium hypochlorite, conc.	X	X
Sodium sulfide, to 10%	250	121
Stannic chloride	200	93
Stannous chloride	160	71
Sulfuric acid, 10%	140	60
Sulfuric acid, 50%	110	43

(continued)

TABLE 4.5 *Continued*

Chemical	Maximum Temperature	
	°F	°C
Sulfuric acid, 70%	110	43
Sulfuric acid, 90%	X	X
Sulfuric acid, 98%	X	X
Sulfuric acid, 100%	X	X
Sulfuric acid, fuming	X	X
Sulfurous acid, 20%	240	116
Thionyl chloride	X	X
Toluene	150	66
Trichloroacetic acid	X	X
White liquor	90	32
Zinc chloride	250	121

The chemicals listed are in the pure state or in a saturated solution unless otherwise indicated. Compatibility is shown to the maximum allowable temperature for which data is available. Incompatibility is shown by an X. A blank space indicates that the data is unavailable.

Source: From P.A. Schweitzer. 2004. *Corrosion Resistance Tables*, Vols. 1–4, 5th ed., New York: Marcel Dekker.

Furan polymers are noted for their excellent resistance to solvents and they exhibit excellent resistance to strong concentrated mineral acids, caustics, and combinations of solvents with acids and bases. These furans are subject to many different formulations, making them suitable for specific applications. Consequently, the manufacturer should be consulted for the correct formulation for a specific application. In general, furan formulations are compatible with the following:

Solvents

Acetone	Benzene	Ethyl acetate
Chlorobenzene	Ethanol	Perchloroethylene
Methanol	Methyl ethyl ketone	Toluene
Styrene	Trichloroethylene	
Xylene	Carbon disulfide	

Acids

Acetic	Hydrochloric	Nitric, 5%
Phosphoric	Sulfuric, 60% to 150°F (66°C)	

Bases

Diethylamine	Sodium carbonate
Sodium sulfide	Sodium hydroxide, 50%

Water

Demineralized	Distilled

Others

Pulp mill liquor

The furan resins are not satisfactory for use with oxidizing media, such as chromic or nitric acids, peroxides, hypochlorites, chlorine, phenol, and concentrated sulfuric acid. Refer to Table 4.6 for the compatibility of furan resins with selected corrodents and Reference [1] for a more complete listing.

4.10.4 Vinyl Esters

The vinyl ester class of resins was developed during the late 1950s and early 1960s. Vinyl esters were first used as dental fillings. They had improved toughness and bonding ability over the acrylic materials that were being used at that time. Over the next several years, changes in the molecular structure of the vinyl esters produced resins that found extensive use in corrosion-resistant equipment.

Present-day vinyl esters possess several advantages over unsaturated polyesters. They provide toughness in the cured polymer while maintaining good thermal stability and physical properties at elevated temperatures.

Vinyl esters are available in various formulations. Halogenated modifications are available where fire resistance and ignition resistance are major concerns. The vinyl esters are resistant up to 400°F (204°C).

Vinyl esters can be used to handle most hot, highly chlorinated and acid mixtures at elevated temperatures. They also provide excellent resistance to strong mineral acids and bleaching solutions. Vinyl esters excel in alkaline and bleach environments and are used extensively in the very corrosive conditions found in the pulp and paper industry.

The family of vinyl esters includes a wide variety of formulations. As a result, there can be differences in the compatibility in a reference table. One must keep in mind that all formulations will not act as shown. An indication that a vinyl ester is compatible generally means that at least one formulation is compatible. This is the case in Table 4.7 which shows the compatibility of vinyl ester with selected corrodents. The resin manufacturer must be consulted to verify the resistance.

4.10.5 Epoxy Polyamide

Polyamide resins (nylons) can react with epoxies to form durable protective linings with a temperature resistance of 225°F (107°C) dry and 150°F (66°C) wet. The chemical resistance of the polyamides is inferior to that of the amine-cured epoxies. They are partially resistant to acids, acid salts, alkaline and organic solvents, and are resistant to moisture. Refer to Table 4.8 for the compatibility of epoxy polyamides with selected corrodents and to Reference [1] for a more complete listing. Applications include storage tanks and nuclear containment facilities.

TABLE 4.6

Compatibility of Furans with Selected Corrodents

Chemical	Maximum Temperature	
	°F	°C
Acetaldehyde	X	X
Acetic acid, 10%	212	100
Acetic acid, 50%	160	71
Acetic acid, 80%	80	27
Acetic acid, glacial	80	27
Acetic anhydride	80	27
Acetone	80	27
Acetyl chloride	200	93
Acrylic acid	80	27
Acrylonitrile	80	27
Adipic acid, 25%	280	138
Allyl alcohol	300	149
Allyl chloride	300	149
Alum, 5%	140	60
Aluminum chloride, aq.	300	149
Aluminum chloride, dry	300	149
Aluminum fluoride	280	138
Aluminum hydroxide	260	127
Aluminum sulfate	160	71
Ammonium carbonate	240	116
Ammonium hydroxide, 25%	250	121
Ammonium hydroxide, sat.	200	93
Ammonium nitrate	250	121
Ammonium persulfate	260	127
Ammonium phosphate	260	127
Ammonium sulfate, 10–40%	260	127
Ammonium sulfide	260	127
Ammonium sulfite	240	116
Amyl acetate	260	127
Amyl alcohol	278	137
Amyl chloride	X	X
Aniline	80	27
Antimony trichloride	250	121
Aqua regia, 3:1	X	X
Barium carbonate	240	116
Barium chloride	260	127
Barium hydroxide	260	127
Barium sulfide	260	127
Benzaldehyde	80	27
Benzene	160	71
Benzenesulfonic acid, 10%	160	71
Benzoic acid	260	127
Benzyl alcohol	80	27
Benzyl chloride	140	60
Borax	140	60
Boric acid	300	149
Bromine gas, dry	X	X

(continued)

TABLE 4.6 *Continued*

Chemical	Maximum Temperature	
	°F	°C
Bromine gas, moist	X	X
Bromine, liquid, 3% max	300	149
Butadiene		
Butyl acetate	260	127
Butyl alcohol	212	100
n-Butylamine	X	X
Butyric acid	260	127
Calcium bisulfite	260	127
Calcium chloride	160	71
Calcium hydroxide, sat.	260	127
Calcium hypochlorite	X	X
Calcium nitrate	260	127
Calcium oxide		
Calcium sulfate	260	127
Caprylic acid	250	121
Carbon bisulfide	160	71
Carbon dioxide, dry	90	32
Carbon dioxide, wet	80	27
Carbon disulfide	260	127
Carbon tetrachloride	212	100
Cellosolve	240	116
Chlorine gas, dry	260	127
Chlorine gas, wet	260	127
Chlorine, liquid	X	X
Chloroacetic acid	240	116
Chloroacetic acid, 50% water	100	38
Chlorobenzene	260	127
Chloroform	X	X
Chlorosulfonic acid	260	127
Chromic acid, 10%	X	X
Chromic acid, 50%	X	X
Chromyl chloride	250	121
Citric acid, 15%	250	121
Citric acid, conc.	250	121
Copper acetate	260	127
Copper carbonate		
Copper chloride	260	127
Copper cyanide	240	116
Copper sulfate	300	149
Cresol	260	127
Cupric chloride, 5%	300	149
Cupric chloride, 50%	300	149
Cyclohexane	140	60
Cyclohexanol		
Dichloroacetic acid	X	X
Dichloroethane (ethylene dichloride)	250	121
Ethylene glycol	160	71
Ferric chloride	260	127

(continued)

TABLE 4.6 *Continued*

Chemical	Maximum Temperature	
	°F	°C
Ferric chloride, 50% in water	160	71
Ferric nitrate, 10–50%	160	71
Ferrous chloride	160	71
Ferrous nitrate		
Fluorine gas, dry	X	X
Fluorine gas, moist	X	X
Hydrobromic acid, 20%	212	100
Hydrobromic acid, 50%	212	100
Hydrobromic acid, dil.	212	100
Hydrochloric acid, 20%	212	100
Hydrochloric acid, 38%	80	27
Hydrocyanic acid, 10%	160	71
Hydrofluoric acid, 30%	230	110
Hydrofluoric acid, 70%	140	60
Hydrofluoric acid, 100%	140	60
Hypochlorous acid	X	X
Iodine solution, 10%	X	X
Ketones, general	100	38
Lactic acid, 25%	212	100
Lactic acid, conc.	160	71
Magnesium chloride	260	127
Malic acid, 10%	260	127
Manganese chloride	200	93
Methyl chloride	120	49
Methyl ethyl ketone	80	27
Methyl isobutyl ketone	160	71
Muriatic acid	80	27
Nitric acid, 5%	X	X
Nitric acid, 20%	X	X
Nitric acid, 70%	X	X
Nitric acid, anhydrous	X	X
Nitrous acid, conc.	X	X
Oleum	190	88
Perchloric acid, 10%	X	X
Perchloric acid, 70%	260	127
Phenol	X	X
Phosphoric acid, 50%	212	100
Picric acid		
Potassium bromide, 30%	260	127
Salicylic acid	260	127
Silver bromide, 10%		
Sodium carbonate	212	100
Sodium chloride	260	127
Sodium hydroxide, 10%	X	X
Sodium hydroxide, 50%	X	X
Sodium hydroxide, conc.	X	X
Sodium hypochlorite, 15%	X	X
Sodium hypochlorite, conc.	X	X

(continued)

TABLE 4.6 *Continued*

Chemical	Maximum Temperature	
	°F	°C
Sodium sulfide, to 10%	260	127
Stannic chloride	260	127
Stannous chloride	250	121
Sulfuric acid, 10%	160	71
Sulfuric acid, 100%	X	X
Sulfuric acid, 50%	80	27
Sulfuric acid, 70%	80	27
Sulfuric acid, 90%	X	X
Sulfuric acid, 98%	X	X
Sulfuric acid, fuming	X	X
Sulfurous acid	160	71
Thionyl chloride	X	X
Toluene	212	100
Trichloroaceteic acid, 30%	80	27
White liquor	140	60
Zinc chloride	160	71

The chemicals listed are in the pure state or in a saturated solution unless otherwise indicated. Compatibility is shown to the maximum allowable temperature for which data is available. Incompatibility is shown by an X. A blank space indicates that the data is unavailable.

Source: From P.A. Schweitzer. 2004. *Corrosion Resistance Tables*, Vols. 1–4, 5th ed., New York: Marcel Dekker.

4.10.6 Coal Tar Epoxy

Coal tar epoxies can be applied to bare steel or concrete without a primer. They will not cure below 50°F (10°C) and have a temperature resistance of 225°F (105°C) dry or 150°F (66°C) wet. They have a low cost per unit coverage.

Coal tar epoxies combine the moisture resistance of coal tar with the chemical resistance of epoxy. It possesses excellent resistance to salt water, fresh water, mild acids, and mild alkalies, but has poor solvent resistance. Refer to Table 4.9 for the compatibility of coal tar epoxy with selected corrodents. Reference [1] has a more complete listing.

Coal tar epoxy finds application as a coating for crude oil storage tanks, and in sewage disposal plants and water works.

4.10.7 Coal Tar

Unless cross-linked with another resin, coal tar is thermoplastic and will flow at temperatures of 100°F (38°C) or less. It hardens and embrittles in cold weather.

TABLE 4.7

Compatibility of Vinyl Ester with Selected Corrodents

| | Maximum Temperature | |
Chemical	°F	°C
Acetaldehyde	X	X
Acetamide		
Acetic acid, 10%	200	93
Acetic acid, 50%	180	82
Acetic acid, 80%	150	66
Acetic acid, glacial	150	66
Acetic anhydride	100	38
Acetone	X	X
Acetyl chloride	X	X
Acrylic acid	100	38
Acrylonitrile	X	X
Adipic acid	182	82
Allyl alcohol	90	32
Allyl chloride	90	32
Alum	240	116
Aluminum acetate	210	99
Aluminum chloride, aqueous	260	127
Aluminum chloride, dry	140	60
Aluminum fluoride	100	38
Aluminum hydroxide	200	93
Aluminum nitrate	200	93
Aluminum sulfate	250	121
Ammonia gas	100	38
Ammonium bifluoride	150	66
Ammonium carbonate	150	66
Ammonium chloride, 10%	200	93
Ammonium chloride, 50%	200	93
Ammonium chloride, sat.	200	93
Ammonium fluoride, 10%	140	60
Ammonium fluoride, 25%	140	60
Ammonium hydroxide, 25%	100	38
Ammonium hydroxide, sat.	130	54
Ammonium nitrate	250	121
Ammonium oxychloride		
Ammonium persulfate	180	82
Ammonium phosphate	200	93
Ammonium sulfate, 10–40%	220	104
Ammonium sulfide	120	49
Ammonium sulfite	220	104
Amyl acetate	110	38
Amyl alcohol	210	99
Amyl chloride	120	49
Aniline	X	X
Antimony trichloride	160	71
Aqua regia, 3:1	X	X
Barium carbonate	260	127
Barium chloride	200	93

(continued)

TABLE 4.7 *Continued*

Chemical	Maximum Temperature	
	°F	°C
Barium hydroxide	150	66
Barium sulfate	200	93
Barium sulfide	180	82
Benzaldehyde	X	X
Benzene	X	X
Benzenesulfonic acid, 10%	200	93
Benzoic acid	180	82
Benzyl alcohol	100	38
Benzyl chloride	90	32
Borax	210	99
Boric acid	200	93
Bromine gas, dry	100	38
Bromine gas, moist	100	38
Bromine, liquid	X	X
Butadiene		
n-Butylamine	X	X
Butyl acetate	80	27
Butyl alcohol	120	49
Butyric acid	130	54
Calcium bisulfide		
Calcium bisulfite	180	82
Calcium carbonate	180	82
Calcium chlorate	260	127
Calcium chloride	180	82
Calcium hydroxide, 10%	180	82
Calcium hydroxide, sat.	180	82
Calcium hypochlorite	180	82
Calcium nitrate	210	99
Calcium oxide	160	71
Calcium sulfate	250	116
Caprylic acid	220	104
Carbon bisulfide	X	X
Carbon dioxide, dry	200	93
Carbon dioxide, wet	220	104
Carbon disulfide	X	X
Carbon monoxide	350	177
Carbon tetrachloride	180	82
Carbonic acid	120	49
Cellosolve	140	60
Chlorine gas, dry	250	121
Chlorine gas, wet	250	121
Chlorine, liquid	X	X
Chloroacetic acid	200	93
Chloroacetic acid, 50% water	150	66
Chlorobenzene	110	43
Chloroform	X	X
Chlorosulfonic acid	X	X
Chromic acid, 10%	150	66

(continued)

TABLE 4.7 *Continued*

	Maximum Temperature	
Chemical	°F	°C
Chromic acid, 50%	X	X
Chromyl chloride	210	99
Citric acid, 15%	210	99
Citric acid, conc.	210	99
Copper acetate	210	99
Copper carbonate		
Copper chloride	220	104
Copper cyanide	210	99
Copper sulfate	240	116
Cresol	X	X
Cupric chloride, 5%	260	127
Cupric chloride, 50%	220	104
Cyclohexane	150	66
Cyclohexanol	150	66
Dibutyl phthalate	200	93
Dichloroacetic acid	100	38
Dichloroethane (ethylene dichloride)	110	43
Ethylene glycol	210	99
Ferric chloride	210	99
Ferric chloride, 50% in water	210	99
Ferric nitrate, 10–50%	200	93
Ferrous chloride	200	93
Ferrous nitrate	200	93
Fluorine gas, dry	X	X
Fluorine gas, moist	X	X
Hydrobromic acid, 20%	180	82
Hydrobromic acid, 50%	200	93
Hydrobromic acid, dil.	180	82
Hydrochloric acid, 20%	220	104
Hydrochloric acid, 38%	180	82
Hydrocyanic acid, 10%	160	71
Hydrofluoric acid, 30%	X	X
Hydrofluoric acid, 70%	X	X
Hydrofluoric acid, 100%	X	X
Hypochlorous acid	150	66
Iodine solution, 10%	150	66
Ketones, general	X	X
Lactic acid, 25%	210	99
Lactic acid, conc.	200	93
Magnesium chloride	260	127
Malic acid, 10%	140	60
Manganese chloride	210	99
Methyl chloride		
Methyl ethyl ketone	X	X
Methyl isobutyl ketone	X	X
Muriatic acid	180	82
Nitric acid, 20%	150	66
Nitric acid, 5%	180	82

(continued)

TABLE 4.7 *Continued*

Chemical	Maximum Temperature	
	°F	°C
Nitric acid, 70%	X	X
Nitrous acid, 10%	150	66
Nitrous acid, anhydrous	X	X
Oleum	X	X
Perchloric acid, 10%	150	66
Perchloric acid, 70%	X	X
Phenol	X	X
Phosphoric acid, 50–80%	210	99
Picric acid	200	93
Potassium bromide, 30%	160	71
Salicylic acid	150	66
Silver bromide, 10%		
Sodium carbonate	180	82
Sodium chloride	180	82
Sodium hydroxide, 10%	170	77
Sodium hydroxide, 50%	220	104
Sodium hydroxide, conc.		
Sodium hypochlorite, 20%	180	82
Sodium hypochlorite, conc.	100	38
Sodium sulfide, to 50%	220	104
Stannic chloride	210	99
Stannous chloride	200	93
Sulfuric acid, 10%	200	93
Sulfuric acid, 100%	X	X
Sulfuric acid, 50%	210	99
Sulfuric acid, 70%	180	82
Sulfuric acid, 90%	X	X
Sulfuric acid, 98%	X	X
Sulfuric acid, fuming	X	X
Sulfurous acid, 10%	120	49
Thionyl chloride	X	X
Toluene	120	49
Trichloroacetic acid, 50%	210	99
White liquor	180	82
Zinc chloride	180	82

The chemicals listed are in the pure state or in a saturated solution unless otherwise indicated. Compatibility is shown to the maximum allowable temperature for which data is available. Incompatibility is shown by an X. A blank space indicates that the data is unavailable.

Source: From P.A. Schweitzer. 2004. *Corrosion Resistance Tables*, Vols. 1–4, 5th ed., New York: Marcel Dekker.

Coal tar exhibits excellent water resistance, good resistance to acids, alkalies, minerals, animal and vegetable oils, and salts. Table 4.10 provides the compatibility of coal tar with selected corrodents. Reference [1] provides a more extensive listing.

TABLE 4.8

Compatibility of Epoxy Polyamides with Selected Corrodents

Chemical	Maximum Temperature	
	°F	°C
Acetaldehyde	X	X
Acetic acid, all conc.	X	X
Acetic acid vapors	X	X
Acetone	X	X
Aluminum chloride, dry	100	38
Aluminum fluoride	X	X
Ammonium chloride, all	100	38
Ammonium hydroxide, 25%	100	38
Aqua regia, 3:1	X	X
Benzene	X	X
Boric acid	140	60
Bromine gas, dry	X	X
Bromine gas, moist	X	X
Calcium chloride	110	43
Calcium hydroxide, all	140	60
Citric acid, all conc.	100	38
Diesel fuels	100	38
Ethanol	100	38
Ferric chloride	100	38
Formaldehyde, to 50%	100	38
Formic acid	X	X
Glucose	100	38
Green liquor	100	38
Hydrobromic acid	X	X
Hydrochloric acid, dil.	100	38
Hydrochloric acid, 20%	X	X
Hydrofluoric acid, dil.	100	38
Hydrofluoric acid, 30%	X	X
Hydrofluoric acid, vapors	100	38
Hydrogen sulfide, dry	100	38
Hydrogen sulfide, wet	100	38
Iodine	X	X
Lactic acid	X	X
Lard oil	X	X
Lauric acid	X	X
Linseed oil	100	38
Magnesium chloride, 50%	100	38
Mercuric chloride	100	38
Mercuric nitrate	100	38
Methyl alcohol	100	38
Methyl sulfate	X	X
Methylene chloride	X	X
Mineral oil	100	38
Nitric acid	X	X
Oil, vegetable	100	38
Oleum	X	X
Oxalic acid, all conc.	100	38

(continued)

TABLE 4.8 *Continued*

Chemical	Maximum Temperature	
	°F	°C
Perchloric acid	X	X
Petroleum oils, sour	100	38
Phenol	X	X
Phosphoric acid	X	X
Potassium chloride, 30%	100	38
Potassium hydroxide, 50%	100	38
Propylene glycol	100	38
Sodium chloride	110	43
Sodium hydroxide, to 50%	100	38
Sulfur dioxide, wet	100	38
Sulfuric acid	X	X
Water, demineralized	110	43
Water, distilled	130	54
Water, salt	130	54
Water, sea	110	43
Water, sewage	100	38
White liquor	150	66
Wines	100	38
Xylene	X	X

The chemicals listed are in the pure state or in a saturated solution unless otherwise indicated. Compatibility is shown to the maximum allowable temperature for which data is available. Incompatibility is shown by an X. A blank space indicates that the data is unavailable.

Source: From P.A. Schweitzer. 2004. *Corrosion Resistance Tables*, Vols. 1–4, 5th ed., New York: Marcel Dekker.

Coal tar finds application as a coating both for interior and exterior of underground pipelines.

4.10.8 Urethanes

Polyurethane-resin-based coatings are extremely versatile. They are priced higher than alkyds but lower than epoxies. Polyurethane resins are available as oil-modified, moisture-curing, bleached, two-component, and lacquers. Because of the versatility of the isocyanate reaction, wide diversity exists in specific coating properties. Exposure to the isocyanate should be minimized to avoid sensitivity that might result in an asthmatic-like breathing condition. Continued exposure to humidity may result in gassing or bubbling of the coating in humid conditions.

The urethane coatings have a maximum operating temperature of 250°F (121°C) dry and 150°F (66°C) wet. These coatings are resistant to most mineral and vegetable oils, greases, fuels, and to aliphatic and chlorinated hydrocarbons. Aromatic hydrocarbons, polar solvents, esters, ethers, and ketones will attack urethanes and alcohols will soften urethanes.

TABLE 4.9

Compatibility of Coal Tar Epoxy with Selected Corrodents

Chemical	Maximum Temperature	
	°F	°C
Acetaldehyde	X	X
Acetic acid, to 20%	100	38
Acetic acid, vapors	100	38
Acetone	X	X
Aluminum chloride, dry	100	38
Aluminum fluoride	120	49
Ammonium chloride, dry	100	88
Ammonium hydroxide, 25%	110	43
Aqua regia, 3:1	X	X
Benzene	X	X
Boric acid	100	38
Bromine gas, dry	100	38
Bromine gas, moist	X	X
Calcium chloride	100	38
Calcium hydroxide, all	100	38
Citric acid, all conc.	100	38
Diesel fuels	100	38
Ethanol	100	38
Ferric chloride	100	38
Formaldehyde, to 50%	100	38
Formic acid	X	X
Glucose	100	38
Green liquor	100	38
Hydrobromic acid	X	X
Hydrochloric acid, dil.	100	38
Hydrochloric acid, 20%	X	X
Hydrofluoric acid, dil.	100	38
Hydrofluoric acid, 30%	X	X
Hydrofluoric acid, vapors	110	43
Hydrogen sulfide, dry	100	38
Hydrogen sulfide, wet	100	38
Iodine	X	X
Lactic acid	X	X
Lard oil	X	X
Lauric acid	X	X
Linseed oil	100	38
Magnesium chloride, 50%	90	32
Mercuric chloride	100	38
Mercuric nitrate	100	38
Methyl alcohol	100	38
Methyl sulfate	X	X
Methylene chloride	X	X
Mineral oil	100	38
Nitric acid	X	X
Oil vegetable	100	38
Oleum	X	X
Oxalic acid, all conc.	100	38

(continued)

TABLE 4.9 *Continued*

Chemical	Maximum Temperature	
	°F	°C
Perchloric acid	X	X
Petroleum oils, sour	100	38
Phenol	X	X
Phosphoric acid	X	X
Potassium chloride, 30%	100	38
Potassium hydroxide, 50%	100	38
Propylene glycol	100	38
Sodium chloride	110	43
Sodium hydroxide, to 50%	100	38
Sulfur dioxide, wet	100	38
Sulfuric acid	X	X
Water, demineralized	100	38
Water, distilled	100	38
Water, salt	130	54
Water, sea	90	32
Water, sewage	100	38
White liquor	100	38
Wines	100	38
Xylene	X	X

The chemicals listed are in the pure state or in a saturated solution unless otherwise indicated. Compatibility is shown to the maximum allowable temperature for which data is available. Incompatibility is shown by an X. A blank space indicates that the data is unavailable.

Source: From P.A. Schweitzer. 2004. *Corrosion Resistance Tables*, Vols. 1–4, 5th ed., New York: Marcel Dekker.

Urethane finds limited service in weak acid solutions and cannot be used in concentrated acids. Urethanes are not resistant to steam or caustics, but they are resistant to deteriorating effects of being immersed in water. Refer to Table 4.11 for the compatibility of urethane with selected corrodents and Reference [1] for a more comprehensive listing.

It is possible to apply uniform coatings or films of urethane to a variety of substrate materials including glass, metal, wood, fabric, and paper. Urethane coatings are often applied to the interior of pipes and tanks.

Filtration units, clarifiers, holding tanks, and treatment sumps constructed of reinforced concrete are widely used in the treatment of municipal, industrial, and thermal generating station wastewater. In many cases, particularly in anaerobic, industrial and thermal generating systems, urethane linings are used to protect the concrete from severe chemical attack and prevent seepage into the concrete of chemicals that can attack the reinforcing steel. These linings provide protection against abrasion and erosion, and act as a waterproofing system to combat leakage of the equipment resulting from concrete movement and shrinkage.

TABLE 4.10

Compatibility of Coal Tar with Selected Corrodents

Chemical	Maximum Temperature	
	°F	°C
Acetaldehyde	X	X
Acetic acid, all conc.		
Acetic acid, vapors		
Acetone	X	X
Aluminum chloride, dry		
Aluminum fluoride		
Ammonium chloride, all		
Ammonium hydroxide, 25%		
Aqua regia, 3:1	X	X
Benzene	X	X
Boric acid		
Bromine gas, dry	X	X
Bromine gas, moist	X	X
Calcium chloride		
Calcium hydroxide, all conc.		
Citric acid, all conc.		
Diesel fuels		
Ethanol		
Ferric chloride		
Formaldehyde, to 50%	X	X
Formic acid	X	X
Glucose		
Green liquor	X	X
Hydrobromic acid	X	X
Hydrochloric acid, dil.	X	X
Hydrochloric acid, 20%	X	X
Hydrofluoric acid, dil.	X	X
Hydrofluoric acid, 30%	X	X
Hydrofluoric acid, vapors		
Hydrogen sulfide, dry		
Hydrogen sulfide, wet		
Iodine	X	X
Lactic acid	X	X
Lard oil	X	X
Lauric acid	X	X
Linseed oil		
Magnesium chloride, 50%		
Mercuric chloride		
Mercuric nitrate		
Methyl alcohol		
Methyl sulfate		
Methylene chloride	X	X
Mineral oil		
Nitric acid	X	X
Oil vegetable		
Oleum	X	X

(continued)

TABLE 4.10 *Continued*

	Maximum Temperature	
Chemical	°F	°C
Oxalic acid, all conc.		
Perchloric acid	X	X
Petroleum oils, sour		
Phenol	X	X
Phosphoric acid	X	X
Potassium chloride, 30%		
Potassium hydroxide, 50%		
Propylene glycol		
Sodium chloride		
Sodium hydroxide, to 50%		
Sulfur dioxide, wet		
Sulfuric acid	X	X
Water, demineralized	90	32
Water, distilled		
Water, salt		
Water, sea	90	32
Water, sewage	90	32
White liquor		
Wines		
Xylene	X	X

The chemicals listed are in the pure state or in a saturated solution unless otherwise indicated. Compatibility is shown to the maximum allowable temperature for which data is available. Incompatibility is shown by an X. A blank space indicates that the data is unavailable.

Source: From P.A. Schweitzer. 2004. *Corrosion Resistance Tables*, Vols. 1–4, 5th ed., New York: Marcel Dekker.

4.10.9 Neoprene

Neoprene is one of the oldest and most versatile of the synthetic rubbers. Chemically, it is polychloroprene. Its basic unit is a chlorinated butadiene whose formula is:

$$CH_2-\underset{\underset{Cl}{|}}{C}-CH-CH_2 \cdot$$

The raw material is acetylene, which makes this product more expensive than some of the other elastomeric materials.

As with other lining materials, neoprene is available in a variety of formulations. Depending on the compounding procedure, material can be produced to impart specific properties to meet application needs.

TABLE 4.11

Compatibility of Urethanes with Selected Corrodents

Chemical	Maximum Temperature	
	°F	°C
Acetaldehyde	X	X
Acetic acid, all conc.	90	32
Acetic acid, vapors	90	32
Acetone	90	32
Aluminum chloride, dry		
Aluminum fluoride		
Ammonium chloride, all	90	32
Ammonium hydroxide, 25%	90	32
Aqua regia, 3:1	X	X
Benzene	X	X
Boric acid	90	32
Bromine gas, dry		
Bromine gas, moist		
Calcium chloride	80	27
Calcium hydroxide, all	90	32
Citric acid, all conc.		
Diesel fuels		
Ethanol	90	32
Ferric chloride	90	32
Formaldehyde, to 50%		
Formic acid	X	X
Glucose	X	X
Green liquor		
Hydrobromic acid		
Hydrochloric acid, dil.	X	X
Hydrochloric acid, 20%	X	X
Hydrofluoric acid, dil.		
Hydrofluoric acid, 30%		
Hydrofluoric acid vapors		
Hydrogen sulfide, dry		
Hydrogen sulfide, wet		
Iodine		
Lactic acid		
Lard oil	90	32
Lauric acid		
Linseed oil	90	32
Magnesium chloride, 50%	90	32
Mercuric chloride		
Mercuric nitrate		
Methyl alcohol	90	32
Methyl sulfate		
Methylene chloride	X	X
Mineral oil	90	32
Nitric acid	X	X
Oil vegetable		

(continued)

TABLE 4.11 *Continued*

	Maximum Temperature	
Chemical	°F	°C
Oleum	X	X
Oxalic acid, all conc.		
Perchloric acid	X	X
Petroleum oils, sour		
Phenol	X	X
Phosphoric acid		
Potassium chloride, 30%	90	32
Potassium hydroxide, 50%	90	32
Propylene glycol		
Sodium chloride	80	27
Sodium hydroxide, to 50%	90	32
Sulfur dioxide, wet		
Sulfuric acid, 10%	90	32
Water, demineralized		
Water, distilled	90	32
Water, salt	X	X
Water, sea	80	27
Water, sewage		
White liquor		
Wines	X	X
Xylene	X	X

The chemicals listed are in the pure state or in a saturated solution unless otherwise indicated. Compatibility is shown to the maximum allowable temperature for which data is available. Incompatibility is shown by an X. A blank space indicates that the data is unavailable.

Source: From P.A. Schweitzer. 2004. *Corrosion Resistance Tables*, Vols. 1–4, 5th ed., New York: Marcel Dekker.

Neoprene is also available in a variety of forms. In addition to a neoprene latex that is similar to natural rubber latex, neoprene is produced in fluid form as either a compounded latex dispersion or solvent solution. After these materials have solidified or cured, they have the same physical and mechanical properties as the solid or cellular forms of neoprene.

Neoprene solvent solutions are prepared by dissolving neoprene in standard rubber solvents. These solutions can be formulated in a range of viscosities suitable for application by brush, spray, or roller. Major areas of application include linings for storage tanks, industrial equipment, and chemical processing equipment. These coatings protect the vessels from corrosion by acids, salts, oils, alkalies, and most hydrocarbons.

Neoprene possesses excellent resistance to attack from solvents, waxes, fats, oils, greases, and many other petroleum-based products. It also exhibits excellent service in contact with aliphatic compounds (methyl and ethyl alcohols, ethylene glycols, etc.) and aliphatic hydrocarbons. It is also resistant to dilute mineral acids, inorganic salt solutions, and alkalies.

Chlorinated and aromatic hydrocarbons, organic esters, aromatic hydroxy compounds, and certain ketones will attack neoprene. Refer to Table 4.12 for the compatibility of neoprene with selected corrodents and Reference [1] for a more complete listing.

4.10.10 Polysulfide Rubber

Polysulfide rubbers are manufactured by combining ethylene ($CH_2=CH_2$) with an alkaline polysulfide. Morton Thiokol, Inc. markets a series of liquid polysulfides that can be oxidized to rubbers.

The polysulfide rubbers possess outstanding resistance to solvents. They exhibit excellent resistance to oils, gasoline and aliphatic and aromatic hydrocarbon solvents, very good resistance to water, good alkali resistance, and fair acid resistance. Contact with strong concentrated inorganic acids such as sulfuric, nitric, or hydrochloric, should be avoided. Refer to Table 4.13 for the compatibility of polysulfides with selected corrodents and Reference [1] for a more comprehensive listing.

4.10.11 Hypalon

Chlorosulfonated polyethylene synthetic rubber is manufactured by DuPont under the trade name Hypalon. In many respects, it is similar to neoprene but it does possess some advantages over neoprene in certain types of service. It has better heat and ozone resistance and better chemical resistance.

Hypalon has a broad range of service temperatures with excellent thermal properties. General purpose compounds can operate continuously at temperatures of 248–275°F (120–135°C). Special compounds can be formulated that can be used intermittently up to 302°F (150°C). On the low temperature side, conventional compounds can be used continuously down to 0 to −20° F (−18 to −28°C).

When properly compounded, Hypalon is resistant to attack by hydrocarbon oils and fuels, even at elevated temperatures. It is also resistant to such oxidizing chemicals as sodium hypochlorite, sodium peroxide, ferric chlorides, and sulfuric, chromic and hydrofluoric acids. Concentrated hydrochloric acid (37%) at elevated temperatures (above 158°F (70°C)) will attack hypalon, but can be handled with no adverse effects at all concentrations below that temperature. Nitric acid up to 60% concentration at room temperature can also be handled without adverse effect. Hypalon is also resistant to salt solutions, alcohols, and both weak and concentrated alkalies. Long-term contact with water has little effect on Hypalon.

Hypalon has poor resistance to aliphatic, aromatic, and chlorinated hydrocarbons, aldehydes, and ketones. Refer to Table 4.14 for the

TABLE 4.12

Compatibility of Neoprene with Selected Corrodents

Chemical	Maximum Temperature	
	°F	°C
Acetaldehyde	200	93
Acetamide	200	93
Acetic acid, 10%	160	71
Acetic acid, 50%	160	71
Acetic acid, 80%	160	71
Acetic acid, glacial	X	X
Acetic anhydride	X	X
Acetone	X	X
Acetyl chloride	X	X
Acrylic acid	X	X
Acrylonitrile	140	60
Adipic acid	160	71
Allyl alcohol	120	49
Allyl chloride	X	X
Alum	200	93
Aluminum acetate		
Aluminum chloride, aq.	150	66
Aluminum chloride, dry		
Aluminum fluoride	200	93
Aluminum hydroxide	180	82
Aluminum nitrate	200	93
Aluminum sulfate	200	93
Ammonia gas	140	60
Ammonium bifluoride	X	X
Ammonium carbonate	200	93
Ammonium chloride, 10%	150	66
Ammonium chloride, 50%	150	66
Ammonium chloride, sat.	150	66
Ammonium fluoride, 10%	200	93
Ammonium fluoride, 25%	200	93
Ammonium hydroxide, 25%	200	93
Ammonium hydroxide, sat.	200	93
Ammonium nitrate	200	93
Ammonium oxychloride		
Ammonium persulfate	200	93
Ammonium phosphate	150	66
Ammonium sulfate, 10–40%	150	66
Ammonium sulfide	160	71
Ammonium sulfite		
Amyl acetate	X	X
Amyl alcohol	200	93
Amyl chloride	X	X
Aniline	X	X
Antimony trichloride	140	60
Aqua regia, 3:1	X	X
Barium carbonate	150	66

(continued)

TABLE 4.12 *Continued*

Chemical	Maximum Temperature	
	°F	°C
Barium chloride	150	66
Barium hydroxide	230	110
Barium sulfate	200	93
Barium sulfide	200	93
Benzaldehyde	X	X
Benzene	X	X
Benzene sulfonic acid, 10%	100	38
Benzoic acid	150	66
Benzyl alcohol	X	X
Benzyl chloride	X	X
Borax	200	93
Boric acid	150	66
Bromine gas, dry	X	X
Bromine gas, moist	X	X
Bromine, liquid	X	X
Butadiene	140	60
Butyl acetate	60	16
Butyl alcohol	200	93
Butyl phthalate		
n-Butylamine		
Butyric acid	X	X
Calcium bisulfide		
Calcium bisulfite	X	X
Calcium carbonate	200	93
Calcium chlorate	200	93
Calcium chloride	150	66
Calcium hydroxide, 10%	230	110
Calcium hydroxide, sat.	230	110
Calcium hypochlorite	X	X
Calcium nitrate	150	66
Calcium oxide	200	93
Calcium sulfate	150	66
Caprylic acid		
Carbon bisulfide	X	X
Carbon dioxide, dry	200	93
Carbon dioxide, wet	200	93
Carbon disulfide	X	X
Carbon monoxide	X	X
Carbon tetrachloride	X	X
Carbonic acid	150	66
Cellosolve	X	X
Chlorine gas, dry	X	X
Chlorine gas, wet	X	X
Chlorine, liquid	X	X
Chloroacetic acid	X	X
Chloroacetic acid, 50% water	X	X
Chlorobenzene	X	X

(continued)

TABLE 4.12 *Continued*

Chemical	Maximum Temperature	
	°F	°C
Chloroform	X	X
Chlorosulfonic acid	X	X
Chromic acid, 10%	140	60
Chromic acid, 50%	100	38
Chromyl chloride		
Citric acid, 15%	150	66
Citric acid, conc.	150	66
Copper acetate	160	71
Copper carbonate		
Copper chloride	200	93
Copper cyanide	160	71
Copper sulfate	200	93
Cresol	X	X
Cupric chloride, 5%	200	93
Cupric chloride, 50%	160	71
Cyclohexane	X	X
Cyclohexanol	X	X
Dichloroacetic acid	X	X
Dichloroethane (ethylene dichloride)	X	X
Ethylene glycol	100	38
Ferric chloride	160	71
Ferric chloride, 50% in water	160	71
Ferric nitrate, 10–50%	200	93
Ferrous chloride	90	32
Ferrous nitrate	200	93
Fluorine gas, dry	X	X
Fluorine gas, moist	X	X
Hydrobromic acid, 20%	X	X
Hydrobromic acid, 50%	X	X
Hydrobromic acid, dil.	X	X
Hydrochloric acid, 20%	X	X
Hydrochloric acid, 38%	X	X
Hydrocyanic acid, 10%	X	X
Hydrofluoric acid, 30%	X	X
Hydrofluoric acid, 70%	X	X
Hydrofluoric acid, 100%	X	X
Hypochlorous acid	X	X
Iodine solution, 10%	80	27
Ketones, general	X	X
Lactic acid, 25%	140	60
Lactic acid, conc.	90	32
Magnesium chloride	200	93
Malic acid		
Manganese chloride	200	93
Methyl chloride	X	X
Methyl ethyl ketone	X	X
Methyl isobutyl ketone	X	X

(continued)

TABLE 4.12   *Continued*

Chemical	Maximum Temperature	
	°F	°C
Muriatic acid	X	X
Nitric acid, 5%	X	X
Nitric acid, 20%	X	X
Nitric acid, 70%	X	X
Nitric acid, anhydrous	X	X
Nitrous acid, conc.	X	X
Oleum	X	X
Perchloric acid, 10%		
Perchloric acid, 70%	X	X
Phenol	X	X
Phosphoric acid, 50–80%	150	66
Picric acid	200	93
Potassium bromide, 30%	160	71
Salicylic acid		
Silver bromide, 10%		
Sodium carbonate	200	93
Sodium chloride	200	93
Sodium hydroxide, 10%	230	110
Sodium hydroxide, 50%	230	110
Sodium hydroxide, conc.	230	110
Sodium hypochlorite, 20%	X	X
Sodium hypochlorite, conc.	X	X
Sodium sulfide, to 50%	200	93
Stannic chloride	200	93
Stannous chloride	X	X
Sulfuric acid, 10%	150	66
Sulfuric acid, 50%	100	38
Sulfuric acid, 70%	X	X
Sulfuric acid, 90%	X	X
Sulfuric acid, 98%	X	X
Sulfuric acid, 100%	X	X
Sulfuric acid, fuming	X	X
Sulfurous acid	100	38
Thionyl chloride	X	X
Toluene	X	X
Trichloroacetic acid	X	X
White liquor	140	60
Zinc chloride	160	71

The chemicals listed are in the pure state or in a saturated solution unless otherwise indicated. Compatibility is shown to the maximum allowable temperature for which data is available. Incompatibility is shown by an X. A blank space indicates that the data is unavailable.

Source: From P.A. Schweitzer. 2004. *Corrosion Resistance Tables*, Vols. 1–4, 5th ed., New York: Marcel Dekker.

TABLE 4.13

Compatibility of Polysulfides with Selected Corrodents

Chemical	Maximum Temperature	
	°F	°C
Acetaldehyde		
Acetic acid, all conc.	80	27
Acetic acid vapors	90	32
Acetone	80	27
Aluminum chloride, dry		
Aluminum fluoride		
Ammonium chloride, all	140	66
Ammonium hydroxide, 25%	X	X
Aqua regia, 3:1		
Benzene	X	X
Boric acid		
Bromine gas, dry		
Bromine gas, moist		
Calcium chloride	150	66
Calcium hydroxide, all	X	X
Citric acid, all conc.	X	X
Diesel fuels	80	27
Ethanol	150	66
Ferric chloride		
Formaldehyde, to 50%	80	27
Formic acid		
Glucose		
Green liquor		
Hydrobromic acid		
Hydrochloric acid, dil.	X	X
Hydrochloric acid, 20%	X	X
Hydrofluoric acid, dil.	X	X
Hydrofluoric acid, 30%	X	X
Hydrofluoric acid, vapors		
Hydrogen sulfide, dry		
Hydrogen sulfide, wet		
Iodine		
Lactic acid	X	X
Lard oil		
Lauric acid		
Linseed oil	150	66
Magnesium chloride, 50%		
Mercuric chloride		
Mercuric nitrate		
Methyl alcohol	80	27
Methyl sulfate		
Methylene chloride		
Mineral oil	80	27
Nitric acid	X	X
Oil vegetable	X	X
Oleum	X	X

(continued)

TABLE 4.13 *Continued*

Chemical	Maximum Temperature	
	°F	°C
Oxalic acid, all conc.	X	X
Perchloric acid		
Petroleum oils, sour		
Phenol	X	X
Phosphoric acid	X	X
Potassium chloride, 30%		
Potassium hydroxide, 50%	80	27
Propylene glycol		
Sodium chloride	80	27
Sodium hydroxide, to 50%	X	X
Sulfur dioxide, wet		
Sulfuric acid	X	X
Water, demineralized	80	27
Water, distilled	80	27
Water salt	80	27
Water, sea	80	27
Water, sewage	80	27
White liquor		
Wines	X	X
Xylene	80	27

The chemicals listed are in the pure state or in a saturated solution unless otherwise indicated. Compatibility is shown to the maximum allowable temperature for which data is available. Incompatibility is shown by an X. A blank space indicates that the data is unavailable.

Source: From P.A. Schweitzer. 2004. *Corrosion Resistance Tables*, Vols. 1–4, 5th ed., New York: Marcel Dekker.

compatibility of Hypalon with selected corrodents and Reference [1] for a more complete listing.

Hypalon finds useful applications in many industries and many fields. Because of its outstanding resistance to oxidizing acids, it is used to line railroad tank cars and other tanks containing oxidizing chemicals and acids.

4.10.12 Plastisols

PVC plastisols are dispersions of PVC and/or PVC copolymer resins in compatible plasticizers. These liquids can vary in viscosity from thin, milk-like fluids to heavy pastes having the consistency of molasses. The lowest viscosity products are generally used for spray coating. PVC plastisol coatings have limited adhesion and require primers; the coatings must be heat cured.

TABLE 4.14

Compatibility of Hypalon with Selected Corrodents

| | Maximum Temperature | |
Chemical	°F	°C
Acetaldehyde	60	16
Acetamide	X	X
Acetic acid, 10%	200	93
Acetic acid, 50%	200	93
Acetic acid, 80%	200	93
Acetic acid, glacial	X	X
Acetic anhydride	200	93
Acetone	X	X
Acetyl chloride	X	X
Acrylonitrile	140	60
Adipic acid	140	60
Allyl alcohol	200	93
Aluminum fluoride	200	93
Aluminum hydroxide	200	93
Aluminum nitrate	200	93
Aluminum sulfate	180	82
Ammonia carbonate	140	60
Ammonia gas	90	32
Ammonium chloride, 10%	190	88
Ammonium chloride, 10%	200	93
Ammonium chloride, 50%	190	88
Ammonium fluoride, sat.	190	88
Ammonium hydroxide, 25%	200	93
Ammonium hydroxide, sat.	200	93
Ammonium nitrate	200	93
Ammonium persulfate	80	27
Ammonium phosphate	140	60
Ammonium sulfate, 10–40%	200	93
Ammonium sulfide	200	93
Amyl acetate	60	16
Amyl alcohol	200	93
Amyl chloride	X	X
Aniline	140	60
Antimony trichloride	140	60
Barium carbonate	200	93
Barium chloride	200	93
Barium hydroxide	200	93
Barium sulfate	200	93
Barium sulfide	200	93
Benzaldehyde	X	X
Benzene	X	X
Benzene sulfonic acid, 10%	X	X
Benzoic acid	200	93
Benzyl alcohol	140	60
Benzyl chloride	X	X
Borax	200	93

(continued)

TABLE 4.14 *Continued*

Chemical	Maximum Temperature	
	°F	°C
Boric acid	200	93
Bromine gas, dry	60	16
Bromine gas, moist	60	16
Bromine liquid	60	16
Butadiene	X	X
Butyl acetate	60	16
Butyl alcohol	200	93
Butyric acid	X	X
Calcium bisulfite	200	93
Calcium carbonate	90	32
Calcium chlorate	90	32
Calcium chloride	200	93
Calcium hydroxide, 10%	200	93
Calcium hydroxide, sat.	200	93
Calcium hypochlorite	200	93
Calcium nitrate	100	38
Calcium oxide	200	93
Calcium sulfate	200	93
Caprylic acid	X	X
Carbon dioxide, dry	200	93
Carbon dioxide, wet	200	93
Carbon disulfide	200	93
Carbon monoxide	X	X
Carbon tetrachloride	200	93
Carbonic acid	X	X
Chloracetic acid	X	X
Chlorine gas, dry	X	X
Chlorine gas, wet	90	32
Chlorobenzene	X	X
Chloroform	X	X
Chlorosulfonic acid	X	X
Chromic acid, 10%	150	66
Chromic acid, 50%	150	66
Chromyl chloride		
Citric acid, 15%	200	93
Citric acid, conc.	200	93
Copper acetate	X	X
Copper chloride	200	93
Copper cyanide	200	93
Copper sulfate	200	93
Cresol	X	X
Cupric chloride, 5%	200	93
Cupric chloride, 50%	200	93
Cyclohexane	X	X
Cyclohexanol	X	X
Dichloroethane (ethylene dichloride)	X	X

(continued)

TABLE 4.14 *Continued*

Chemical	Maximum Temperature	
	°F	°C
Ethylene glycol	200	93
Ferric chloride	200	93
Ferric chloride, 50% in water	200	93
Ferric nitrate, 10–50%	200	93
Ferrous chloride	200	93
Fluorine gas, dry	140	60
Hydrobromic acid, 20%	100	38
Hydrobromic acid, 50%	100	38
Hydrobromic acid, dil.	90	32
Hydrochloric acid, 20%	160	71
Hydrochloric acid, 38%	140	60
Hydrocyanic acid, 10%	90	32
Hydrofluoric acid, 30%	90	32
Hydrofluoric acid, 70%	90	32
Hydrofluoric acid, 100%	90	32
Hypochlorous acid	X	X
Ketones, general	X	X
Lactic acid, 25%	140	60
Lactic acid, conc.	80	27
Magnesium chloride	200	93
Manganese chloride	180	82
Methyl chloride	X	X
Methyl ethyl ketone	X	X
Methyl isobutyl ketone	X	X
Muriatic acid	140	60
Nitric acid, 5%	100	38
Nitric acid, 20%	100	38
Nitric acid, 70%	X	X
Nitric acid, anhydrous	X	X
Oleum	X	X
Perchloric acid, 10%	100	38
Perchloric acid, 70%	90	32
Phenol	X	X
Phosphoric acid, 50–80%	200	93
Picric acid	80	27
Potassium bromide, 30%	200	93
Sodium carbonate	200	93
Sodium chloride	200	93
Sodium hydroxide, 10%	200	93
Sodium hydroxide, 50%	200	93
Sodium hydroxide, conc.	200	93
Sodium hypochlorite, 20%	200	93
Sodium hypochlorite, conc.		
Sodium sulfide, to 50%	200	93
Stannic chloride	90	32
Stannous chloride	200	93

(continued)

TABLE 4.14 *Continued*

| | Maximum Temperature | |
Chemical	°F	°C
Sulfuric acid, 10%	200	93
Sulfuric acid, 50%	200	93
Sulfuric acid, 70%	160	71
Sulfuric acid, 90%	X	X
Sulfuric acid, 98%	X	X
Sulfuric acid, 100%	X	X
Sulfurous acid	160	71
Toluene	X	X
Zinc chloride	200	93

The chemicals listed are in the pure state or in a saturated solution unless otherwise indicated. Compatibility is shown to the maximum allowable temperature for which data is available. Incompatibility is shown by an X.

Source: From P.A. Schweitzer. 2004. *Corrosion Resistance Tables*, Vols. 1–4, 5th ed., New York: Marcel Dekker.

The viscosity of plastisol is controlled by formulator techniques, and is often kept low by the addition of inactive diluents such as odorless mineral spirits. If more than minor amounts of diluents are used, the product is often referred to as an organsol.

These compounds share common compounding technology. The primary components are the dispersion grade resin, plasticizers, PVC stabilizers (which are common to all PVC), and assorted fillers, pigments, and a wide variety of additives to control properties of the product in storage, during processing, and in the finished state.

Plasticizers are liquids that provide mobility to the plastisol system. They are of primary significance, and they are selected first when formulating. Plasticizers differ in the performance characteristics that they impart to the finished product. The blend of plasticizers will also assist in the control of viscosity and its stability, and in the fusion characteristics of the finished plastisol.

Because plasticized PVC is compounded of a PVC dispersion of high molecular weight vinyl chloride polymers in a suitable liquid plasticizer, formulations can be made for special applications. By selective compounding, both physical and corrosion-resistant properties can be modified. For certain applications, this feature can be advantageous.

Two types of PVC resin are produced: normal impact (type 1) and high impact (type 2). Type 1 is unplasticized PVC having normal impact and optimum chemical resistance. Type 2 is a plasticized PVC and has optimum impact strength and reduced chemical resistance. Plastisol PVC is the latter type.

Type 1 PVC resists attack by most acids and strong alkalies, gasoline, kerosene, aliphatic alcohols, and hydrocarbons. It is particularly useful in the handling of hydrochloric acid. The chemical resistance of type 2 PVC (plastisol) to oxidizing and highly alkaline chemicals is reduced. Plastisol can be attacked by aromatics, chlorinated organic compounds, and lacquer solvents.

In addition to handling highly corrosive and abrasive chemicals, many applications have also been found in marine environments. Table 4.15 lists the compatibility of plastisols with selected corrodents and Reference [1] provides additional listings.

Vinyl plastisols are popular for use as an acid-resisting coating. Plastisol has a maximum operating temperature of 140°F (60°C).

4.10.13 Perfluoroalkoxy

Perfluoroalkoxy (PFA) is manufactured by DuPont. It is not degraded by systems commonly encountered in chemical processes. It is inert to strong mineral acids, inorganic bases, inorganic oxidizing agents, salt solutions, and such organic compounds as organic acids, aldehydes, anhydrides, aliphatic hydrocarbons, alcohols, esters, ethers, chlorocarbons, fluorocarbons, and mixtures of the above. Refer to Table 4.16 for the compatibility of PFA with selected corrodents and Reference [1] for additional listings.

PFA will be attacked by certain halogen complexes containing fluorine. These include chlorine trifluoride, bromine trifluoride, iodine pentafluoride, and fluorine itself. It is also attacked by such metals as sodium or potassium, particularly in their molten states.

Standard lining thickness is nominally 0.040 in. on interior and wetted surfaces. When abrasion is a problem, a thickness of 0.090 in. is available. Coatings applied on carbon steel or stainless steel have a continuous service temperature of between −60°F (−51°C) and 400°F (204°C).

A primer is required prior to applying the coating. If damaged, the coating cannot be repaired. Heat is required to cure the coating.

Applications are used to provide corrosion protection, nonstick surfaces, and purity protection of chemicals being handled.

4.10.14 Fluorinated Ethylene Propylene

Fluorinated Ethylene Propylene (FEP) is a fluorinated thermoplast but is less expensive than PTFE (Teflon). With few exceptions, FEP exhibits the same corrosion resistance as PTFE but at a lower temperature. It is resistant to practically all chemicals, the exception being extremely potent oxidizers such as chlorine trifluoride, and related compounds. Some chemicals in high concentrations will attack FEP when at or near the service temperature limit. Refer to Table 4.17 for the compatibility of FEP with selected corrodents and Reference [1] for additional listings.

TABLE 4.15

Compatibility of Plastisols with Selected Corrodents

Chemical	Maximum Temperature	
	°F	°C
Acetaldehyde	X	X
Acetamide	X	X
Acetic acid, 10%	100	38
Acetic acid, 50%	90	32
Acetic acid, 80%	X	X
Acetic acid, glacial	X	X
Acetic anhydride	X	X
Acetone	X	X
Acetyl chloride	X	X
Acrylic acid	X	X
Acrylonitrile	X	X
Adipic acid	140	60
Allyl alcohol	90	32
Allyl chloride	X	X
Alum	140	60
Aluminum acetate	100	38
Aluminum chloride, aq.	140	60
Aluminum fluoride	140	60
Aluminum hydroxide	140	60
Aluminum nitrate	140	60
Aluminum oxychloride	140	60
Aluminum sulfate	140	60
Ammonia gas	140	60
Ammonium bifluoride	90	32
Ammonium carbonate	140	60
Ammonium chloride, 10%	140	60
Ammonium chloride, 50%	140	60
Ammonium chloride, sat.	140	60
Ammonium fluoride, 10%	90	32
Ammonium fluoride, 25%	90	32
Ammonium hydroxide, 25%	140	60
Ammonium hydroxide, sat.	140	60
Ammonium nitrate	140	60
Ammonium persulfate	140	60
Ammonium phosphate	140	60
Ammonium sulfate, 10–40%	140	60
Ammonium sulfide	140	60
Amyl acetate	X	X
Amyl alcohol	X	X
Amyl chloride	X	X
Aniline	X	X
Antimony trichloride	140	60
Aqua regia, 3:1	X	X
Barium carbonate	140	60
Barium chloride	140	60
Barium hydroxide	140	60

(continued)

TABLE 4.15 *Continued*

Chemical	Maximum Temperature	
	°F	°C
Barium sulfate	140	60
Barium sulfide	140	60
Benzaldehyde	X	X
Benzene	X	X
Benzene sulfonic acid, 10%	140	60
Benzoic acid	140	60
Benzyl alcohol	X	X
Borax	140	60
Boric acid	140	60
Bromine gas, dry	X	X
Bromine gas, moist	X	X
Bromine liquid	X	X
Butadiene	60	16
Butyl acetate	X	X
Butyl alcohol	X	X
n-Butylamine	X	X
Butyric acid	X	X
Calcium bisulfide	140	60
Calcium bisulfide	X	X
Calcium bisulfite	140	60
Calcium carbonate	140	60
Calcium chlorate	140	60
Calcium chloride	140	60
Calcium hydroxide, 10%	140	60
Calcium hydroxide, sat.	140	60
Calcium hypochlorite	140	60
Calcium nitrate	140	60
Calcium oxide	140	60
Calcium sulfate	140	60
Carbon dioxide, dry	140	60
Carbon dioxide, wet	140	60
Carbon disulfide	X	X
Carbon monoxide	140	60
Carbon tetrachloride	X	X
Carbonic acid	140	60
Cellosolve	X	X
Chloracetic acid	105	40
Chlorine gas, dry	140	60
Chlorine gas, wet	X	X
Chlorine, liquid	X	X
Chlorobenzene	X	X
Chloroform	X	X
Chlorosulfonic acid	60	16
Chromic acid, 10%	140	60
Chromic acid, 50%	X	X
Citric acid, 15%	140	60
Citric acid, conc.	140	60

(continued)

TABLE 4.15 *Continued*

Chemical	Maximum Temperature	
	°F	°C
Copper carbonate	140	60
Copper chloride	140	60
Copper cyanide	140	60
Copper sulfate	140	60
Cresol	X	X
Cyclohexanol	X	X
Dichloroacetic acid	120	49
Dichloroethane (ethylene dichloride)	X	X
Ethylene glycol	140	60
Ferric chloride	140	60
Ferric nitrate, 10–50%	140	60
Ferrous chloride	140	60
Ferrous nitrate	140	60
Fluorine gas, dry	X	X
Fluorine gas, moist	X	X
Hydrobromic acid, 20%	140	60
Hydrobromic acid, 50%	140	60
Hydrobromic acid, dil.	140	60
Hydrochloric acid, 20%	140	60
Hydrochloric acid, 38%	140	60
Hydrocyanic acid, 10%	140	60
Hydrofluoric acid, 30%	120	49
Hydrofluoric acid, 70%	68	20
Hypochlorous acid	140	60
Ketones, general	X	X
Lactic acid, 25%	140	60
Lactic acid, conc.	80	27
Magnesium chloride	140	60
Malic acid	140	60
Methyl chloride	X	X
Methyl ethyl ketone	X	X
Methyl isobutyl ketone	X	X
Muriatic acid	140	60
Nitric acid, 5%	100	38
Nitric acid, 20%	140	60
Nitric acid, 70%	70	23
Nitric acid, anhydrous	X	X
Nitrous acid, conc.	60	16
Oleum	X	X
Perchloric acid, 10%	60	16
Perchloric acid, 70%	60	16
Phenol	X	X
Phosphoric acid, 50–80%	140	60
Picric acid	X	X
Potassium bromide, 30%	140	60
Salicylic acid	X	X

(continued)

TABLE 4.15 *Continued*

Chemical	Maximum Temperature	
	°F	°C
Silver bromide, 10%	105	40
Sodium carbonate	140	60
Sodium chloride	140	60
Sodium hydroxide, 10%	140	60
Sodium hydroxide, 50%	140	60
Sodium hydroxide, conc.	140	60
Sodium hypochlorite, 20%	140	60
Sodium hypochlorite, conc.	140	60
Sodium sulfide, to 50%	140	60
Stannic chloride	140	60
Stannous chloride	140	60
Sulfuric acid, 10%	140	60
Sulfuric acid, 50%	140	60
Sulfuric acid, 70%	140	60
Sulfuric acid, 90%	X	X
Sulfuric acid, 98%	X	X
Sulfuric acid, 100%	X	X
Sulfuric acid, fuming	X	X
Sulfurous acid	140	60
Thionyl chloride	X	X
Toluene	X	X
Trichloroacetic acid	X	X
White liquor	140	60
Zinc chloride	140	60

The chemicals listed are in the pure state or in a saturated solution unless otherwise indicated. Compatibility is shown to the maximum allowable temperature for which data is available. Incompatibility is shown by an X. A blank space indicates that the data is unavailable.

Source: From P.A. Schweitzer. 2004. *Corrosion Resistance Tables*, Vols. 1–4, 5th ed., New York: Marcel Dekker.

Coating thickness range from 0.010 to 0.060 in. with a maximum service temperature of 390°F (199°C). Damage to these coatings cannot be repaired. The coating is a fusion from a water or solvent dispersion and requires a heat cure temperature of 500–660°F (260–315°C). Application is by spray. Previously glass-linked tanks can be refurbished with this lining.

4.10.15 PTFE (Teflon)

PTFE coatings are spray applied as water or solvent dispersions. They require heat curing at approximately 750°F (399°C). Their maximum service temperature is 500°F (260°C).

PTFE is chemically inert in the presence of most corrodents. There are very few chemicals that will attack PTFE within normal use temperatures.

TABLE 4.16

Compatibility of PFA with Selected Corrodents

Chemical	Maximum Temperature	
	°F	°C
Acetaldehyde	450	232
Acetamide	450	232
Acetic acid, 10%	450	232
Acetic acid, 50%	450	232
Acetic acid, 80%	450	232
Acetic acid, glacial	450	232
Acetic anhydride	450	232
Acetone	450	232
Acetyl chloride	450	232
Acrylonitrile	450	232
Adipic acid	450	232
Allyl alcohol	450	232
Allyl chloride	450	232
Alum	450	232
Aluminum chloride, aq.	450	232
Aluminum fluoride	450	232
Aluminum hydroxide	450	232
Aluminum nitrate	450	232
Aluminum oxychloride	450	232
Aluminum sulfate	450	232
Ammonia gas[a]	450	232
Ammonium bifluoride[a]	450	232
Ammonium carbonate	450	232
Ammonium chloride, 10%	450	232
Ammonium chloride, 50%	450	232
Ammonium chloride, sat.	450	232
Ammonium fluoride, 10%[a]	450	232
Ammonium fluoride, 25%[a]	450	232
Ammonium hydroxide, 25%	450	232
Ammonium hydroxide, sat.	450	232
Ammonium nitrate	450	232
Ammonium persulfate	450	232
Ammonium phosphate	450	232
Ammonium sulfate, 10–40%	450	232
Ammonium sulfide	450	232
Amyl acetate	450	232
Amyl alcohol	450	232
Amyl chloride	450	232
Aniline[b]	450	232
Antimony trichloride	450	232
Aqua regia, 3:1	450	232
Barium carbonate	450	232
Barium chloride	450	232
Barium hydroxide	450	232
Barium sulfate	450	232
Barium sulfide	450	232

(continued)

TABLE 4.16 *Continued*

Chemical	Maximum Temperature	
	°F	°C
Benzaldehyde[b]	450	232
Benzene sulfonic acid, 10%	450	232
Benzene[a]	450	232
Benzoic acid	450	232
Benzyl alcohol[b]	450	232
Benzyl chloride[a]	450	232
Borax	450	232
Boric acid	450	232
Bromine gas, dry[a]	450	232
Bromine, liquid[a,b]	450	232
Butadiene[a]	450	232
Butyl acetate	450	232
Butyl alcohol	450	232
Butyl phthalate	450	232
n-Butylamine[b]	450	232
Butyric acid	450	232
Calcium bisulfide	450	232
Calcium bisulfite	450	232
Calcium carbonate	450	232
Calcium chlorate	450	232
Calcium chloride	450	232
Calcium hydroxide, 10%	450	232
Calcium hydroxide, sat.	450	232
Calcium hypochlorite	450	232
Calcium nitrate	450	232
Calcium oxide	450	232
Calcium sulfate	450	232
Caprylic acid	450	232
Carbon bisulfide[a]	450	232
Carbon dioxide, dry	450	232
Carbon dioxide, wet	450	232
Carbon disulfide[a]	450	232
Carbon monoxide	450	232
Carbon tetrachloride[a,b,c]	450	232
Carbonic acid	450	232
Chloracetic acid	450	232
Chloracetic acid, 50% water	450	232
Chlorine gas, dry	X	X
Chlorine gas, wet[a]	450	232
Chlorine, liquid[b]	X	X
Chlorobenzene[a]	450	232
Chloroform[a]	450	232
Chlorosulfonic acid[b]	450	232
Chromic acid, 10%	450	232
Chromic acid, 50%[b]	450	232
Chromyl chloride	450	232

(continued)

TABLE 4.16 *Continued*

Chemical	Maximum Temperature	
	°F	°C
Citric acid, 15%	450	232
Citric acid, conc.	450	232
Copper carbonate	450	232
Copper chloride	450	232
Copper cyanide	450	232
Copper sulfate	450	232
Cresol	450	232
Cupric chloride, 5%	450	232
Cupric chloride, 50%	450	232
Cyclohexane	450	232
Cyclohexanol	450	232
Dichloroacetic acid	450	232
Dichloroethane (ethylene dichloride)[a]	450	232
Ethylene glycol	450	232
Ferric chloride	450	232
Ferric chloride, 50% in water[b]	450	232
Ferric nitrate, 10–50%	450	232
Ferrous chloride	450	232
Ferrous nitrate	450	232
Fluorine gas, dry	X	X
Fluorine gas, moist	X	X
Hydrobromic acid, 20%[a,c]	450	232
Hydrobromic acid, 50%[a,c]	450	232
Hydrobromic acid, dil.[a,c]	450	232
Hydrochloric acid, 20%[a,c]	450	232
Hydrochloric acid, 38%[a,c]	450	232
Hydrocyanic acid, 10%	450	232
Hydrofluoric acid, 30%[a]	450	232
Hydrofluoric acid, 70%[a]	450	232
Hydrofluoric acid, 100%[a]	450	232
Hypochlorous acid	450	232
Iodine solution, 10%[a]	450	232
Ketones, general	450	232
Lactic acid, 25%	450	232
Lactic acid, conc.	450	232
Magnesium chloride	450	232
Malic acid	450	232
Methyl chloride[a]	450	232
Methyl ethyl ketone[a]	450	232
Methyl isobutyl ketone[a]	450	232
Muriatic acid[a]	450	232
Nitric acid, 5%[a]	450	232
Nitric acid, 20[a]	450	232
Nitric acid, 70%[a]	450	232
Nitric acid, anhydrous[a]	450	232
Nitrous acid, 10%	450	232

(continued)

TABLE 4.16 *Continued*

Chemical	Maximum Temperature	
	°F	°C
Oleum	450	232
Perchloric acid, 10%	450	232
Perchloric acid, 70%	450	232
Phenol[a]	450	232
Phosphoric acid, 50–80%[b]	450	232
Picric acid	450	232
Potassium bromide, 30%	450	232
Salicylic acid	450	232
Sodium carbonate	450	232
Sodium chloride	450	232
Sodium hydroxide, 10%	450	232
Sodium hydroxide, 50%	450	232
Sodium hydroxide, conc.	450	232
Sodium hypochlorite, 20%	450	232
Sodium hypochlorite, conc.	450	232
Sodium sulfide, to 50%	450	232
Stannic chloride	450	232
Stannous chloride	450	232
Sulfuric acid, 10%	450	232
Sulfuric acid, 50%	450	232
Sulfuric acid, 70%	450	232
Sulfuric acid, 90%	450	232
Sulfuric acid, 98%	450	232
Sulfuric acid, 100%	450	232
Sulfuric acid, fuming[a]	450	232
Sulfurous acid	450	232
Thionyl chloride[a]	450	232
Toluene[a]	450	232
Trichloroacetic acid	450	232
White liquor	450	232
Zinc chloride[b]	450	232

The chemicals listed are in the pure state or in a saturated solution unless otherwise indicated. Compatibility is shown to the maximum allowable temperature for which data is available. Incompatibility is shown by an X. A blank space indicates that the data is unavailable.

[a] Material will permeate.
[b] Material will be absorbed.
[c] Material will cause stress cracking.

Source: From P.A. Schweitzer. 2004. *Corrosion Resistance Tables*, Vols. 1–4, 5th ed., New York: Marcel Dekker.

These reactants are among the most violent oxidizing and reducing agents known. Elemental sodium in contact with fluorocarbons removes fluorine from the polymer molecule. The other alkali metals (potassium, lithium, etc.) react in a similar manner.

TABLE 4.17

Compatibility of FEP with Selected Corrodents

| | Maximum Temperature | |
Chemical	°F	°C
Acetaldehyde	200	93
Acetamide	400	204
Acetic acid, 10%	400	204
Acetic acid, 50%	400	204
Acetic acid, 80%	400	204
Acetic acid, glacial	400	204
Acetic anhydride	400	204
Acetone[a]	400	204
Acetyl chloride	400	204
Acrylic acid	200	93
Acrylonitrile	400	204
Adipic acid	400	204
Allyl alcohol	400	204
Allyl chloride	400	204
Alum	400	204
Aluminum acetate	400	204
Aluminum chloride, aq.	400	204
Aluminum chloride, dry	300	149
Aluminum fluoride[b]	400	204
Aluminum hydroxide	400	204
Aluminum nitrate	400	204
Aluminum oxychloride	400	204
Aluminum sulfate	400	204
Ammonia gas[b]	400	204
Ammonium bifluoride[b]	400	204
Ammonium carbonate	400	204
Ammonium chloride, 10%	400	204
Ammonium chloride, 50%	400	204
Ammonium chloride, sat.	400	204
Ammonium fluoride, 10%[b]	400	204
Aluminum fluoride, 25%[b]	400	204
Ammonium hydroxide, 25%	400	204
Ammonium hydroxide, sat.	400	204
Ammonium nitrate	400	204
Ammonium persulfate	400	204
Ammonium phosphate	400	204
Ammonium sulfate, 10–40%	400	204
Ammonium sulfide	400	204
Ammonium sulfite	400	204
Amyl acetate	400	204
Amyl alcohol	400	204
Amyl chloride	400	204
Aniline[a]	400	204
Antimony trichloride	250	121
Aqua regia, 3:1	400	204
Barium carbonate	400	204
Barium chloride	400	204

(continued)

TABLE 4.17 *Continued*

Chemical	Maximum Temperature	
	°F	°C
Barium hydroxide	400	204
Barium sulfate	400	204
Barium sulfide	400	204
Benzaldehyde[a]	400	204
Benzene[a,b]	400	204
Benzenesulfonic acid, 10%	400	204
Benzoic acid	400	204
Benzyl alcohol	400	204
Benzyl chloride	400	204
Borax	400	204
Boric acid	400	204
Bromine gas, dry[b]	200	93
Bromine gas, moist[b]	200	93
Bromine liquid[a,b]	400	204
Butadiene[b]	400	204
Butyl acetate	400	204
Butyl alcohol	400	204
n-Butylamine[a]	400	204
Butyl phthalate	400	204
Butyric acid	400	204
Calcium bisulfide	400	204
Calcium bisulfide[b]	400	204
Calcium bisulfite	400	204
Calcium carbonate	400	204
Calcium chlorate	400	204
Calcium chloride	400	204
Calcium hydroxide, 10%	400	204
Calcium hydroxide, sat.	400	204
Calcium hypochlorite	400	204
Calcium nitrate	400	204
Calcium oxide	400	204
Calcium sulfate	400	204
Caprylic acid	400	204
Carbon dioxide, dry	400	204
Carbon dioxide, wet	400	204
Carbon disulfide	400	204
Carbon monoxide	400	204
Carbon tetrachloride[a,b,c]	400	204
Carbonic acid	400	204
Cellosolve	400	204
Chloracetic acid	400	204
Chloracetic acid, 50% water	400	204
Chlorine gas, dry	X	X
Chlorine gas, wet[b]	400	204
Chlorine, liquid[a]	400	204
Chlorobenzene[b]	400	204
Chloroform[b]	400	204
Chlorosulfonic acid[a]	400	204

(continued)

TABLE 4.17 *Continued*

Chemical	Maximum Temperature	
	°F	°C
Chromic acid, 10%	400	204
Chromic acid, 50%[a]	400	204
Chromyl chloride	400	204
Citric acid, 15%	400	204
Citric acid, conc.	400	204
Copper acetate	400	204
Copper carbonate	400	204
Copper chloride	400	204
Copper cyanide	400	204
Copper sulfate	400	204
Cresol	400	204
Cupric chloride, 5%	400	204
Cupric chloride, 50%	400	204
Cyclohexane	400	204
Cyclohexanol	400	204
Dichloroacetic acid	400	204
Dichloroethane (ethylene dichloride)[b]	400	204
Ethylene glycol	400	204
Ferric chloride	400	204
Ferric chloride, 50% in water[a]	260	127
Ferric nitrate, 10–50%	260	127
Ferrous chloride	400	204
Ferrous nitrate	400	204
Fluorine gas, dry	200	93
Fluorine gas, moist	X	X
Hydrobromic acid, 20%[b,c]	400	204
Hydrobromic acid. 50%[b,c]	400	204
Hydrobromic acid, dil.	400	204
Hydrochloric acid, 20%[b,c]	400	204
Hydrochloric acid, 38%[b,c]	400	204
Hydrocyanic acid, 10%[b]	400	204
Hydrofluoric acid, 30%[b]	400	204
Hydrofluoric acid, 70%[b]	400	204
Hydrofluoric acid, 100%[b]	400	204
Hypochlorous acid	400	204
Iodine solution, 10%[b]	400	204
Ketones, general	400	204
Lactic acid, 25%	400	204
Lactic acid, conc.	400	204
Magnesium chloride	400	204
Malic acid	400	204
Manganese chloride	300	149
Methyl chloride[b]	400	204
Methyl ethyl ketone[b]	400	204
Methyl isobutyl ketone[b]	400	204
Muriatic acid[b]	400	204
Nitric acid, 5%[b]	400	204

(continued)

TABLE 4.17   *Continued*

Chemical	Maximum Temperature	
	°F	°C
Nitric acid, 20%[b]	400	204
Nitric acid. 70%[b]	400	204
Nitric acid, anhydrous[b]	400	204
Nitrous acid, conc.	400	204
Oleum	400	204
Perchloric acid, 10%	400	204
Perchloric acid, 70%	400	204
Phenol[b]	400	204
Phosphoric acid, 50–80%	400	204
Picric acid	400	204
Potassium bromide, 30%	400	204
Salicylic acid	400	204
Silver bromide, 10%	400	204
Sodium carbonate	400	204
Sodium chloride	400	204
Sodium hydroxide, 10%[a]	400	204
Sodium hydroxide, 50%	400	204
Sodium hydroxide, conc.	400	204
Sodium hypochlorite, 20%	400	204
Sodium hypochlorite, conc.	400	204
Sodium sulfide, to 50%	400	204
Stannic chloride	400	204
Stannous chloride	400	204
Sulfuric acid, 10%	400	204
Sulfuric acid, 50%	400	204
Sulfuric acid, 70%	400	204
Sulfuric acid, 90%	400	204
Sulfuric acid, 98%	400	204
Sulfuric acid, 100%	400	204
Sulfuric acid, fuming[b]	400	204
Sulfurous acid	400	204
Thionyl chloride[b]	400	204
Toluene[b]	400	204
Trichloroacetic acid	400	204
White liquor	400	204
Zinc chloride[c]	400	204

The chemicals listed are in the pure state or in a saturated solution unless otherwise indicated. Compatibility is shown to the maximum allowable temperature for which data is available. Incompatibility is shown by an X. A blank space indicates that the data is unavailable.

[a] Material will be absorbed.
[b] Material will permeate.
[c] Material will cause stress cracking.

Source: From P.A. Schweitzer. 2004. *Corrosion Resistance Tables*, Vols. 1–4, 5th ed., New York: Marcel Dekker.

Fluorine and related compounds (e.g., chlorine trifluoride) are absorbed into the PTFE resin with such intimate contact that the mixture becomes sensitive to a source of ignition, such as impact.

The handling of 80% sodium hydroxide, aluminum chloride, ammonia, and certain amines at high temperature may produce the same effect as elemental sodium. Also, slow oxidative attack can be produced by 70% nitric acid under pressure at 488°F (250°C).

Refer to Table 4.18 for the compatibility of PTFE with selected corrodents and Reference [1] for additional listings.

4.10.16   Tefzel

Tefzel (ETFE) is a trademark of DuPont. ETFE is a modified, partially fluorinated copolymer of ethylene and polytetrafluoroethylene (PTFE). Because it contains more than 75% by weight PTFE it has better resistance to abrasion and cut-through than PTFE, while retaining most of the corrosion-resistant properties.

The typical Tefzel coating thickness is nominally 0.040 in. thick on all interior surfaces and flange faces. Coating thickness from 0.020 to 0.090 in. are available, depending on application requirements and part geometry. For coatings or linings applied on carbon steel or stainless steel, the continuous service temperature range is from −25°F (−32°C) to 225°F (107°C).

Tefzel is inert to strong mineral acids, inorganic bases, halogens, and strong metal salt solutions. Even carboxylic acids, aromatic and aliphatic hydrocarbons, alcohols, ketones, aldehydes, ethers, chlorocarbons, and classic polymer solvents have little effect on Tefzel. Very strong oxidizing acids near their boiling points, such as nitric acid at high concentrations, will effect ETFE to varying degrees as will strong organic bases such as amines and sulfonic acids. Refer to Table 4.19 for the compatibility of ETFE with selected corrodents and Reference [1] for additional listings.

4.10.17   Halar

Halar (ECTFE) is manufactured under the trade name of Halar by Ausimont. Ethylene–chlorotrifluoroethylene is a 1:1 alternating copolymer of ethylene and chlorotrifluoroethylene. This chemical structure gives the polymer a unique combination of properties. It possesses excellent chemical resistance and, of all the fluoropolymers, ECTFE ranks among the best for abrasion resistance.

The resistance to permeation by oxygen, carbon dioxide, chlorine gas, or hydrochloric acid is superior to that of PTFE or FEP, being 10–100 times better. Water absorption is less than 0.1%.

Halar exhibits outstanding chemical resistance. It is virtually unaffected by all corrosive chemicals commonly encountered in industry, including strong mineral and oxidizing acids, alkalies, metal etchants, liquid oxygen, and essentially all organic solvents except hot amines (e.g., aniline

TABLE 4.18

Compatibility of PTFE with Selected Corrodents

| | Maximum Temperature | |
Chemical	°F	°C
Acetaldehyde	450	232
Acetamide	450	232
Acetic acid, 10%	450	232
Acetic acid, 50%	450	232
Acetic acid, 80%	450	232
Acetic acid, glacial	450	232
Acetic anhydride	450	232
Acetone	450	232
Acetyl chloride	450	232
Acrylonitrile	450	232
Adipic acid	450	232
Allyl alcohol	450	232
Allyl chloride	450	232
Alum	450	232
Aluminum chloride, aq.	450	232
Aluminum fluoride	450	232
Aluminum hydroxide	450	232
Aluminum nitrate	450	232
Aluminum oxychloride	450	232
Aluminum sulfate	450	232
Ammonia gas[a]	450	232
Ammonium bifluoride	450	232
Ammonium carbonate	450	232
Ammonium chloride, 10%	450	232
Ammonium chloride, 50%	450	232
Ammonium chloride, sat.	450	232
Ammonium fluoride, 10%	450	232
Ammonium fluoride, 25%	450	232
Ammonium hydroxide, 25%	450	232
Ammonium hydroxide, sat.	450	232
Ammonium nitrate	450	232
Ammonium persulfate	450	232
Ammonium phosphate	450	232
Ammonium sulfate, 10–40%	450	232
Ammonium sulfide	450	232
Amyl acetate	450	232
Amyl alcohol	450	232
Amyl chloride	450	232
Aniline	450	232
Antimony trichloride	450	232
Aqua regia, 3:1	450	232
Barium carbonate	450	232
Barium chloride	450	232
Barium hydroxide	450	232
Barium sulfate	450	232
Barium sulfide	450	232

(continued)

TABLE 4.18 *Continued*

Chemical	Maximum Temperature	
	°F	°C
Benzaldehyde	450	232
Benzene sulfonic acid, 10%	450	232
Benzene[a]	450	232
Benzoic acid	450	232
Benzyl alcohol	450	232
Benzyl chloride	450	232
Borax	450	232
Boric acid	450	232
Bromine gas, dry[a]	450	232
Bromine liquid[a]	450	232
Butadiene[a]	450	232
Butyl acetate	450	232
Butyl alcohol	450	232
n-Butylamine	450	232
Butyl phthalate	450	232
Butyric acid	450	232
Calcium bisulfide	450	232
Calcium bisulfite	450	232
Calcium carbonate	450	232
Calcium chlorate	450	232
Calcium chloride	450	232
Calcium hydroxide, 10%	450	232
Calcium hydroxide, sat.	450	232
Calcium hypochlorite	450	232
Calcium nitrate	450	232
Calcium oxide	450	232
Calcium sulfate	450	232
Caprylic acid	450	232
Carbon dioxide, dry	450	232
Carbon dioxide, wet	450	232
Carbon disulfide	450	232
Carbon disulfide[a]	450	232
Carbon monoxide	450	232
Carbon tetrachloride[b]	450	232
Carbonic acid	450	232
Chloracetic acid	450	232
Chloracetic acid, 50% water	450	232
Chlorine gas, dry	X	X
Chlorine gas, wet[a]	450	232
Chlorine, liquid	X	X
Chlorobenzene[a]	450	232
Chloroform[a]	450	232
Chlorosulfonic acid	450	232
Chromic acid, 10%	450	232
Chromic acid, 50%	450	232
Chromyl chloride	450	232
Citric acid, 15%	450	232

(continued)

TABLE 4.18 *Continued*

Chemical	Maximum Temperature	
	°F	°C
Citric acid, conc.	450	232
Copper carbonate	450	232
Copper chloride	450	232
Copper cyanide, 10%	450	232
Copper sulfate	450	232
Cresol	450	232
Cupric chloride, 5%	450	232
Cupric chloride, 50%	450	232
Cyclohexane	450	232
Cyclohexanol	450	232
Dichloroethane (ethylene dichloride)[a]	450	232
Dichloroethane acid	450	232
Ethylene glycol	450	232
Ferric chloride	450	232
Ferric chloride, 50% in water	450	232
Ferric nitrate, 10–50%	450	232
Ferrous chloride	450	232
Ferrous nitrate	450	232
Fluorine gas, dry	X	X
Fluorine gas, moist	X	X
Hydrobromic acid, 20%[b]	450	232
Hydrobromic acid, 50%[b]	450	232
Hydrobromic acid, dil.[a,b]	450	232
Hydrochloric acid, 20%[b]	450	232
Hydrochloric acid, 38%[b]	450	232
Hydrocyanic acid, 10%[b]	450	232
Hydrofluoric acid, 30%[b]	450	232
Hydrofluoric acid, 70%[b]	450	232
Hydrofluoric acid, 100%[a]	450	232
Hypochlorous acid	450	232
Iodine solution, 10%[a]	450	232
Ketones, general	450	232
Lactic acid, 25%	450	232
Lactic acid, conc.	450	232
Magnesium chloride	450	232
Malic acid	450	232
Methyl chloride[a]	450	232
Methyl ethyl ketone[a]	450	232
Methyl isobutyl ketone[b]	450	232
Muriatic acid[a]	450	232
Nitric acid, 20%[a]	450	232
Nitric acid, 5%[a]	450	232
Nitric acid, 70%[a]	450	232
Nitric acid, anhydrous[a]	450	232
Nitrous acid, 10%	450	232
Oleum	450	232

(continued)

TABLE 4.18 *Continued*

Chemical	Maximum Temperature	
	°F	°C
Perchloric acid, 10%	450	232
Perchloric acid, 70%	450	232
Phenol[a]	450	232
Phosphoric acid, 50–80%	450	232
Picric acid	450	232
Potassium bromide, 30%	450	232
Salicylic acid	450	232
Sodium carbonate	450	232
Sodium chloride	450	232
Sodium hydroxide, 10%	450	232
Sodium hydroxide, 50%	450	232
Sodium hydroxide, conc.	450	232
Sodium hypochlorite, 20%	450	232
Sodium hypochlorite, conc.	450	232
Sodium sulfide, to 50%	450	232
Stannic chloride	450	232
Stannous chloride	450	232
Sulfuric acid, 10%	450	232
Sulfuric acid, 50%	450	232
Sulfuric acid, 70%	450	232
Sulfuric acid, 90%	450	232
Sulfuric acid, 98%	450	232
Sulfuric acid, 100%	450	232
Sulfurous acid	450	232
Sulfurous acid, fuming[a]	450	232
Thionyl chloride	450	232
Toluene[a]	450	232
Trichloroacetic acid	450	232
White liquor	450	232
Zinc chloride[c]	450	232

The chemicals listed are in the pure state or in a saturated solution unless otherwise indicated. Compatibility is shown to the maximum allowable temperature for which data is available. Incompatibility is shown by an X. A blank space indicates that the data is unavailable.

[a] Material will permeate.
[b] Material will cause stress cracking.
[c] Material will be absorbed.

Source: From P.A. Schweitzer. 2004. *Corrosion Resistance Tables*, Vols. 1–4, 5th ed., New York: Marcel Dekker.

dimethylamine). As with other fluorocarbons, Halar will be attacked by metallic sodium and potassium. Refer to Table 4.20 for the compatibility of ECTFE with selected corrodents and Reference [1] for additional listings.

Coating thickness range from 0.010 to 0.040 in. with a temperature range from cryogenic to 320°F (160°C). In addition to its corrosion resistance,

TABLE 4.19

Compatibility of ETFE with Selected Corrodents

Chemical	Maximum Temperature	
	°F	°C
Acetaldehyde	200	93
Acetamide	250	121
Acetic acid, 10%	250	121
Acetic acid, 50%	250	121
Acetic acid, 80%	230	110
Acetic acid, glacial	230	110
Acetic anhydride	300	149
Acetone	150	66
Acetyl chloride	150	66
Acrylonitrile	150	66
Adipic acid	280	138
Allyl alcohol	210	99
Allyl chloride	190	88
Alum	300	149
Aluminum chloride, aq.	300	149
Aluminum chloride, dry	300	149
Aluminum fluoride	300	149
Aluminum hydroxide	300	149
Aluminum nitrate	300	149
Aluminum oxychloride	300	149
Aluminum sulfate	300	149
Ammonium bifluoride	300	149
Ammonium carbonate	300	149
Ammonium chloride, 10%	300	149
Ammonium chloride, 50%	290	143
Ammonium chloride, sat.	300	149
Ammonium fluoride, 10%	300	149
Ammonium fluoride, 25%	300	149
Ammonium hydroxide, 25%	300	149
Ammonium hydroxide, sat.	300	149
Ammonium nitrate	230	110
Ammonium persulfate	300	149
Ammonium phosphate	300	149
Ammonium sulfate, 10–40%	300	149
Ammonium sulfide	300	140
Amyl acetate	250	121
Amyl alcohol	300	149
Amyl chloride	300	149
Aniline	230	110
Antimony trichloride	210	99
Aqua regia, 3:1	210	99
Barium carbonate	300	149
Barium chloride	300	149
Barium hydroxide	300	149
Barium sulfate	300	149
Barium sulfide	300	149

(continued)

TABLE 4.19 *Continued*

Chemical	Maximum Temperature °F	°C
Benzaldehyde	210	99
Benzene	210	99
Benzene sulfonic acid, 10%	210	99
Benzoic acid	270	132
Benzyl alcohol	300	149
Benzyl chloride	300	149
Borax	300	149
Boric acid	300	149
Bromine gas, dry	150	66
Bromine water, 10%	230	110
Butadiene	250	121
Butyl acetate	230	110
Butyl alcohol	300	149
Butyl phthalate	150	66
n-Butylamine	120	49
Butyric acid	250	121
Calcium bisulfide	300	149
Calcium carbonate	300	149
Calcium chlorate	300	149
Calcium chloride	300	149
Calcium hydroxide, 10%	300	149
Calcium hydroxide, sat.	300	149
Calcium hypochlorite	300	149
Calcium nitrate	300	149
Calcium oxide	260	127
Calcium sulfate	300	149
Caprylic acid	210	99
Carbon bisulfide	150	66
Carbon dioxide, dry	300	149
Carbon dioxide, wet	300	149
Carbon disulfide	150	66
Carbon monoxide	300	149
Carbon tetrachloride	270	132
Carbonic acid	300	149
Cellosolve	300	149
Chloracetic acid, 50%	230	110
Chloracetic acid, 50% water	230	110
Chlorine gas, dry	210	99
Chlorine gas, wet	250	121
Chlorine, water	100	38
Chlorobenzene	210	99
Chloroform	230	110
Chlorosulfonic acid	80	27
Chromic acid, 10%	150	66
Chromic acid, 50%	150	66
Chromyl chloride	210	99
Citric acid, 15%	120	49

(continued)

TABLE 4.19 *Continued*

Chemical	Maximum Temperature	
	°F	°C
Copper chloride	300	149
Copper cyanide	300	149
Copper sulfate	300	149
Cresol	270	132
Cupric chloride, 5%	300	149
Cyclohexane	300	149
Cyclohexanol	250	121
Dichloroacetic acid	150	66
Ethylene glycol	300	149
Ferric chloride, 50% in water	300	149
Ferric nitrate, 10–50%	300	149
Ferrous chloride	300	149
Ferrous nitrate	300	149
Fluorine gas, dry	100	38
Fluorine gas, moist	100	38
Hydrobromic acid, 20%	300	149
Hydrobromic acid, 50%	300	149
Hydrobromic acid, dil.	300	149
Hydrochloric acid, 20%	300	149
Hydrochloric acid, 38%	300	149
Hydrocyanic acid, 10%	300	149
Hydrofluoric acid, 30%	270	132
Hydrofluoric acid, 70%	250	121
Hydrofluoric acid, 100%	230	110
Hypochlorous acid	300	149
Lactic acid, 25%	250	121
Lactic acid, conc.	250	121
Magnesium chloride	300	149
Malic acid	270	132
Manganese chloride	120	49
Methyl chloride	300	149
Methyl ethyl ketone	230	110
Methyl isobutyl ketone	300	149
Muriatic acid	300	149
Nitric acid, 5%	150	66
Nitric acid, 20%	150	66
Nitric acid, 70%	80	27
Nitric acid, anhydrous	X	X
Nitrous acid, conc.	210	99
Oleum	150	66
Perchloric acid, 10%	230	110
Perchloric acid, 70%	150	66
Phenol	210	99
Phosphoric acid, 50–80%	270	132
Picric acid	130	54
Potassium bromide, 30%	300	149
Salicylic acid	250	121

(continued)

TABLE 4.19 *Continued*

Chemical	Maximum Temperature	
	°F	°C
Sodium carbonate	300	149
Sodium chloride	300	149
Sodium hydroxide, 10%	230	110
Sodium hydroxide, 50%	230	110
Sodium hypochlorite, 20%	300	149
Sodium hypochlorite, conc.	300	149
Sodium sulfide, to 50%	300	149
Stannic chloride	300	149
Stannous chloride	300	149
Sulfuric acid, 10%	300	149
Sulfuric acid, 50%	300	149
Sulfuric acid, 70%	300	149
Sulfuric acid. 90%	300	149
Sulfuric acid, 98%	300	149
Sulfuric acid, 100%	300	149
Sulfuric acid, fuming	120	49
Sulfurous acid	210	99
Thionyl chloride	210	99
Toluene	250	121
Trichloroacetic acid	210	99
Zinc chloride	300	149

The chemicals listed are in the pure state or in a saturated solution unless otherwise indicated. Compatibility is shown to the maximum allowable temperature for which data is available. Incompatibility is shown by an X. A blank space indicates that the data is unavailable.

Source: From P.A. Schweitzer. 2004. *Corrosion Resistance Tables*, Vols. 1–4, 5th ed., New York: Marcel Dekker.

the material has excellent impact strength and abrasion resistance. Previously glass-lined vessels can be refurbished using ECTFE.

4.10.18 Fluoroelastomers

Fluoroelastomers (FKM) are fluorine-containing hydrocarbon polymers with a saturated structure obtained by polymerizing fluorinated monomers such as vinylidene fluoride, hexafluoropropene, and tetrafluoroethylene. They are manufactured under trade names such as Viton by DuPont, Technoflon by Ausimont, and Fluorel by 3M.

The FKM have been approved by the U.S. Food and Drug Administration for use in repeated contact with food products. More details are available in the *Federal Register* (Vol. 33, No. 5, January 9, 1968), Part 121—Food Additives, Subpart F—Food Additives Resulting from Contact with Containers or Equipment and Food Additives Otherwise Affecting Food-Rubber Articles Intended for Repeated Use.

TABLE 4.20

Compatibility of ECTFE with Selected Corrodents

Chemical	Maximum Temperature	
	°F	°C
Acetic acid, 10%	250	121
Acetic acid, 50%	250	121
Acetic acid, 80%	150	66
Acetic acid, glacial	200	93
Acetic anhydride	100	38
Acetone	150	66
Acetyl chloride	150	66
Acrylonitrile	150	66
Adipic acid	150	66
Allyl chloride	300	149
Alum	300	149
Aluminum chloride, aq.	300	149
Aluminum chloride, dry		
Aluminum fluoride	300	149
Aluminum hydroxide	300	149
Aluminum nitrate	300	149
Aluminum oxychloride	150	66
Aluminum sulfate	300	149
Ammonia gas	300	149
Ammonium bifluoride	300	149
Ammonium carbonate	300	149
Ammonium chloride, 10%	290	143
Ammonium chloride, 50%	300	149
Ammonium chloride, sat.	300	149
Ammonium fluoride, 10%[a]	300	149
Ammonium fluoride. 25%[a]	300	149
Ammonium hydroxide, 25%	300	149
Ammonium hydroxide, sat.	300	149
Ammonium nitrate	300	149
Ammonium persulfate	150	66
Ammonium phosphate	300	149
Ammonium sulfate, 10–40%	300	149
Ammonium sulfide	300	149
Amyl acetate	160	71
Amyl alcohol	300	149
Amyl chloride	300	149
Aniline[b]	90	32
Antimony trichloride	100	38
Aqua regia, 3:1	250	121
Barium carbonate	300	149
Barium chloride	300	149
Barium hydroxide	300	149
Barium sulfate	300	149
Barium sulfide	300	149
Benzaldehyde	150	66

(continued)

TABLE 4.20 *Continued*

Chemical	Maximum Temperature	
	°F	°C
Benzene	150	66
Benzene sulfonic acid, 10%	150	66
Benzoic acid	250	121
Benzyl alcohol	300	149
Benzyl chloride	300	149
Borax	300	149
Boric acid	300	149
Bromine gas, dry	X	X
Bromine liquid	150	66
Butadiene	250	121
Butyl acetate	150	66
Butyl alcohol	300	149
Butyric acid	250	121
Calcium bisulfide	300	149
Calcium bisulfite	300	149
Calcium carbonate	300	149
Calcium chlorate	300	149
Calcium chloride	300	149
Calcium hydroxide, 10%	300	149
Calcium hydroxide, sat.	300	149
Calcium hypochlorite	300	149
Calcium nitrate	300	149
Calcium oxide	300	149
Calcium sulfate	300	149
Caprylic acid	220	104
Carbon bisulfide	80	27
Carbon dioxide, dry	300	149
Carbon dioxide, wet	300	149
Carbon disulfide	80	27
Carbon monoxide	150	66
Carbon tetrachloride	300	149
Carbonic acid	300	149
Cellosolve	300	149
Chloracetic acid	250	121
Chloracetic acid, 50% water	250	121
Chlorine gas, dry	150	66
Chlorine gas, wet	250	121
Chlorine, liquid	250	121
Chlorobenzene	150	66
Chloroform	250	121
Chlorosulfonic acid	80	27
Chromic acid, 10%	250	121
Chromic acid, 50%[b]	250	121
Citric acid, 15%	300	149
Citric acid, conc.	300	149
Copper carbonate	150	66

(continued)

TABLE 4.20 *Continued*

Chemical	Maximum Temperature	
	°F	°C
Copper chloride	300	149
Copper cyanide	300	149
Copper sulfate	300	149
Cresol	300	149
Cupric chloride, 5%	300	149
Cupric chloride, 50%	300	149
Cyclohexane	300	149
Cyclohexanol	300	149
Ethylene glycol	300	149
Ferric chloride	300	149
Ferric chloride, 50% in water	300	149
Ferric nitrate, 10–50%	300	149
Ferrous chloride	300	149
Ferrous nitrate	300	149
Fluorine gas, dry	X	X
Fluorine gas, moist	80	27
Hydrobromic acid, 20%	300	149
Hydrobromic acid, 50%	300	149
Hydrobromic acid, dil.	300	149
Hydrochloric acid, 20%	300	149
Hydrochloric acid, 38%	300	149
Hydrocyanic acid, 10%	300	149
Hydrofluoric acid, 30%	250	121
Hydrofluoric acid, 70%	240	116
Hydrofluoric acid, 100%	240	116
Hypochlorous acid	300	149
Iodine solution, 10%	250	121
Lactic acid, 25%	150	66
Lactic acid, conc.	150	66
Magnesium chloride	300	149
Malic acid	250	121
Methyl chloride	300	149
Methyl ethyl ketone	150	66
Methyl isobutyl ketone	150	66
Muriatic acid	300	149
Nitric acid, 5%	300	149
Nitric acid, 20%[a]	250	121
Nitric acid, 70%[a]	150	66
Nitric acid, anhydrous	150	66
Nitrous acid, conc.	250	121
Oleum	X	X
Perchloric acid, 10%	150	66
Perchloric acid, 70%	150	66
Phenol	150	66
Phosphoric acid, 50–80%	250	121
Picric acid	88	27

(continued)

TABLE 4.20   *Continued*

Chemical	Maximum Temperature	
	°F	°C
Potassium bromide, 30%	300	149
Salicylic acid	250	121
Sodium carbonate	300	149
Sodium chloride	300	149
Sodium hydroxide, 10%	300	149
Sodium hydroxide, 50%	250	121
Sodium hydroxide, conc.	150	66
Sodium hypochlorite, 20%	300	149
Sodium hypochlorite, conc.	300	149
Sodium sulfide, to 50%	300	149
Stannic chloride	300	149
Stannous chloride	300	149
Sulfuric acid, 10%	250	121
Sulfuric acid, 50%	250	121
Sulfuric acid, 70%	250	121
Sulfuric acid, 90%	150	66
Sulfuric acid, 98%	150	66
Sulfuric acid, 100%	80	27
Sulfuric acid, fuming	300	149
Sulfurous acid	250	121
Thionyl chloride	150	66
Toluene	150	66
Trichloroacetic acid	150	66
White liquor	250	121
Zinc chloride	300	149

The chemicals listed are in the pure state or in a saturated solution unless otherwise indicated. Compatibility is shown to the maximum allowable temperature for which data is available. Incompatibility is shown by an X. A blank space indicates that the data is unavailable.

[a] Material will permeate.
[b] Material will be absorbed.

Source: From P.A. Schweitzer. 2004. *Corrosion Resistance Tables*, Vols. 1–4, 5th ed., New York: Marcel Dekker.

As with other rubbers, fluoroelastomers are capable of being compounded with various additives to enhance specific properties for particular applications. FKM coatings have an allowable temperature range of −40°F (−40°C) to 400°F (204°C).

Fluoroelastomers provide excellent resistance to oils, fuels, lubricants, most mineral acids, many aliphatic and aromatic hydrocarbons (carbon tetrachloride, benzene, toluene, and xylene) that act as solvents for chlorinated solvents, and pesticides. Special formulations can be produced to obtain resistance to hot mineral acids, steam, and hot water.

These elastomers are not suitable for use with low-molecular-weight esters and ketones, certain amines, and hot anhydrous hydrofluoric or chlorosulfonic acids. Their solubility in low-molecular-weight ketones is an advantage in producing solution coatings of fluoroelastomers. Refer to Table 4.21 for the compatibility of fluoroelastomers with selected corrodents and Reference [1] for additional listings.

The chemical stability of these elastomers is an important property for their use as protective linings. Applications include linings for power station stacks operated with high sulfur fuels, and tank linings for the chemical industry.

4.10.19 Polyvinylidene Fluoride

Polyvinylidene Fluoride (PVDF) is a homopolymer of 1:1 difluoroethane with alternating CH_2 and CF_2 groups along the polymer chain. These groups impart a unique polarity that influences its solubility. The polymer has the characteristic stability of fluoropolymers when exposed to aggressive chemical and thermal conditions.

PVDF is manufactured under the tradename of Kynar by Elf Atochem, Solef by Solvay, Hylar by Ausimont U.S.A., and Super Pro and ISO by Asahi/Amarica.

PVDF can be used in applications intended for repeated contact with food per Title 21, Code of Federal Regulations Chapter 1. Part 177.2520. It is also permitted for use in processing or storage areas in contact with meat or poultry food products prepared under federal inspection according to the U.S. Department of Agriculture (U.S.D.A.). Use is also permitted under "3-A Sanitary Standards for Multiple-Use Plastic Materials Used as Product Contact Surfaces for Dairy Equipment" Serial No. 2000.

PVDF linings have an operating temperature range from $-4°F$ ($-20°C$) to $280°F$ ($138°C$). Coating thickness range from 0.010 to 0.040 in.

Polyvinylidene fluoride is resistant to most acids, alkalies, aliphatic and aromatic hydrocarbons, alcohols, and strong oxidizing agents. Highly polar solvents, such as acetone, and ethyl acetate may cause swelling. When used with strong alkalies, stress cracking results. Refer to Table 4.22 for the compatibility of PVDF with selected corrodents and Reference [1] for a more complete listing.

Typical applications include lining vessels, agitators, pump housings, centrifuge housings, piping, and dust collectors. Previously glass-lined tanks and accessories can be refurbished with PVDF.

4.10.20 Isophthalic Polyester

The isophthalic polyesters use isophthalic acid in place of phthalic anhydrides as the saturated monomer. This increases the cost of production but improves the chemical resistance.

The standard corrosion-grade isophthalic polyesters are made with a 1:1 molar ratio of maleic anhydride or fumaric acid with propylene glycol.

TABLE 4.21

Compatibility of Fluoroelastomers with Selected Corrodents

	Maximum Temperature	
Chemical	°F	°C
Acetaldehyde	X	X
Acetamide	210	99
Acetic acid, 10%	190	88
Acetic acid, 50%	180	82
Acetic acid, 80%	180	82
Acetic acid, glacial	X	X
Acetic anhydride	X	X
Acetone	X	X
Acetyl chloride	400	204
Acrylic acid	X	X
Acrylonitrile	X	X
Adipic acid	190	88
Allyl alcohol	190	88
Allyl chloride	100	38
Alum	190	88
Aluminum acetate	180	82
Aluminum chloride, aq.	400	204
Aluminum fluoride	400	204
Aluminum hydroxide	190	88
Aluminum nitrate	400	204
Aluminum oxychloride	X	X
Aluminum sulfate	390	199
Ammonia gas	X	X
Ammonium bifluoride	140	60
Ammonium carbonate	190	88
Ammonium chloride, 10%	400	204
Ammonium chloride, 50%	300	149
Ammonium chloride, sat.	300	149
Ammonium fluoride, 10%	140	60
Ammonium fluoride, 25%	140	60
Ammonium hydroxide, 25%	190	88
Ammonium hydroxide, sat.	190	88
Ammonium nitrate	X	X
Ammonium persulfate	140	60
Ammonium phosphate	180	82
Ammonium sulfate, 10–40%	180	82
Ammonium sulfide	X	X
Amyl acetate	X	X
Amyl alcohol	200	93
Amyl chloride	190	88
Aniline	230	110
Antimony trichloride	190	88
Aqua regia, 3:1	190	88
Barium carbonate	250	121
Barium chloride	400	204
Barium hydroxide	400	204

(continued)

TABLE 4.21 *Continued*

Chemical	Maximum Temperature	
	°F	°C
Barium sulfate	400	204
Barium sulfide	400	204
Benzaldehyde	X	X
Benzene	400	204
Benzene sulfonic acid, 10%	190	88
Benzoic acid	400	204
Benzyl alcohol	400	204
Benzyl chloride	400	204
Borax	190	88
Boric acid	400	204
Bromine gas, dry, 25%	180	82
Bromine gas, moist, 25%	180	82
Bromine liquid	350	177
Butadiene	400	204
Butyl acetate	X	X
Butyl alcohol	400	204
n-Butylamine	X	X
Butyl phthalate	80	27
Butyric acid	120	49
Calcium bisulfide	400	204
Calcium bisulfite	400	204
Calcium carbonate	190	88
Calcium chlorate	190	88
Calcium chloride	300	149
Calcium hydroxide, 10%	300	149
Calcium hydroxide, sat.	400	204
Calcium hypochlorite	400	204
Calcium nitrate	400	204
Calcium sulfate	200	93
Carbon bisulfide	400	204
Carbon dioxide, dry	80	27
Carbon dioxide, wet	X	X
Carbon disulfide	400	204
Carbon monoxide	400	204
Carbon tetrachloride	350	177
Carbonic acid	400	204
Cellosolve	X	X
Chloracetic acid	X	X
Chloracetic acid, 50% water	X	X
Chlorine gas, dry	190	88
Chlorine gas, wet	190	88
Chlorine, liquid	190	88
Chlorobenzene	400	204
Chloroform	400	204
Chlorosulfonic acid	X	X
Chromic acid, 10%	350	177
Chromic acid, 50%	350	177

(continued)

TABLE 4.21 *Continued*

Chemical	Maximum Temperature	
	°F	°C
Citric acid, 15%	300	149
Citric acid, conc.	400	204
Copper acetate	X	X
Copper carbonate	190	88
Copper chloride	400	204
Copper cyanide	400	204
Copper sulfate	400	204
Cresol	X	X
Cupric chloride, 5%	180	82
Cupric chloride, 50%	180	82
Cyclohexane	400	204
Cyclohexanol	400	204
Dichloroethane (ethylene dichloride)	190	88
Ethylene glycol	400	204
Ferric chloride	400	204
Ferric chloride, 50% in water	400	204
Ferric nitrate, 10–50%	400	204
Ferrous chloride	180	82
Ferrous nitrate	210	99
Fluorine gas, dry	X	X
Fluorine gas, moist	X	X
Hydrobromic acid, 20%	400	204
Hydrobromic acid, 50%	400	204
Hydrobromic acid, dil.	400	204
Hydrochloric acid, 20%	350	177
Hydrochloric acid, 38%	350	177
Hydrocyanic acid, 10%	400	204
Hydrofluoric acid, 30%	210	99
Hydrofluoric acid, 70%	350	177
Hydrofluoric acid, 100%	X	X
Hypochlorous acid	400	204
Iodine solution, 10%	190	88
Ketones, general	X	X
Lactic acid, 25%	300	149
Lactic acid, conc.	400	204
Magnesium chloride	390	199
Malic acid	390	199
Manganese chloride	180	82
Methyl chloride	190	88
Methyl ethyl ketone	X	X
Methyl isobutyl ketone	X	X
Muriatic acid	350	149
Nitric acid, 5%	400	204
Nitric acid, 20%	400	204
Nitric acid, 70%	190	88
Nitric acid, anhydrous	190	88
Nitrous acid, conc.	90	32

(continued)

TABLE 4.21 *Continued*

Chemical	Maximum Temperature	
	°F	°C
Oleum	190	88
Perchloric acid, 10%	400	204
Perchloric acid, 70%	400	204
Phenol	210	99
Phosphoric acid, 50–80%	300	149
Picric acid	400	204
Potassium bromide, 30%	190	88
Salicylic acid	300	149
Sodium carbonate	190	88
Sodium chloride	400	204
Sodium hydroxide, 10%	X	X
Sodium hydroxide, 50%	X	X
Sodium hydroxide, conc.	X	X
Sodium hypochlorite, 20%	400	204
Sodium hypochlorite, conc.	400	204
Sodium sulfide, to 50%	190	88
Stannic chloride	400	204
Stannous chloride	400	204
Sulfuric acid, 10%	350	149
Sulfuric acid, 50%	350	149
Sulfuric acid, 70%	350	149
Sulfuric acid, 90%	350	149
Sulfuric acid, 98%	350	149
Sulfuric acid, 100%	180	82
Sulfuric acid, fuming	200	93
Sulfurous acid	400	204
Thionyl chloride	X	X
Toluene	400	204
Trichloroacetic acid	190	88
White liquor	190	88
Zinc chloride	400	204

The chemicals listed are in the pure state or in a saturated solution unless otherwise indicated. Compatibility is shown to the maximum allowable temperature for which data is available. Incompatibility is shown by an X. A blank space indicates that the data is unavailable.

Source: From P.A. Schweitzer. 2004. *Corrosion Resistance Tables*, Vols. 1–4, 5th ed., New York: Marcel Dekker.

The isophthalic polyesters are the most common type used for chemical service applications. They have a wide range of corrosion resistance, being satisfactory for use up to 125°F (52°C) in such acids as 10% acetic, benzoic, citric, oleic, 25% phosphoric, 10 to 25% sulfuric, and fatty acids. Most inorganic salts are compatible with isophthalic polyesters. Solvents such as amyl alcohol, ethylene glycol, formaldehyde, gasoline, kerosene, and naphtha are also compatible.

TABLE 4.22

Compatibility of PVDF with Selected Corrodents

	Maximum Temperature	
Chemical	°F	°C
Acetaldehyde	150	66
Acetamide	90	32
Acetic acid 10%	300	149
Acetic acid, 50%	300	149
Acetic acid, 80%	190	88
Acetic acid, glacial	190	88
Acetic anhydride	100	38
Acetone	X	X
Acetyl chloride	120	49
Acrylic acid	150	66
Acrylonitrile	130	54
Adipic acid	280	138
Allyl alcohol	200	93
Allyl chloride	200	93
Alum	180	82
Aluminum acetate	250	121
Aluminum chloride, aq.	300	149
Aluminum chloride, dry	270	132
Aluminum fluoride	300	149
Aluminum hydroxide	260	127
Aluminum nitrate	300	149
Aluminum oxychloride	290	143
Aluminum sulfate	300	149
Ammonia gas	270	132
Ammonium bifluoride	250	121
Ammonium carbonate	280	138
Ammonium chloride, 10%	280	138
Ammonium chloride, 50%	280	138
Ammonium chloride, sat.	280	138
Ammonium fluoride, 10%	280	138
Ammonium fluoride, 25%	280	138
Ammonium hydroxide, 25%	280	138
Ammonium hydroxide, sat.	280	138
Ammonium nitrate	280	138
Ammonium persulfate	280	138
Ammonium phosphate	280	138
Ammonium sulfate, 10–40%	280	138
Ammonium sulfide	280	138
Ammonium sulfite	280	138
Amyl acetate	190	88
Amyl alcohol	280	138
Amyl chloride	280	138
Aniline	200	93
Antimony trichloride	150	66
Aqua regia, 3:1	130	54
Barium carbonate	280	138
Barium chloride	280	138
Barium hydroxide	280	138

(continued)

TABLE 4.22 *Continued*

Chemical	Maximum Temperature	
	°F	°C
Barium sulfate	280	138
Barium sulfide	280	138
Benzaldehyde	120	49
Benzene	150	66
Benzene sulfonic acid, 10%	100	38
Benzoic acid	250	121
Benzyl alcohol	280	138
Benzyl chloride	280	138
Borax	280	138
Boric acid	280	138
Bromine gas, dry	210	99
Bromine gas, moist	210	99
Bromine liquid	140	60
Butadiene	280	138
Butyl acetate	140	60
Butyl alcohol	280	138
n-Butylamine	X	X
Butyl phthalate	80	27
Butyric acid	230	110
Calcium bisulfide	280	138
Calcium bisulfite	280	138
Calcium carbonate	280	138
Calcium chlorate	280	138
Calcium chloride	280	138
Calcium hydroxide, 10%	270	132
Calcium hydroxide, sat.	280	138
Calcium hypochlorite	280	138
Calcium nitrate	280	138
Calcium oxide	250	121
Calcium sulfate	280	138
Caprylic acid	220	104
Carbon bisulfide	80	27
Carbon dioxide, dry	280	138
Carbon dioxide, wet	280	138
Carbon disulfide	80	27
Carbon monoxide	280	138
Carbon tetrachloride	280	138
Carbonic acid	280	138
Cellosolve	280	138
Chloracetic acid	200	93
Chloracetic acid, 50% water	210	99
Chloride gas, dry	210	99
Chlorine gas, wet, 10%	210	99
Chlorine, liquid	210	99
Chlorobenzene	220	104
Chloroform	250	121
Chlorosulfonic acid	110	43
Chromic acid, 10%	220	104
Chromic acid, 50%	250	121

(continued)

TABLE 4.22 *Continued*

Chemical	Maximum Temperature	
	°F	°C
Chromyl chloride	110	43
Citric acid, 15%	250	121
Citric acid, conc.	250	121
Copper acetate	250	121
Copper carbonate	250	121
Copper chloride	280	138
Copper cyanide	280	138
Copper sulfate	280	138
Cresol	210	99
Cupric chloride, 5%	270	132
Cupric chloride, 50%	270	132
Cyclohexane	250	121
Cyclohexanol	210	99
Dichlorethane (ethylene dichloride)	280	138
Dichloroacetic acid	120	49
Ethylene glycol	280	138
Ferric chloride	280	138
Ferric chloride, 50% in water	280	138
Ferrous chloride	280	138
Ferrous nitrate	280	138
Ferrous nitrate, 10–50%	280	138
Fluorine gas, dry	80	27
Fluorine gas, moist	80	27
Hydrobromic acid, 20%	280	138
Hydrobromic acid, 50%	280	138
Hydrobromic acid, dil.	260	127
Hydrochloric acid, 20%	280	138
Hydrochloric acid, 38%	280	138
Hydrocyanic acid, 10%	280	138
Hydrofluoric acid, 30%	260	127
Hydrofluoric acid, 70%	200	93
Hydrofluoric acid, 100%	200	93
Hypochlorous acid	280	138
Iodine solution	250	121
Ketones, general	110	43
Lactic acid, 25%	130	54
Lactic acid, conc.	110	43
Magnesium chloride	280	138
Malic acid	250	121
Manganese chloride	280	138
Methyl chloride	X	X
Methyl ethyl ketone	X	X
Methyl isobutyl ketone	110	43
Muriatic acid	280	138
Nitric acid, 20%	180	82
Nitric acid, 5%	200	93
Nitric acid, 70%	120	49
Nitric acid, anhydrous	150	66
Nitrous acid, conc.	210	99

(continued)

TABLE 4.22 *Continued*

Chemical	Maximum Temperature	
	°F	°C
Oleum	X	X
Perchloric acid, 10%	210	99
Perchloric acid, 70%	120	49
Phenol	200	93
Phosphoric acid, 50–80%	220	104
Picric acid	80	27
Potassium bromide, 30%	280	138
Salicylic acid	220	104
Silver bromide, 10%	250	121
Sodium carbonate	280	138
Sodium chloride	280	138
Sodium hydroxide, 10%	230	110
Sodium hydroxide, 50%	220	104
Sodium hydroxide, conc.[a]	150	66
Sodium hypochlorite, 20%	280	138
Sodium hypochlorite, conc.	280	138
Sodium sulfide, to 50%	280	138
Stannic chloride	280	138
Stannous chloride	280	138
Sulfuric acid, 10%	250	121
Sulfuric acid, 50%	220	104
Sulfuric acid, 70%	220	104
Sulfuric acid, 90%	210	99
Sulfuric acid, 98%	140	60
Sulfuric acid. 100%	X	X
Sulfuric acid, fuming	X	X
Sulfurous acid	220	104
Thionyl chloride	X	X
Toluene	X	X
Trichloroacetic acid	130	54
White liquor	80	27
Zinc chloride	260	127

The chemicals listed are in the pure state or in a saturated solution unless otherwise indicated. Compatibility is shown to the maximum allowable temperature for which data is available. Incompatibility is shown by an X. A blank space indicates that the data is unavailable.

[a] Subject to stress corrosion cracking.

Source: From P.A. Schweitzer. 2004. *Corrosion Resistance Tables*, Vols. 1–4, 5th ed., New York: Marcel Dekker.

The isophthalic polyesters are not resistant to acetone, amyl acetate, benzene, carbon disulfide, solutions of alkaline salts of potassium and sodium, hot distilled water, or higher concentrations of oxidizing acids. Refer to Table 4.23 for the compatibility of isophthalic polyesters with selected corrodents and Reference [1] for a more complete listing.

TABLE 4.23

Compatibility of Isophthalic Polyester with Selected
Corrodents

Chemical	Maximum Temperature	
	°F	°C
Acetaldehyde	X	X
Acetic acid, 10%	180	82
Acetic acid, 50%	110	43
Acetic acid, 80%	X	X
Acetic acid, glacial	X	X
Acetic anhydride	X	X
Acetone	X	X
Acetyl chloride	X	X
Acrylic acid	X	X
Acrylonitrile	X	X
Adipic acid	220	104
Allyl alcohol	X	X
Allyl chloride	X	X
Alum	250	121
Aluminum chloride, aq.	180	82
Aluminum chloride, dry	170	77
Aluminum fluoride, 10%	140	60
Aluminum hydroxide	160	71
Aluminum nitrate	160	71
Aluminum sulfate	180	82
Ammonia gas	90	32
Ammonium carbonate	X	X
Ammonium chloride, 10%	160	71
Ammonium chloride, 50%	160	71
Ammonium chloride, sat.	180	82
Ammonium fluoride, 10%	90	32
Ammonium fluoride, 25%	90	32
Ammonium hydroxide, 25%	X	X
Ammonium hydroxide, sat.	X	X
Ammonium nitrate	160	11
Ammonium persulfate	160	71
Ammonium phosphate	160	71
Ammonium sulfate, 10%	180	82
Ammonium sulfide	X	X
Ammonium sulfite	X	X
Amyl acetate	X	X
Amyl alcohol	160	71
Amyl chloride	X	X
Aniline	X	X
Antimony trichloride	160	71
Aqua regia, 3:1	X	X
Barium carbonate	190	88
Barium chloride	140	60
Barium hydroxide	X	X
Barium sulfate	160	71

(continued)

TABLE 4.23 *Continued*

Chemical	Maximum Temperature °F	°C
Barium sulfide	90	32
Benzaldehyde	X	X
Benzene	X	X
Benzene sulfonic acid, 10%	180	82
Benzoic acid	180	82
Benzyl alcohol	X	X
Benzyl chloride	X	X
Borax	140	60
Boric acid	180	82
Bromine gas, dry	X	X
Bromine gas, moist	X	X
Bromine liquid	X	X
Butyl acetate	X	X
Butyl alcohol	80	27
n-Butylamine	X	X
Butyric acid, 25%	129	49
Calcium bisulfide	160	71
Calcium bisulfite	150	66
Calcium carbonate	160	71
Calcium chlorate	160	71
Calcium chloride	180	82
Calcium hydroxide, 10%	160	71
Calcium hydroxide, sat.	160	71
Calcium hypochlorite, 10%	120	49
Calcium nitrate	140	60
Calcium oxide	160	71
Calcium sulfate	160	71
Caprylic acid	160	71
Carbon bisulfide	X	X
Carbon dioxide, dry	160	71
Carbon dioxide, wet	160	71
Carbon disulfide	X	X
Carbon monoxide	160	71
Carbon tetrachloride	X	X
Carbonic acid	160	71
Cellosolve	X	X
Chloracetic acid, 50% water	X	X
Chloride gas, dry	160	71
Chlorine gas, wet	160	71
Chlorine, liquid	X	X
Chloroacetic acid, 25%	150	66
Chlorobenzene	X	X
Chloroform	X	X
Chlorosulfonic acid	X	X
Chromic acid, 10%	X	X
Chromic acid, 50%	X	X
Chromyl chloride	140	60

(continued)

TABLE 4.23 *Continued*

Chemical	Maximum Temperature	
	°F	°C
Citric acid, 15%	160	71
Citric acid, conc.	200	93
Copper acetate	160	71
Copper chloride	180	82
Copper cyanide	160	71
Copper sulfate	200	93
Cresol	X	X
Cupric chloride, 5%	170	77
Cupric chloride, 50%	170	77
Cyclohexane	80	27
Dichloroacetic acid	X	X
Dichloroethane (ethylene dichloride)	X	X
Ethylene glycol	120	49
Ferric chloride	180	82
Ferric chloride, 50% in water	160	71
Ferric nitrate, 10–50%	180	82
Ferrous chloride	180	82
Ferrous nitrate	160	71
Fluorine gas, dry	X	X
Fluorine gas, moist	X	X
Hydrobromic acid, dil.	120	49
Hydrobromic acid, 20%	140	60
Hydrobromic acid, 50%	140	60
Hydrochloric acid, 20%	160	71
Hydrochloric acid, 38%	160	71
Hydrocyanic acid, 10%	90	32
Hydrofluoric acid, 30%	X	X
Hydrofluoric acid, 70%	X	X
Hydrofluoric acid, 100%	X	X
Hypochlorous acid	90	32
Ketones, general	X	X
Lactic acid, 25%	160	71
Magnesium chloride	180	82
Malic acid	90	32
Methyl ethyl ketone	X	X
Methyl isobutyl ketone	X	X
Muriatic acid	160	71
Nitric acid, 5%	120	49
Nitric acid, 20%	X	X
Nitric acid, 70%	X	X
Nitric acid, anhydrous	X	X
Nitrous acid, conc.	120	49
Oleum	X	X
Perchloric acid, 10%	X	X
Perchloric acid, 70%	X	X
Phenol	X	X
Phosphoric acid, 50–80%	180	82

(continued)

TABLE 4.23 *Continued*

Chemical	Maximum Temperature	
	°F	°C
Picric acid	X	X
Potassium bromide, 30%	160	71
Salicylic acid	100	38
Sodium carbonate, 20%	90	32
Sodium chloride	200	93
Sodium hydroxide, 10%	X	X
Sodium hydroxide, 50%	X	X
Sodium hydroxide, conc.	X	X
Sodium hypochlorite, 20%	X	X
Sodium hypochlorite, conc.	X	X
Sodium sulfide, to 50%	X	X
Stannic chloride	180	82
Stannous chloride	180	82
Sulfuric acid, 10%	160	71
Sulfuric acid, 50%	150	66
Sulfuric acid, 70%	X	X
Sulfuric acid, 90%	X	X
Sulfuric acid, 98%	X	X
Sulfuric acid, 100%	X	X
Sulfuric acid, fuming	X	X
Sulfurous acid	X	X
Thionyl chloride	X	X
Toluene	110	43
Trichloroacetic acid, 50%	170	77
White liquor	X	X
Zinc chloride	180	82

The chemicals listed are in the pure state or in a saturated solution unless otherwise indicated. Compatibility is shown to the maximum allowable temperature for which data is available. Incompatibility is shown by an X. A blank space indicates that the data is unavailable.

Source: From P.A. Schweitzer. 2004. *Corrosion Resistance Tables*, Vols. 1–4, 5th ed., New York: Marcel Dekker.

Applications include coating of chemical storage tanks. In food contact applications these resins withstand acids and corrosive salts encountered in foods and food handling.

4.10.21 Bisphenol-A Fumurate Polyesters

This is a premium-grade corrosion resistant resin. It costs approximately one-third more than an isophthalic resin.

Standard bisphenol-A fumurate resins are derived from the propylene glycol or oxide diether of bisphenol-A and fumuric acid. The aromatic structure contributed by the bisphenol-A provides several benefits. Thermal stability is improved, and because the number of interior chain ester groups is reduced, the resistance to hydrolysis and saponification increases. Bisphenol-A polyesters have the best hydrolysis resistance of any commercial unsaturated polyester.

The bisphenol polyesters are superior in their corrosion resistant properties to the isophthalic polyesters. They show good performance with moderate alkaline solutions, and excellent resistance to the various categories of bleaching agents. The bisphenol polyesters will break down in highly concentrated alkalies. These resins can be used in the handling of the following materials.

Acids, to 200°F (93°C)

Acetic	Fatty acids	Stearic
Benzoic	Hydrochloric, 10%	Sulfonic, 30%
Boric	Lactic	Tannic
Butyric	Maleic	Tartaric
Chloroacetic, 15%	Oleic	Trichloroacetic, 50%
Chromic, 5%	Oxalic	Rayon spin bath
Citric	Phosphoric, 80%	

Salt solutions to 200°F (93°C)

All aluminum salts	Cadmium salts	Copper salts
Most ammonium salts	Most plating solutions	Iron salts
Zinc salts		

Alkalies

Ammonium hydroxide, 5% to 160°F (71°C)	Calcium hypochlorite, 20°F to 200°F (93°C)	Potassium hydroxide, 25% to 160°F (71°C)
Calcium hydroxide, 25% to 160°F (71°C)	Chlorine dioxide, 15% to 200°F (93°C)	Sodium chlorite, to 200°F (93°C)
Sodium hydrosulfite, to 200°F (93°C)		

Solvents

Sour crude oil	Glycerine	Alcohols at ambient temperatures
Linseed oil		

Gases, to 200°F (93°C)

Carbon dioxide	Chlorine, wet	Rayon waste gases, 150°F (65°C)
Sulfur trioxide	Carbon monoxide	
Sulfur dioxide, dry		

Solvents such as benzene, carbon disulfide, ether methyl ethyl ketone, toluene, xylene, trichloroethylene, and trichloroethane will attack the resin. Sulfuric acid above 70% sodium hydroxide, and 30% chromic acid will also attack the resin. Refer to Table 4.24 for the compatibility of bisphenol-A fumarate resin with selected corrodents and Table 4.25 for the compatibility of hydrogenated bisphenol-A fumarate polyester resin. Reference [1] provides additional listings for both resins.

TABLE 4.24

Compatibility of Bisphenol-A Fumarate Polyester with
Selected Corrodents

	Maximum Temperature	
Chemical	°F	°C
Acetaldehyde	X	X
Acetic acid, 10%	220	104
Acetic acid, 50%	160	171
Acetic acid, 80%	160	171
Acetic acid, glacial	X	X
Acetic anhydride	110	43
Acetone	X	X
Acetyl chloride	X	X
Acrylic acid	100	38
Acrylonitrile	X	X
Adipic acid	220	104
Allyl alcohol	X	X
Allyl chloride	X	X
Alum	220	104
Aluminum chloride, aq.	200	93
Aluminum fluoride, 10%	90	32
Aluminum hydroxide	160	71
Aluminum nitrate	200	93
Aluminum sulfate	200	93
Ammonia gas	200	93
Ammonium carbonate	90	32
Ammonium chloride 10%	200	93
Ammonium chloride, 50%	220	104
Ammonium chloride, sat.	220	104
Ammonium fluoride, 10%	180	82
Ammonium fluoride, 25%	120	49
Ammonium hydroxide, 20%	140	60
Ammonium hydroxide, 25%	100	38
Ammonium nitrate	220	104
Ammonium persulfate	180	82
Ammonium phosphate	80	27
Ammonium sulfate, 10–40%	220	104
Ammonium sulfide	110	43
Ammonium sulfite	80	27
Amyl acetate	80	27
Amyl alcohol	200	93
Amyl chloride	X	X
Aniline	X	X
Antimony trichloride	220	104
Aqua regia, 3:1	X	X
Barium carbonate	200	93
Barium chloride	220	104
Barium hydroxide	150	66
Barium sulfate	220	104
Barium sulfide	140	60

(continued)

TABLE 4.24 *Continued*

Chemical	Maximum Temperature	
	°F	°C
Benzaldehyde	X	X
Benzene	X	X
Benzene sulfonic acid, 10%	200	93
Benzoic acid	180	82
Benzyl alcohol	X	X
Benzyl chloride	X	X
Borax	220	104
Boric acid	220	104
Bromine gas, dry	90	32
Bromine gas, moist	100	38
Bromine, liquid	X	X
Butyl acetate	80	27
Butyl alcohol	80	27
n-Butylamine	X	X
Butyric acid	220	93
Calcium bisulfite	180	82
Calcium carbonate	210	99
Calcium chlorate	200	93
Calcium chloride	220	104
Calcium hydroxide, 10%	180	82
Calcium hydroxide, sat.	160	71
Calcium hypochlorite, 10%	80	27
Calcium nitrate	220	93
Calcium sulfate	220	93
Caprylic acid	160	71
Carbon bisulfide	X	X
Carbon dioxide, dry	350	177
Carbon dioxide, wet	210	99
Carbon disulfide	X	X
Carbon monoxide	350	177
Carbon tetrachloride	110	43
Carbonic acid	90	32
Cellosolve	140	60
Chlorine gas, dry	200	93
Chlorine gas, wet	200	93
Chlorine, liquid	X	X
Chloroacetic acid, 50% water	140	60
Chloroacetic acid, to 25%	80	27
Chlorobenzene	X	X
Chloroform	X	X
Chlorosulfonic acid	X	X
Chromic acid, 10%	X	X
Chromic acid, 50%	X	X
Chromyl chloride	150	66
Citric acid, 15%	220	104
Citric acid, conc.	220	104
Copper acetate	180	82

(continued)

TABLE 4.24 *Continued*

Chemical	Maximum Temperature	
	°F	°C
Copper chloride	220	104
Copper cyanide	220	104
Copper sulfate	220	104
Cresol	X	X
Cyclohexane	X	X
Dichloroacetic acid	100	38
Dichloroethane (ethylene dichloride)	X	X
Ethylene glycol	220	104
Ferric chloride	220	104
Ferric chloride, 50% in water	220	104
Ferric nitrate, 10–50%	220	104
Ferrous chloride	220	104
Ferrous nitrate	220	104
Fluorine gas, moist	220	104
Hydrobromic acid, 20%	220	104
Hydrobromic acid, 50%	160	71
Hydrobromic acid, dil.	220	104
Hydrochloric acid, 20%	190	88
Hydrochloric acid, 38%	X	X
Hydrocyanic acid, 10%	200	93
Hydrofluoric acid, 30%	90	32
Hypochlorous acid, 20%	90	32
Iodine solution, 10%	200	104
Lactic acid, 5%	210	99
Lactic acid, conc.	220	104
Magnesium chloride	220	104
Malic acid	160	71
Methyl ethyl ketone	X	X
Methyl isobutyl ketone	X	X
Muriatic acid	130	54
Nitric acid, 5%	160	71
Nitric acid, 20%	100	38
Nitric acid, 70%	X	X
Nitric acid, anhydrous	X	X
Oleum	X	X
Phenol	X	X
Phosphoric acid, 50–80%	220	104
Picric acid	110	43
Potassium bromide, 30%	200	93
Salicylic acid	150	66
Sodium carbonate	160	71
Sodium chloride	220	104
Sodium hydroxide, 10%	130	54
Sodium hydroxide, 50%	220	104
Sodium hydroxide, conc.	200	93
Sodium hypochlorite, 20%	X	X
Sodium sulfide, to 50%	210	99

(continued)

TABLE 4.24 *Continued*

Chemical	Maximum Temperature	
	°F	°C
Stannic chloride	200	93
Stannous chloride	220	104
Sulfuric acid, 10%	220	104
Sulfuric acid, 50%	220	104
Sulfuric acid, 70%	160	71
Sulfuric acid, 90%	X	X
Sulfuric acid, 98%	X	X
Sulfuric acid, 100%	X	X
Sulfuric acid, fuming	X	X
Sulfurous acid	110	43
Thionyl chloride	X	X
Toluene	X	X
Trichloroacetic acid, 50%	180	82
White liquor	180	82
Zinc chloride	250	121

The chemicals listed are in the pure state or in a saturated solution unless otherwise indicated. Compatibility is shown to the maximum allowable temperature for which data is available. Incompatibility is shown by an X. A blank space indicates that the data is unavailable.

Source: From P.A. Schweitzer. 2004. *Corrosion Resistance Tables*, Vols. 1–4, 5th ed., New York: Marcel Dekker.

4.10.22 Halogenated Polyesters

Halogenated resins consist of chlorinated or brominated polyesters. The chlorinated polyester resins cured at room temperature are also known as chlorendic polyesters. These resins have the highest heat resistance of any of the polyesters. They are also inherently fire retardant. A combustible rating of 20 can be achieved, making this the safest possible polyester for stacks, hoods, or wherever a fire hazard exists.

Refer to Table 4.26 for the performance of chlorinated polyesters at elevated temperatures. This permits them to survive high-temperature upsets in flue gas desulfurization scrubbers, some of which can reach a temperature of 400°F (204°C).

The halogenated polyesters exhibit excellent resistance in contact with oxidizing acids and solutions such as 35% nitric acid at room temperature, 40% chromic acid, chlorine water, wet chlorine, and 15% hypochlorites. They also resist neutral and acid salts, nonoxidizing acids, organic acids, mercaptans, ketones, aldehydes, alcohols, organic esters, and fats and oils.

These polyesters are not resistant to highly alkaline solutions of sodium hydroxide, concentrated sulfuric acid, alkaline solutions with a pH greater than 10, aliphatic, primary, and aromatic amines, amides and other alkaline

TABLE 4.25

Compatibility of Hydrogenated Bisphenol A Fumarate
Polyester with Selected Corrodents

	Maximum Temperature	
Chemical	°F	°C
Acetic acid, 10%	200	93
Acetic acid, 50%	160	71
Acetic anhydride	X	X
Acetone	X	X
Acetyl chloride	X	X
Acrylonitrile	X	X
Aluminum acetate		
Aluminum chloride, aq.	200	93
Aluminum fluoride	X	X
Aluminum sulfate	200	93
Ammonium chloride, sat.	200	93
Ammonium nitrate	200	93
Ammonium persulfate	200	93
Ammonium sulfide	100	38
Amyl acetate	X	X
Amyl alcohol	200	93
Amyl chloride	90	32
Aniline	X	X
Antimony trichloride	80	27
Aqua regia, 3:1	X	X
Barium carbonate	180	82
Barium chloride	200	93
Benzaldehyde	X	X
Benzene	X	X
Benzoic acid	210	99
Benzyl alcohol	X	X
Benzyl chloride	X	X
Boric acid	210	99
Bromine, liquid	X	X
Butyl acetate	X	X
n-Butylamine	X	X
Butyric acid	X	X
Calcium bisulfide	120	49
Calcium chlorate	210	99
Calcium chloride	210	99
Calcium hypochlorite, 10%	180	82
Carbon bisulfide	X	X
Carbon disulfide	X	X
Carbon tetrachloride	X	X
Chlorine gas, dry	210	99
Chlorine gas, wet	210	99
Chloroacetic acid, 50% water	90	32
Chloroform	X	X
Chromic acid, 50%	X	X

(continued)

TABLE 4.25 *Continued*

Chemical	Maximum Temperature	
	°F	°C
Citric acid, 15%	200	93
Citric acid, conc.	210	99
Copper acetate	210	99
Copper chloride	210	99
Copper cyanide	210	99
Copper sulfate	210	99
Cresol	X	X
Cyclohexane	210	99
Dichloroethane (ethylene dichloride)	X	X
Ferric chloride	210	99
Ferric chloride, 50% in water	200	93
Ferric nitrate, 10–50%	200	93
Ferrous chloride	210	99
Ferrous nitrate	210	99
Hydrobromic acid, 20%	90	32
Hydrobromic acid, 50%	90	32
Hydrochloric acid, 20%	180	82
Hydrochloric acid, 38%	190	88
Hydrocyanic acid, 10%	X	X
Hydrofluoric acid, 30%	X	X
Hydrofluoric acid, 50%	210	99
Hydrofluoric acid, 70%	X	X
Hydrofluoric acid, 100%	X	X
Lactic acid, 25%	210	99
Lactic acid, conc.	210	99
Magnesium chloride	210	99
Methyl ethyl ketone	X	X
Methyl isobutyl ketone	X	X
Muriatic acid	190	88
Nitric acid, 5%	90	32
Oleum	X	X
Perchloric acid, 10%	X	X
Perchloric acid, 70%	X	X
Phenol	X	X
Phosphoric acid, 50–80%	210	99
Sodium carbonate, 10%	100	38
Sodium chloride	210	99
Sodium hydroxide, 10%	100	38
Sodium hydroxide, 50%	X	X
Sodium hydroxide, conc.	X	X
Sodium hypochlorite, 10%	160	71
Sulfuric acid, 10%	210	99
Sulfuric acid, 50%	210	99
Sulfuric acid, 70%	90	32
Sulfuric acid, 90%	X	X
Sulfuric acid, 98%	X	X

(continued)

TABLE 4.25 *Continued*

Chemical	Maximum Temperature	
	°F	°C
Sulfuric acid, 100%	X	X
Sulfuric acid, fuming	X	X
Sulfurous acid, 25%	210	99
Toluene	90	32
Trichloroacetic acid	90	32
Zinc chloride	200	93

The chemicals listed are in the pure state or in a saturated solution unless otherwise indicated. Compatibility is shown to the maximum allowable temperature for which data is available. Incompatibility is shown by an X. A blank space indicates that the data is unavailable.

Source: From P.A. Schweitzer. 2004. *Corrosion Resistance Tables*, Vols. 1–4, 5th ed., New York: Marcel Dekker.

TABLE 4.26

General Application Guide for Chlorinated Polyesters

Environment	Comments
Acid halides	Not recommended
Acids, mineral nonoxidizing	Resistant to 250°F/121°C
Acids, organic	Resistant to 250°F/121°C; glacial acetic acid to 120°F/49°C
Alcohols	Resistant to 180°F/82°C
Aldehydes	Resistant to 180°F/82°C
Alkaline solutions pH>10	Note recommended for continuous exposure
Amines, aliphatic, primary aromatic	Can cause severe attack
Amides, other alkaline organics	Can cause severe attack
Esters, organic	Resistant to 180°F/82°C
Fats and oils	Resistant to 200°F/95°C
Glycols	Resistant to 180°F/82°C
Ketones	Resistant to 180°F/82°C
Mercaptans	Resistant to 180°F/82°C
Phenol	Not recommended
Salts, acid	Resistant to 250°F/121°C
Salts, neutral	Resistant to 250°F/121°C
Water, demineralized, distilled, deionized, steam and condensate	Resistant to 212°F/100°C; lowest absorption of any polyester

TABLE 4.27

Compatibility of Halogenated Polyesters with Selected
Corrodents

Chemical	Maximum Temperature	
	°F	°C
Acetaldehyde	X	X
Acetic acid, 10%	140	60
Acetic acid, 50%	90	32
Acetic acid, glacial	110	43
Acetic anhydride	100	38
Acetone	X	X
Acetyl chloride	X	X
Acrylic acid	X	X
Acrylonitrile	X	X
Adipic acid	220	104
Allyl alcohol	X	X
Allyl chloride	X	X
Alum, 10%	200	93
Aluminum chloride, aq.	120	49
Aluminum fluoride, 10%	90	32
Aluminum hydroxide	170	77
Aluminum nitrate	160	71
Aluminum oxychloride		
Aluminum sulfate	250	121
Ammonia gas	150	66
Ammonium carbonate	140	60
Ammonium chloride, 10%	200	93
Ammonium chloride, 50%	200	93
Ammonium chloride, sat.	200	93
Ammonium fluoride, 10%	140	60
Ammonium fluoride, 25%	140	60
Ammonium hydroxide, 25%	90	32
Ammonium hydroxide, sat.	90	32
Ammonium nitrate	200	93
Ammonium persulfate	140	60
Ammonium phosphate	150	66
Ammonium sulfate, 10–40%	200	93
Ammonium sulfide	120	49
Ammonium sulfite	100	38
Amyl acetate	190	85
Amyl alcohol	200	93
Amyl chloride	X	X
Aniline	120	49
Antimony trichloride, 50%	200	93
Aqua regia, 3:1	X	X
Barium carbonate	250	121
Barium chloride	250	121
Barium hydroxide	X	X
Barium sulfate	180	82
Barium sulfide	X	X

(continued)

TABLE 4.27 *Continued*

Chemical	Maximum Temperature	
	°F	°C
Benzaldehyde	X	X
Benzene	90	32
Benzene sulfonic acid, 10%	120	49
Benzoic acid	250	121
Benzyl alcohol	X	X
Benzyl chloride	X	X
Borax	190	88
Boric acid	180	82
Bromine gas, dry	100	38
Bromine gas, moist	100	38
Bromine, liquid	X	X
Butyl acetate	80	27
Butyl alcohol	100	38
n-Butylamine	X	X
Butyric acid, 20%	200	93
Calcium bisulfide	X	X
Calcium bisulfite	150	66
Calcium carbonate	210	99
Calcium chlorate	250	121
Calcium chloride	250	121
Calcium hydroxide, sat.	X	X
Calcium hypochlorite, 20%	80	27
Calcium nitrate	220	104
Calcium oxide	150	66
Calcium sulfate	250	121
Caprylic acid	140	60
Carbon bisulfide	X	X
Carbon dioxide, dry	250	121
Carbon dioxide, wet	250	121
Carbon disulfide	X	X
Carbon monoxide	170	77
Carbon tetrachloride	120	49
Carbonic acid	160	71
Cellosolve	80	27
Chlorine gas, dry	200	93
Chlorine gas, wet	220	104
Chlorine, liquid	X	X
Chloroacetic acid, 25%	90	32
Chloroacetic acid, 50% water	100	38
Chlorobenzene	X	X
Chloroform	X	X
Chlorosulfonic acid	X	X
Chromic acid, 10%	180	82
Chromic acid, 50%	140	60
Chromyl chloride	210	99
Citric acid, 15%	250	121
Citric acid, conc.	250	121

(continued)

TABLE 4.27   *Continued*

Chemical	Maximum Temperature	
	°F	°C
Copper acetate	210	99
Copper chloride	250	121
Copper cyanide	250	121
Copper sulfate	250	121
Cresol	X	X
Cyclohexane	140	60
Dibutyl phthalate	100	38
Dichloroacetic acid	100	38
Dichloroethane (ethylene dichloride)	X	X
Ethylene glycol	250	121
Ferric chloride	250	121
Ferric chloride, 50% in water	250	121
Ferric nitrate, 10–50%	250	121
Ferrous chloride	250	121
Ferrous nitrate	160	71
Hydrobromic acid, 20%	160	71
Hydrobromic acid, 50%	200	93
Hydrobromic acid, dil.	200	93
Hydrochloric acid, 20%	230	110
Hydrochloric acid, 38%	180	82
Hydrocyanic acid, 10%	150	66
Hydrofluoric acid, 10%	100	38
Hydrofluoric acid, 30%	120	49
Lactic acid, 25%	200	93
Lactic acid, conc.	200	93
Magnesium chloride	250	121
Malic acid, 10%	90	32
Methyl chloride	80	27
Methyl ethyl ketone	X	X
Methyl isobutyl ketone	80	27
Muriatic acid	190	88
Nitric acid, 20%	80	27
Nitric acid, 5%	210	99
Nitric acid, 70%	80	27
Nitrous acid, conc.	90	32
Oleum	X	X
Perchloric acid, 10%	90	32
Perchloric acid, 70%	90	32
Phenol, 5%	90	32
Phosphoric acid, 50–80%	250	121
Picric acid	100	38
Potassium bromide, 30%	230	110
Salicylic acid	130	54
Sodium carbonate, 10%	190	88
Sodium chloride	250	121
Sodium hydroxide, 10%	110	43
Sodium hydroxide, 50%	X	X

(continued)

TABLE 4.27 *Continued*

Chemical	Maximum Temperature	
	°F	°C
Sodium hypochlorite, 20%	X	X
Sodium hypochlorite, conc.	X	X
Sodium hypochlorite, conc.	X	X
Sodium sulfide, to 50%	X	X
Stannic chloride	80	27
Stannous chloride	250	121
Sulfuric acid, 10%	260	127
Sulfuric acid, 100%	X	X
Sulfuric acid, 50%	200	93
Sulfuric acid, 70%	190	88
Sulfuric acid, 90%	X	X
Sulfuric acid, 98%	X	X
Sulfuric acid, fuming	X	X
Sulfurous acid, 10%	80	27
Thionyl chloride	X	X
Toluene	110	43
Trichloroacetic acid, 50%	200	93
White liquor	X	X
Zinc chloride	200	93

The chemicals listed are in the pure state or in a saturated solution unless otherwise indicated. Compatibility is shown to the maximum allowable temperature for which data is available. Incompatibility is shown by an X. A blank space indicates that the data is unavailable.

Source: From P.A. Schweitzer. 2004. *Corrosion Resistance Tables*, Vols. 1–4, 5th ed., New York: Marcel Dekker.

organics, phenol, and acid halides. Table 4.27 lists the compatibility of halogenated polyesters with selected corrodents. Reference [1] provides additional information.

Halogenated polyesters are widely used in the pulp and paper industry in bleach atmospheres.

4.10.23 Silicones

The silicone systems are quite expensive, being based on organic silicone compounds (which have silicon rather than carbon linkages in the structure). They are primarily used for high-temperature service, where carbon-based linings would oxidize.

Typically, the silicon atoms will have one or more side groups attached to them, generally phenol ($C_6H_5^-$), methyl (CH_3^-), or vinyl ($CH_2=CH^-$) units. These groups impart properties such as solvent resistance, lubricity, and reactivity with organic chemicals and polymers. Because these side groups affect the corrosion resistance of the resin, it is necessary to verify with the supplier the properties of the resin being supplied.

The maximum allowable operating temperature is 572°F (300°C). These resins are available for operation at cryogenic temperatures.

A second high-temperature formulation with aluminum can be operated up to 1200°F (649°C). This high-temperature type requires baking for a good cure. It is also water repellent.

The silicone resins can be used in contact with dilute acids and alkalies, alcohols, animal and vegetable oils, and lubrication oils. They are also resistant to aliphatic hydrocarbons, but aromatic solvents such as benzene, toluene, gasoline, and chlorinated solvents will cause excessive swelling.

Table 4.28 lists the corrosion resistance of methyl appended silicone with selected corrodents. Reference [1] provides additional listings.

TABLE 4.28

Compatibility of Methyl-Appended Silicone with Selected Corrodents

| | Maximum Temperature | |
Chemical	°F	°C
Acetic acid, 10%	90	32
Acetic acid, 50%	90	32
Acetic acid, 80%	90	32
Acetic acid, glacial	90	32
Acetone	100	43
Acrylic acid, 75%	80	27
Acrylonitrile	X	X
Alum	220	104
Aluminum sulfate	410	210
Ammonium chloride, 10%	X	X
Ammonium chloride, 50%	80	27
Ammonium chloride, sat.	80	27
Ammonium fluoride, 25%	80	27
Ammonium hydroxide, 25%	X	X
Ammonium nitrate	210	99
Amyl acetate	80	27
Amyl alcohol	X	X
Amyl chloride	X	X
Aniline	X	X
Antimony trichloride	80	27
Aqua regia, 3:1	X	X
Benzene	X	X
Benzyl chloride	X	X
Boric acid	390	189
Butyl alcohol	80	27
Calcium bisulfide	400	204
Calcium chloride	300	149
Calcium hydroxide, 30%	200	93
Calcium hydroxide, sat.	400	204
Carbon bisulfide	X	X

(continued)

TABLE 4.28 *Continued*

Chemical	Maximum Temperature	
	°F	°C
Carbon disulfide	X	X
Carbon monoxide	400	204
Carbonic acid	400	204
Chlorobenzene	X	X
Chlorosulfonic acid	X	X
Ethylene glycol	400	204
Ferric chloride	400	204
Hydrobromic acid, 50%	X	X
Hydrochloric acid, 20%	90	32
Hydrochloric acid, 38%	X	X
Hydrofluoric acid, 30%	X	X
Lactic acid, all conc.	80	27
Lactic acid, conc.	80	27
Magnesium chloride	400	204
Methyl alcohol	410	210
Methyl ethyl ketone	X	X
Methyl isobutyl ketone	X	X
Nitric acid, 20%	X	X
Nitric acid, 5%	80	27
Nitric acid, 70%	X	X
Nitric acid, anhydrous	X	X
Oleum	X	X
Phenol	X	X
Phosphoric acid, 50–80%	X	X
Propyl alcohol	400	204
Sodium carbonate	300	149
Sodium chloride, 10%	400	204
Sodium hydroxide, 10%	90	27
Sodium hydroxide, 50%	90	27
Sodium hydroxide, conc.	90	27
Sodium hypochlorite, 20%	X	X
Sodium sulfate	400	204
Stannic chloride	80	27
Sulfuric acid, 10%	X	X
Sulfuric acid, 50%	X	X
Sulfuric acid, 70%	X	X
Sulfuric acid, 90%	X	X
Sulfuric acid, 98%	X	X
Sulfuric acid, 100%	X	X
Sulfuric acid, fuming	X	X
Sulfurous acid	X	X
Tartaric acid	400	204
Tetrahydrofuran	X	X
Toluene	X	X
Tributyl phosphate	X	X
Turpentine	X	X
Vinegar	400	204

(continued)

TABLE 4.28 *Continued*

| | Maximum Temperature | |
Chemical	°F	°C
Water, acid mine	210	99
Water, demineralized	210	99
Water, distilled	210	99
Water, salt	210	99
Water, sea	210	99
Xylene	X	X
Zinc chloride	400	204

The chemicals listed are in the pure state or in a saturated solution unless otherwise indicated. Compatibility is shown to the maximum allowable temperature. Incompatibility is shown by an X.

Source: From P.A. Schweitzer. 2004. *Corrosion Resistance Tables*, Vols. 1–4, 5th ed., New York: Marcel Dekker.

The silicones are used primarily as linings for high-temperature exhaust stacks, ovens, and space heaters.

References

1. P.A. Schweitzer. 2004. *Corrosion Resistance Tables*, 5th ed., vols. 1–4, New York: Marcel Dekker.
2. J.S. Fry, C.N. Merriam, and W.H. Boyd. 1985. Chemistry and technology of phenolic resins and coatings, in *Symposium Series Applied Polymer Science*, Washington, DC: American Chemical Society.
3. R.T. Morrison and R.N. Boyd. 1973. *Organic Chemistry*, 3rd ed., Boston: Allyn Bacon.
4. P.A. Schweitzer. 2000. *Mechanical and Corrosion Resistant Properties of Plastics and Elastomers*, New York: Marcel Dekker.

5

Comparative Resistance of Organic Linings

Following is a series of tables comparing the corrosion resistance of various organic lining materials in contact with commonly used corrodents.

The chemicals listed are in the pure state or in a saturated solution unless otherwise indicated. Compatibility is shown to the maximum allowable temperature for which data is available. An "X" shows incompatibility. A blank space indicates that data is unavailable.

One must keep in mind that most of the lining materials listed are capable of being formulated to meet specific conditions. In the tables, when a lining material is listed as satisfactory, it means that at least one formulation is acceptable. Therefore, before being used the manufacturer should be contacted to be sure that his formulation will be compatible with the application. More extensive listings will be found in Reference [1].

Acetic Acid, 10%

	Maximum Temperature °F (°C)
Phenolics	212 (100)
Epoxy	190 (88)
Furans	212 (100)
Vinyl ester	200 (93)
Epoxy polyamide	X
Coal tar epoxy	100 (38)
Coal tar	
Urethanes	90 (32)
Neoprene	160 (71)
Polysulfides	80 (27)
Hypalon	200 (93)
Plastisols	100 (38)
PFA	450 (232)
FEP	400 (204)
PTFE	450 (232)
ETFE	250 (121)
ECTFE	250 (121)
Fluoroelastomers	190 (88)
PVDF	300 (149)
Isophthalic polyesters	180 (82)
Bisphenol A fumurate	220 (104)
Hydrogenated polyester	200 (93)
Halogenated polyester	140 (60)
Methyl-appended silicone	90 (32)

Acetic Acid, 50%

	Maximum Temperature °F (°C)
Phenolics	
Epoxy	110 (43)
Furans	160 (71)
Vinyl ester	180 (82)
Epoxy polyamide	X
Coal tar epoxy, 20%	100 (38)
Coal tar	
Urethanes, 20%	90 (32)
Neoprene	160 (71)
Polysulfides	80 (27)
Hypalon	200 (93)
Plastisols	90 (32)
PFA	450 (232)
FEP	400 (204)
PTFE	450 (232)
ETFE	250 (121)
ECTFE	250 (121)
Fluoroelastomers	180 (82)
PVDF	300 (149)
Isophthalic polyesters	110 (43)
Bisphenol A fumurate	160 (71)
Hydrogenated polyester	160 (71)
Halogenated polyester	90 (32)
Methyl-appended silicone	90 (32)

Acetic Acid, 80%

	Maximum Temperature °F (°C)
Phenolics	
Epoxy	110 (43)
Furans	80 (27)
Vinyl ester	150 (66)
Epoxy polyamide	X
Coal tar epoxy	
Coal tar	
Urethanes	
Neoprene	160 (71)
Polysulfides	80 (27)
Hypalon	200 (93)
Plastisols	X
PFA	450 (232)
FEP	400 (204)
PTFE	450 (232)
ETFE	230 (110)
ECTFE	150 (66)
Fluoroelastomers	180 (82)
PVDF	190 (88)
Isophthalic polyesters	X
Bisphenol A fumurate	160 (71)
Hydrogenated polyester	
Halogenated polyester	
Methyl-appended silicone	90 (32)

Acetic Acid, Glacial

	Maximum Temperature °F (°C)
Phenolics	70 (21)
Epoxy	
Furans	80 (27)
Vinyl ester	150 (66)
Epoxy polyamide	X
Coal tar epoxy	
Coal tar	
Urethanes	
Neoprene	X
Polysulfides	80 (87)
Hypalon	X
Plastisols	X
PFA	450 (232)
FEP	400 (204)
PTFE	450 (232)
ETFE	230 (110)
ECTFE	200 (93)
Fluoroelastomers	X
PVDF	190 (88)
Isophthalic polyesters	X
Bisphenol A fumurate	X
Hydrogenated polyester	
Halogenated polyester	110 (43)
Methyl-appended silicone	90 (32)

Acetic Acid, Vapors

	Maximum Temperature °F (°C)
Phenolics	110 (43)
Epoxy	X
Furans	
Vinyl ester	
Epoxy polyamide	X
Coal tar epoxy	100 (38)
Coal tar	
Urethanes	90 (32)
Neoprene	90 (32)
Polysulfides	90 (32)
Hypalon	90 (32)
Plastisols	X
PFA	200 (93)
FEP	400 (204)
PTFE	400 (204)
ETFE	
ECTFE	200 (93)
Fluoroelastomers	90 (32)
PVDF	180 (82)
Isophthalic polyesters, 50%	110 (43)
Bisphenol A fumurate	
Hydrogenated polyester	
Halogenated polyester, 25%	180 (82)
Methyl-appended silicone	

Acetic Anhydride

	Maximum Temperature °F (°C)
Phenolics	X
Epoxy	X
Furans	200 (93)
Vinyl ester	100 (38)
Epoxy polyamide	X
Coal tar epoxy	X
Coal tar	
Urethanes	X
Neoprene	200 (93)
Polysulfides	
Hypalon	200 (93)
Plastisols	X
PFA	200 (93)
FEP	400 (202)
PTFE	450 (232)
ETFE	300 (149)
ECTFE	100 (38)
Fluoroelastomers	X
PVDF	100 (38)
Isophthalic polyesters	X
Bisphenol A fumurate	100 (38)
Hydrogenated polyester	X
Halogenated polyester	100 (38)
Methyl-appended silicone	

Acetone

	Maximum Temperature °F (°C)
Phenolics	X
Epoxy	110 (43)
Furans	80 (27)
Vinyl ester	X
Epoxy polyamide	X
Coal tar epoxy	X
Coal tar	X
Urethanes	90 (32)
Neoprene	X
Polysulfides	80 (27)
Hypalon	X
Plastisols	X
PFA	450 (232)
FEP	400 (204)
PTFE	450 (232)
ETFE	150 (66)
ECTFE	150 (66)
Fluoroelastomers	X
PVDF	X
Isophthalic polyesters	X
Bisphenol A fumurate	X
Hydrogenated polyester	X
Halogenated polyester	X
Methyl-appended silicone	100 (43)

Ammonium Carbonate

	Maximum Temperature °F (°C)
Phenolics	90 (32)
Epoxy	140 (60)
Furans	240 (116)
Vinyl ester	150 (66)
Epoxy polyamide	
Coal tar epoxy	
Coal tar	
Urethanes	
Neoprene	200 (93)
Polysulfides	
Hypalon	140 (60)
Plastisols	140 (60)
PFA	450 (232)
FEP	400 (204)
PTFE	450 (232)
ETFE	300 (149)
ECTFE	300 (149)
Fluoroelastomers	190 (88)
PVDF	280 (138)
Isophthalic polyesters	X
Bisphenol A fumurate	90 (32)
Hydrogenated polyester	
Halogenated polyester	140 (60)
Methyl-appended silicone	

Ammonium Hydroxide, 25%

	Maximum Temperature °F (°C)
Phenolics	X
Epoxy	140 (60)
Furans	250 (121)
Vinyl ester	100 (38)
Epoxy polyamide	100 (38)
Coal tar epoxy	110 (43)
Coal tar	
Urethanes	90 (32)
Neoprene	200 (93)
Polysulfides	X
Hypalon	200 (93)
Plastisols	140 (60)
PFA	450 (232)
FEP	400 (204)
PTFE	450 (232)
ETFE	300 (149)
ECTFE	300 (149)
Fluoroelastomers	190 (88)
PVDF	280 (138)
Isophthalic polyesters	X
Bisphenol A fumurate, 20%	140 (60)
Hydrogenated polyester	
Halogenated polyester	90 (32)
Methyl-appended silicone	X

Ammonium Hydroxide, Sat.

	Maximum Temperature °F (°C)
Phenolics	X
Epoxy	150 (66)
Furans	200 (93)
Vinyl ester	130 (54)
Epoxy polyamide	
Coal tar epoxy	
Coal tar	
Urethanes	
Neoprene	200 (93)
Polysulfides	X
Hypalon	200 (93)
Plastisols	140 (60)
PFA	450 (232)
FEP	400 (204)
PTFE	450 (232)
ETFE	300 (149)
ECTFE	300 (149)
Fluoroelastomers	190 (88)
PVDF	280 (138)
Isophthalic polyesters	X
Bisphenol A fumurate	
Hydrogenated polyester	
Halogenated polyester	90 (32)
Methyl-appended silicone	X

Aniline

	Maximum Temperature °F (°C)
Phenolics	X
Epoxy	150 (66)
Furans	80 (27)
Vinyl ester	X
Epoxy polyamide	
Coal tar epoxy	
Coal tar	
Urethanes	
Neoprene	X
Polysulfides	
Hypalon	140 (60)
Plastisols	X
PFA	450 (232)
FEP	400 (204)
PTFE	450 (232)
ETFE	230 (110)
ECTFE	90 (32)
Fluoroelastomers	230 (110)
PVDF	200 (93)
Isophthalic polyesters	X
Bisphenol A fumurate	X
Hydrogenated polyester	X
Halogenated polyester	120 (49)
Methyl-appended silicone	X

Benzoic Acid

	Maximum Temperature °F (°C)
Phenolics, 10%	100 (38)
Epoxy	200 (93)
Furans	260 (127)
Vinyl ester	180 (82)
Epoxy polyamide	100 (38)
Coal tar epoxy	100 (38)
Coal tar	
Urethanes	X
Neoprene	150 (66)
Polysulfides	150 (66)
Hypalon	200 (93)
Plastisols	140 (60)
PFA	450 (232)
FEP	400 (204)
PTFE	450 (232)
ETFE	270 (132)
ECTFE	250 (121)
Fluoroelastomers	400 (204)
PVDF	250 (121)
Isophthalic polyesters	180 (82)
Bisphenol A fumurate	180 (82)
Hydrogenated polyester	210 (99)
Halogenated polyester	250 (121)
Methyl-appended silicone	

Bromine Gas, Dry

	Maximum Temperature °F (°C)
Phenolics	
Epoxy	X
Furans	X
Vinyl ester	100 (38)
Epoxy polyamide	X
Coal tar epoxy	100 (38)
Coal tar	X
Urethanes	
Neoprene	X
Polysulfides	
Hypalon	60 (16)
Plastisols	X
PFA	450 (232)
FEP	200 (93)
PTFE	450 (232)
ETFE	150 (66)
ECTFE	X
Fluoroelastomers, 25%	180 (82)
PVDF	210 (99)
Isophthalic polyesters	X
Bisphenol A fumurate	90 (32)
Hydrogenated polyester	
Halogenated polyester	100 (38)
Methyl-appended silicone	

Bromine Gas, Moist

	Maximum Temperature °F (°C)
Phenolics	
Epoxy	X
Furans	X
Vinyl ester	100 (38)
Epoxy polyamide	X
Coal tar epoxy	X
Coal tar	X
Urethanes	
Neoprene	X
Polysulfides	
Hypalon	60 (16)
Plastisols	X
PFA	200 (93)
FEP	200 (93)
PTFE	250 (121)
ETFE	
ECTFE	
Fluoroelastomers, 25%	180 (82)
PVDF	210 (99)
Isophthalic polyesters	X
Bisphenol A fumurate	100 (38)
Hydrogenated polyester	
Halogenated polyester	100 (38)
Methyl-appended silicone	

Bromine, Liquid

	Maximum Temperature °F (°C)
Phenolics	
Epoxy	X
Furans, 3% max	300 (149)
Vinyl ester	X
Epoxy polyamide	
Coal tar epoxy	
Coal tar	
Urethanes	
Neoprene	X
Polysulfides	
Hypalon	60 (16)
Plastisols	X
PFA	450 (232)
FEP	400 (204)
PTFE	450 (232)
ETFE	
ECTFE	150 (66)
Fluoroelastomers	350 (177)
PVDF	140 (60)
Isophthalic polyesters, 50%	X
Bisphenol A fumurate	X
Hydrogenated polyester	X
Halogenated polyester	X
Methyl-appended silicone	

Calcium Hydroxide

	Maximum Temperature °F (°C)
Phenolics	X
Epoxy	180 (32)
Furans	260 (170)
Vinyl ester	180 (82)
Epoxy polyamide	140 (60)
Coal tar epoxy	100 (38)
Coal tar	X
Urethanes	90 (32)
Neoprene	230 (110)
Polysulfides	X
Hypalon	200 (93)
Plastisols	140 (60)
PFA	450 (232)
FEP	400 (204)
PTFE	450 (232)
ETFE	300 (149)
ECTFE	300 (149)
Fluoroelastomers	400 (204)
PVDF	280 (138)
Isophthalic polyesters	160 (71)
Bisphenol A fumurate	160 (71)
Hydrogenated polyester	
Halogenated polyester	X
Methyl-appended silicone	400 (201)

Carbon Tetrachloride

	Maximum Temperature °F (°C)
Phenolics	200 (93)
Epoxy	170 (77)
Furans	212 (100)
Vinyl ester	180 (82)
Epoxy polyamide	212 (100)
Coal tar epoxy	X
Coal tar	X
Urethanes	X
Neoprene	X
Polysulfides	
Hypalon	200 (93)
Plastisols	X
PFA	450 (232)
FEP	400 (204)
PTFE	450 (232)
ETFE	270 (132)
ECTFE	300 (149)
Fluoroelastomers	350 (177)
PVDF	280 (138)
Isophthalic polyesters	X
Bisphenol A fumurate	110 (43)
Hydrogenated polyester	X
Halogenated polyester	120 (49)
Methyl-appended silicone	X

Chlorine Gas, Wet

	Maximum Temperature °F (°C)
Phenolics	X
Epoxy	X
Furans	250 (121)
Vinyl ester	250 (121)
Epoxy polyamide	X
Coal tar epoxy	X
Coal tar	X
Urethanes	X
Neoprene	X
Polysulfides	
Hypalon	90 (32)
Plastisols	X
PFA	450 (232)
FEP	400 (204)
PTFE	450 (232)
ETFE	250 (121)
ECTFE	250 (121)
Fluoroelastomers	190 (88)
PVDF, 10%	210 (99)
Isophthalic polyesters	160 (71)
Bisphenol A fumurate	200 (93)
Hydrogenated polyester	210 (99)
Halogenated polyester	220 (104)
Methyl-appended silicone	X

Chlorine, Liquid

	Maximum Temperature °F (°C)
Phenolics	X
Epoxy	
Furans, 3% max	X
Vinyl ester	X
Epoxy polyamide	X
Coal tar epoxy	
Coal tar	
Urethanes	
Neoprene	X
Polysulfides	
Hypalon	
Plastisols	X
PFA	X
FEP	400 (204)
PTFE	X
ETFE	
ECTFE	250 (121)
Fluoroelastomers	190 (88)
PVDF	210 (99)
Isophthalic polyesters	X
Bisphenol A fumurate	X
Hydrogenated polyester	
Halogenated polyester	X
Methyl-appended silicone	X

Chlorobenzene

	Maximum Temperature °F (°C)
Phenolics	260 (127)
Epoxy	150 (66)
Furans	260 (127)
Vinyl ester	110 (43)
Epoxy polyamide	100 (38)
Coal tar epoxy	100 (38)
Coal tar	X
Urethanes	X
Neoprene	X
Polysulfides	X
Hypalon	X
Plastisols	X
PFA	450 (232)
FEP	400 (204)
PTFE	450 (232)
ETFE	210 (99)
ECTFE	150 (66)
Fluoroelastomers	400 (204)
PVDF	220 (104)
Isophthalic polyesters	X
Bisphenol A fumurate	X
Hydrogenated polyester	
Halogenated polyester	X
Methyl-appended silicone	X

Chloroform

	Maximum Temperature °F (°C)
Phenolics	160 (71)
Epoxy	110 (43)
Furans	X
Vinyl ester	X
Epoxy polyamide	X
Coal tar epoxy	X
Coal tar	X
Urethanes	X
Neoprene	X
Polysulfides	
Hypalon	X
Plastisols	X
PFA	450 (232)
FEP	400 (204)
PTFE	450 (232)
ETFE	230 (110)
ECTFE	250 (121)
Fluoroelastomers	400 (204)
PVDF	250 (121)
Isophthalic polyesters	X
Bisphenol A fumurate	X
Hydrogenated polyester	X
Halogenated polyester	X
Methyl-appended silicone	X

Chlorosulfonic Acid

	Maximum Temperature °F (°C)
Phenolics	140 (60)
Epoxy	X
Furans	260 (127)
Vinyl ester	X
Epoxy polyamide	X
Coal tar epoxy	X
Coal tar	X
Urethanes	X
Neoprene	X
Polysulfides	
Hypalon	X
Plastisols	60 (16)
PFA	450 (232)
FEP	400 (204)
PTFE	450 (232)
ETFE	80 (27)
ECTFE	80 (27)
Fluoroelastomers	X
PVDF	110 (43)
Isophthalic polyesters	X
Bisphenol A fumurate	X
Hydrogenated polyester	
Halogenated polyester	X
Methyl-appended silicone	X

Citric Acid, 10%

	Maximum Temperature °F (°C)
Phenolics	160 (71)
Epoxy	190 (88)
Furans	250 (121)
Vinyl ester	210 (99)
Epoxy polyamide	100 (38)
Coal tar epoxy	100 (38)
Coal tar	
Urethanes	
Neoprene	150 (66)
Polysulfides	X
Hypalon	200 (93)
Plastisols	140 (60)
PFA	450 (232)
FEP	400 (204)
PTFE	450 (232)
ETFE	120 (49)
ECTFE	300 (149)
Fluoroelastomers	300 (149)
PVDF	250 (121)
Isophthalic polyesters	160 (71)
Bisphenol A fumurate	220 (104)
Hydrogenated polyester	200 (93)
Halogenated polyester	250 (121)
Methyl-appended silicone	X

Citric Acid, Conc.

	Maximum Temperature °F (°C)
Phenolics	160 (71)
Epoxy	X
Furans	250 (121)
Vinyl ester	200 (93)
Epoxy polyamide	100 (38)
Coal tar epoxy	100 (38)
Coal tar	
Urethanes	
Neoprene	200 (93)
Polysulfides	X
Hypalon	250 (121)
Plastisols	140 (60)
PFA	370 (188)
FEP	400 (204)
PTFE	450 (232)
ETFE	
ECTFE	300 (149)
Fluoroelastomers	400 (204)
PVDF	250 (121)
Isophthalic polyesters	200 (93)
Bisphenol A fumurate	220 (104)
Hydrogenated polyester	210 (99)
Halogenated polyester	250 (121)
Silicone	390 (199)

Dextrose

	Maximum Temperature °F (°C)
Phenolics	
Epoxy	100 (38)
Furans, 3% max	260 (127)
Vinyl ester	240 (116)
Epoxy polyamide	100 (38)
Coal tar epoxy	100 (38)
Coal tar	
Urethanes	X
Neoprene	200 (93)
Polysulfides	
Hypalon	200 (93)
Plastisols	140 (60)
PFA	200 (93)
FEP	400 (204)
PTFE	450 (232)
ETFE	
ECTFE	240 (116)
Fluoroelastomers	400 (204)
PVDF	280 (138)
Isophthalic polyesters	180 (82)
Bisphenol A fumurate	220 (104)
Hydrogenated polyester	
Halogenated polyester	220 (104)
Methyl-appended silicone	170 (77)

Dichloroacetic Acid

	Maximum Temperature °F (°C)
Phenolics	
Epoxy	X
Furans	X
Vinyl ester	100 (38)
Epoxy polyamide	X
Coal tar epoxy	X
Coal tar	
Urethanes	
Neoprene	X
Polysulfides	
Hypalon	
Plastisols, 20%	100 (38)
PFA	
FEP	400 (204)
PTFE	400 (204)
ETFE	150 (66)
ECTFE	
Fluoroelastomers	
PVDF	120 (49)
Isophthalic polyesters	X
Bisphenol A fumurate	100 (38)
Hydrogenated polyester	
Halogenated polyester	100 (38)
Methyl-appended silicone	

Diesel Fuels

	Maximum Temperature °F (°C)
Phenolics	
Epoxy	100 (38)
Furans	240 (116)
Vinyl ester	220 (104)
Epoxy polyamide	100 (38)
Coal tar epoxy	100 (38)
Coal tar	
Urethanes	
Neoprene	80 (27)
Polysulfides	80 (27)
Hypalon	80 (27)
Plastisols	100 (38)
PFA	200 (93)
FEP	400 (204)
PTFE	400 (204)
ETFE	300 (149)
ECTFE	300 (149)
Fluoroelastomers	400 (204)
PVDF	280 (138)
Isophthalic polyesters	160 (71)
Bisphenol A fumurate	180 (82)
Hydrogenated polyester	
Halogenated polyester	180 (82)
Methyl-appended silicone	

Diethylamine

	Maximum Temperature °F (°C)
Phenolics	X
Epoxy	X
Furans	200 (93)
Vinyl ester	X
Epoxy polyamide	X
Coal tar epoxy	X
Coal tar	X
Urethanes	
Neoprene	120 (49)
Polysulfides	
Hypalon	X
Plastisols	X
PFA	200 (93)
FEP	400 (204)
PTFE	400 (204)
ETFE	200 (93)
ECTFE	X
Fluoroelastomers	X
PVDF	100 (38)
Isophthalic polyesters	120 (49)
Bisphenol A fumurate	X
Hydrogenated polyester	
Halogenated polyester	X
Methyl-appended silicone	

Dimethyl Formamide

	Maximum Temperature °F (°C)
Phenolics	
Epoxy	X
Furans	X
Vinyl ester	X
Epoxy polyamide	X
Coal tar epoxy	X
Coal tar	X
Urethanes	
Neoprene	160 (71)
Polysulfides	
Hypalon	X
Plastisols	X
PFA	200 (93)
FEP	400 (204)
PTFE	450 (232)
ETFE	250 (121)
ECTFE	100 (38)
Fluoroelastomers	X
PVDF	X
Isophthalic polyesters	X
Bisphenol A fumurate	X
Hydrogenated polyester	X
Halogenated polyester	X
Methyl-appended silicone	

Ethyl Acetate

	Maximum Temperature °F (°C)
Phenolics	
Epoxy	X
Furans	200 (93)
Vinyl ester	X
Epoxy polyamide	X
Coal tar epoxy	X
Coal tar	X
Urethanes	90 (32)
Neoprene	X
Polysulfides	80 (27)
Hypalon	140 (60)
Plastisols	X
PFA	200 (93)
FEP	400 (204)
PTFE	400 (204)
ETFE	150 (66)
ECTFE	150 (66)
Fluoroelastomers	X
PVDF	160 (71)
Isophthalic polyesters	X
Bisphenol A fumurate	X
Hydrogenated polyester	X
Halogenated polyester	X
Methyl-appended silicone	

Ethyl Alcohol

	Maximum Temperature °F (°C)
Phenolics	110 (43)
Epoxy	140 (60)
Furans	140 (60)
Vinyl ester	100 (38)
Epoxy polyamide	100 (38)
Coal tar epoxy	100 (38)
Coal tar	
Urethanes	90 (32)
Neoprene	200 (93)
Polysulfides	150 (66)
Hypalon	200 (93)
Plastisols	140 (60)
PFA	200 (93)
FEP	200 (93)
PTFE	400 (204)
ETFE	300 (149)
ECTFE	300 (149)
Fluoroelastomers	300 (149)
PVDF	280 (138)
Isophthalic polyesters	80 (27)
Bisphenol A fumurate	90 (32)
Hydrogenated polyester	90 (32)
Halogenated polyester	140 (60)
Silicone	400 (204)

Hydrobromic Acid, Dil.

	Maximum Temperature °F (°C)
Phenolics	200 (93)
Epoxy	180 (82)
Furans	212 (100)
Vinyl ester	180 (82)
Epoxy polyamide	X
Coal tar epoxy	X
Coal tar	X
Urethanes	
Neoprene	X
Polysulfides	
Hypalon	90 (32)
Plastisols	140 (60)
PFA	450 (232)
FEP	400 (204)
PTFE	450 (232)
ETFE	300 (149)
ECTFE	300 (149)
Fluoroelastomers	400 (204)
PVDF	260 (127)
Isophthalic polyesters	120 (49)
Bisphenol A fumurate	220 (104)
Hydrogenated polyester	
Halogenated polyester	200 (93)
Methyl-appended silicone	X

Hydrobromic Acid, 20%

	Maximum Temperature °F (°C)
Phenolics	200 (93)
Epoxy	180 (82)
Furans	212 (100)
Vinyl ester	180 (82)
Epoxy polyamide	X
Coal tar epoxy	X
Coal tar	X
Urethanes	
Neoprene	X
Polysulfides	
Hypalon	100 (38)
Plastisols	190 (60)
PFA	450 (232)
FEP	400 (204)
PTFE	450 (232)
ETFE	300 (149)
ECTFE	300 (149)
Fluoroelastomers	400 (204)
PVDF	280 (138)
Isophthalic polyesters	140 (60)
Bisphenol A fumurate	220 (104)
Hydrogenated polyester	90 (32)
Halogenated polyester	160 (71)
Methyl-appended silicone	X

Hydrobromic Acid, 50%

	Maximum Temperature °F (°C)
Phenolics	200 (93)
Epoxy	110 (43)
Furans	212 (100)
Vinyl ester	200 (93)
Epoxy polyamide	X
Coal tar epoxy	X
Coal tar	X
Urethanes	
Neoprene	X
Polysulfides	
Hypalon	100 (38)
Plastisols	140 (60)
PFA	450 (232)
FEP	400 (204)
PTFE	450 (232)
ETFE	300 (149)
ECTFE	300 (149)
Fluoroelastomers	400 (204)
PVDF	280 (138)
Isophthalic polyesters	140 (60)
Bisphenol A fumurate	160 (71)
Hydrogenated polyester	90 (32)
Halogenated polyester	200 (93)
Methyl-appended silicone	X

Hydrochloric Acid, Dil.

	Maximum Temperature °F (°C)
Phenolics	300 (149)
Epoxy	200 (93)
Furans	212 (100)
Vinyl ester	220 (104)
Epoxy polyamide	100 (38)
Coal tar epoxy	100 (38)
Coal tar	X
Urethanes	X
Neoprene	X
Polysulfides	X
Hypalon	160 (71)
Plastisols	140 (60)
PFA	450 (232)
FEP	400 (204)
PTFE	450 (232)
ETFE	300 (149)
ECTFE	300 (149)
Fluoroelastomers	350 (177)
PVDF	280 (138)
Isophthalic polyesters	160 (71)
Bisphenol A fumurate	190 (88)
Hydrogenated polyester	180 (82)
Halogenated polyester	230 (110)
Methyl-appended silicone	90 (32)

Hydrochloric Acid, 20%

	Maximum Temperature °F (°C)
Phenolics	300 (149)
Epoxy	200 (93)
Furans	212 (100)
Vinyl ester	220 (104)
Epoxy polyamide	X
Coal tar epoxy	X
Coal tar	X
Urethanes	X
Neoprene	X
Polysulfides	X
Hypalon	160 (71)
Plastisols	140 (60)
PFA	450 (232)
FEP	400 (204)
PTFE	450 (232)
ETFE	300 (149)
ECTFE	300 (149)
Fluoroelastomers	350 (177)
PVDF	280 (138)
Isophthalic polyesters	160 (71)
Bisphenol A fumurate	190 (88)
Hydrogenated polyester	180 (82)
Halogenated polyester	230 (110)
Methyl-appended silicone	90 (32)

Hydrochloric Acid, 35%

	Maximum Temperature °F (°C)
Phenolics	300 (149)
Epoxy	140 (60)
Furans	80 (27)
Vinyl ester	180 (82)
Epoxy polyamide	X
Coal tar epoxy	X
Coal tar	X
Urethanes	X
Neoprene	X
Polysulfides	X
Hypalon	140 (60)
Plastisols	140 (60)
PFA	450 (232)
FEP	400 (204)
PTFE	450 (232)
ETFE	300 (149)
ECTFE	300 (149)
Fluoroelastomers	350 (177)
PVDF	280 (138)
Isophthalic polyesters	160 (71)
Bisphenol A fumurate	X
Hydrogenated polyester	190 (88)
Halogenated polyester	180 (82)
Methyl-appended silicone	X

Hydrofluoric Acid, 30%

	Maximum Temperature °F (°C)
Phenolics	X
Epoxy	X
Furans	230 (110)
Vinyl ester	X
Epoxy polyamide	X
Coal tar epoxy	X
Coal tar	X
Urethanes	
Neoprene	X
Polysulfides	X
Hypalon	90 (32)
Plastisols	120 (49)
PFA	450 (232)
FEP	400 (204)
PTFE	450 (232)
ETFE	270 (132)
ECTFE	250 (121)
Fluoroelastomers	210 (99)
PVDF	260 (127)
Isophthalic polyesters	X
Bisphenol A fumurate	X
Hydrogenated polyester	
Halogenated polyester	120 (49)
Methyl-appended silicone	X

Hydrofluoric Acid, 70%

	Maximum Temperature °F (°C)
Phenolics	X
Epoxy	X
Furans	140 (60)
Vinyl ester	X
Epoxy polyamide	X
Coal tar epoxy	X
Coal tar	X
Urethanes	
Neoprene	X
Polysulfides	X
Hypalon	90 (32)
Plastisols	68 (20)
PFA	450 (232)
FEP	400 (204)
PTFE	450 (232)
ETFE	250 (121)
ECTFE	240 (116)
Fluoroelastomers	350 (177)
PVDF	200 (93)
Isophthalic polyesters	X
Bisphenol A fumurate	X
Hydrogenated polyester	X
Halogenated polyester	
Methyl-appended silicone	X

Hydrofluoric Acid, 100%

	Maximum Temperature °F (°C)
Phenolics	X
Epoxy	X
Furans	140 (60)
Vinyl ester	X
Epoxy polyamide	X
Coal tar epoxy	X
Coal tar	X
Urethanes	
Neoprene	X
Polysulfides	X
Hypalon	90 (32)
Plastisols	
PFA	450 (232)
FEP	400 (204)
PTFE	450 (232)
ETFE	230 (110)
ECTFE	240 (116)
Fluoroelastomers	X
PVDF	200 (93)
Isophthalic polyesters	X
Bisphenol A fumurate	X
Hydrogenated polyester	
Halogenated polyester	
Methyl-appended silicone	X

Hypochlorous Acid, 100%

	Maximum Temperature °F (°C)
Phenolics	
Epoxy	200 (93)
Furans	X
Vinyl ester	150 (66)
Epoxy polyamide	X
Coal tar epoxy	X
Coal tar	
Urethanes	
Neoprene	X
Polysulfides	
Hypalon	X
Plastisols	140 (60)
PFA	450 (232)
FEP	400 (204)
PTFE	450 (232)
ETFE	300 (149)
ECTFE	300 (149)
Fluoroelastomers	400 (204)
PVDF	280 (138)
Isophthalic polyesters	90 (32)
Bisphenol A fumurate, 20%	90 (32)
Hydrogenated polyester, 50%	210 (99)
Halogenated polyester, 10%	100 (38)
Methyl-appended silicone	

Lactic Acid, 25%

	Maximum Temperature °F (°C)
Phenolics	160 (71)
Epoxy	220 (104)
Furans	212 (100)
Vinyl ester	210 (99)
Epoxy polyamide	X
Coal tar epoxy	X
Coal tar	X
Urethanes	
Neoprene	140 (60)
Polysulfides	X
Hypalon	140 (60)
Plastisols	140 (60)
PFA	450 (232)
FEP	400 (204)
PTFE	450 (232)
ETFE	250 (121)
ECTFE	150 (66)
Fluoroelastomers	300 (149)
PVDF	130 (54)
Isophthalic polyesters	160 (71)
Bisphenol A fumurate	210 (99)
Hydrogenated polyester	210 (99)
Halogenated polyester	200 (93)
Methyl-appended silicone	X

Lactic Acid, Conc.

	Maximum Temperature °F (°C)
Phenolics	
Epoxy	200 (93)
Furans	160 (71)
Vinyl ester	200 (93)
Epoxy polyamide	X
Coal tar epoxy	X
Coal tar	X
Urethanes	
Neoprene	90 (32)
Polysulfides	X
Hypalon	80 (27)
Plastisols	80 (27)
PFA	450 (232)
FEP	400 (204)
PTFE	450 (232)
ETFE	250 (121)
ECTFE	150 (66)
Fluoroelastomers	400 (204)
PVDF	110 (43)
Isophthalic polyesters	160 (71)
Bisphenol A fumurate	220 (104)
Hydrogenated polyester	210 (99)
Halogenated polyester	200 (93)
Methyl-appended silicone	X

Methyl Alcohol

	Maximum Temperature °F (°C)
Phenolics	140 (60)
Epoxy	X
Furans	160 (71)
Vinyl ester	90 (32)
Epoxy polyamide	100 (38)
Coal tar epoxy	100 (38)
Coal tar	
Urethanes	90 (32)
Neoprene	140 (60)
Polysulfides	80 (27)
Hypalon	200 (93)
Plastisols	140 (60)
PFA	200 (93)
FEP	400 (204)
PTFE	400 (204)
ETFE	300 (149)
ECTFE	300 (149)
Fluoroelastomers	X
PVDF	200 (93)
Isophthalic polyesters	X
Bisphenol A fumurate	140 (60)
Hydrogenated polyester	
Halogenated polyester	140 (60)
Methyl-appended silicone	410 (210)

Methyl Cellosolve

	Maximum Temperature °F (°C)
Phenolics	100 (38)
Epoxy	80 (27)
Furans	X
Vinyl ester	X
Epoxy polyamide	
Coal tar epoxy	
Coal tar	
Urethanes	
Neoprene	200 (93)
Polysulfides	
Hypalon	X
Plastisols	
PFA	190 (88)
FEP	400 (204)
PTFE	400 (204)
ETFE	300 (149)
ECTFE	300 (149)
Fluoroelastomers	X
PVDF	280 (138)
Isophthalic polyesters	
Bisphenol A fumurate	
Hydrogenated polyester	
Halogenated polyester	
Methyl-appended silicone	X

Methyl Chloride

	Maximum Temperature °F (°C)
Phenolics	300 (149)
Epoxy	X
Furans	120 (49)
Vinyl ester	X
Epoxy polyamide	X
Coal tar epoxy	X
Coal tar	
Urethanes	X
Neoprene	X
Polysulfides	140 (60)
Hypalon	X
Plastisols	X
PFA	200 (93)
FEP	400 (204)
PTFE	400 (204)
ETFE	300 (149)
ECTFE	300 (149)
Fluoroelastomers	190 (88)
PVDF	300 (149)
Isophthalic polyesters	
Bisphenol A fumurate	
Hydrogenated polyester	
Halogenated polyester	80 (27)
Methyl-appended silicone	

Methyl Ethyl Ketone

	Maximum Temperature °F (°C)
Phenolics	160 (71)
Epoxy	90 (32)
Furans	80 (27)
Vinyl ester	X
Epoxy polyamide	X
Coal tar epoxy	X
Coal tar	X
Urethanes	
Neoprene	X
Polysulfides	
Hypalon	X
Plastisols	X
PFA	450 (232)
FEP	400 (204)
PTFE	450 (232)
ETFE	230 (110)
ECTFE	150 (66)
Fluoroelastomers	X
PVDF	X
Isophthalic polyesters	X
Bisphenol A fumurate	X
Hydrogenated polyester	X
Halogenated polyester	X
Methyl-appended silicone	X

Methyl Isobutyl Ketone

	Maximum Temperature °F (°C)
Phenolics	X
Epoxy	140 (60)
Furans	160 (71)
Vinyl ester	X
Epoxy polyamide	X
Coal tar epoxy	X
Coal tar	X
Urethanes	X
Neoprene	X
Polysulfides	80 (27)
Hypalon	X
Plastisols	X
PFA	450 (232)
FEP	400 (204)
PTFE	450 (232)
ETFE	300 (149)
ECTFE	150 (66)
Fluoroelastomers	X
PVDF	110 (43)
Isophthalic polyesters	X
Bisphenol A fumurate	X
Hydrogenated polyester	X
Halogenated polyester	80 (27)
Methyl-appended silicone	X

Methylene Chloride

	Maximum Temperature °F (°C)
Phenolics	
Epoxy	X
Furans	280 (138)
Vinyl ester	X
Epoxy polyamide	X
Coal tar epoxy	X
Coal tar	X
Urethanes	X
Neoprene	X
Polysulfides	
Hypalon	X
Plastisols	X
PFA	200 (93)
FEP	400 (204)
PTFE	400 (204)
ETFE	210 (99)
ECTFE	X
Fluoroelastomers	X
PVDF	120 (49)
Isophthalic polyesters	X
Bisphenol A fumurate	X
Hydrogenated polyester	X
Halogenated polyester	X
Methyl-appended silicone	X

Naphtha

	Maximum Temperature °F (°C)
Phenolics	110 (43)
Epoxy	100 (38)
Furans	200 (93)
Vinyl ester	200 (93)
Epoxy polyamide	100 (38)
Coal tar epoxy	100 (38)
Coal tar	
Urethanes	90 (32)
Neoprene	X
Polysulfides	80 (27)
Hypalon	X
Plastisols	140 (60)
PFA	200 (93)
FEP	400 (204)
PTFE	400 (204)
ETFE	300 (149)
ECTFE	300 (149)
Fluoroelastomers	400 (204)
PVDF	280 (138)
Isophthalic polyesters	200 (93)
Bisphenol A fumurate	150 (66)
Hydrogenated polyester	200 (93)
Halogenated polyester	200 (93)
Methyl-appended silicone	X

Nitric Acid, 5%

	Maximum Temperature °F (°C)
Phenolics	X
Epoxy	X
Furans	200 (93)
Vinyl ester	180 (82)
Epoxy polyamide	X
Coal tar epoxy	X
Coal tar	X
Urethanes	X
Neoprene	X
Polysulfides	X
Hypalon	100 (38)
Plastisols	100 (38)
PFA	450 (232)
FEP	400 (204)
PTFE	450 (232)
ETFE	150 (66)
ECTFE	300 (149)
Fluoroelastomers	400 (204)
PVDF	200 (93)
Isophthalic polyesters	120 (49)
Bisphenol A fumurate	160 (71)
Hydrogenated polyester	90 (32)
Halogenated polyester	210 (99)
Methyl-appended silicone	80 (27)

Nitric Acid, 20%

	Maximum Temperature °F (°C)
Phenolics	X
Epoxy	100 (38)
Furans	X
Vinyl ester	150 (66)
Epoxy polyamide	X
Coal tar epoxy	X
Coal tar	X
Urethanes	X
Neoprene	X
Polysulfides	X
Hypalon	100 (38)
Plastisols	140 (60)
PFA	450 (232)
FEP	400 (204)
PTFE	450 (232)
ETFE	150 (66)
ECTFE	250 (121)
Fluoroelastomers	400 (204)
PVDF	180 (82)
Isophthalic polyesters	X
Bisphenol A fumurate	100 (38)
Hydrogenated polyester	
Halogenated polyester	80 (27)
Methyl-appended silicone	X

Nitric Acid, Dil.

Coatings for Immersion Service	Maximum Temperature °F (°C)	Paints (S= Splash Resistant W= Immersion Resistant)	
Phenolics	X	Acrylics	R, S
Epoxy	160 (71)	Alkyds:	
Furans	200 (93)	Long oil	
Vinyl ester	180 (82)	Short oil	
Epoxy polyamide	X	Asphalt	X
Coal tar epoxy	X	Chlorinated rubber	R, W
Coal tar	X	Coal tar	X
Urethanes	X	Coal tar epoxy	X
Neoprene	X	Epoxies:	
Polysulfides	X	Aliphatic polyamine	X
Hypalon	100 (38)	Polyamide	X
Plastisols	100 (38)	Polyamine	X
PFA	450 (232)	Phenolic	X
FEP	400 (204)	Polyesters	R
PTFE	450 (232)	Polyvinyl butyral	
ETFE	150 (66)	Polyvinyl formal	
ECTFE	300 (149)	Silicone (methyl)	R
Fluoroelastomers	400 (204)	Urethanes:	
PVDF	200 (93)	Aliphatic	R
Isophthalic PE	120 (49)	Aromatic	R
Bis. A fum. PE	160 (71)	Vinyls	R, W
Hydrogenated PE	90 (32)	Vinyl ester	R
Halogenated PE	210 (99)	Zinc rioh	R
Silicone (methyl)	80 (27)		
Mortars			
Sodium silicate	450 (232)		
Potassium silicate	450 (232)		
Silica	450 (232)		
Furan	X		
Polyester	140 (60)		
Epoxy	X		
Vinyl ester	180 (82)		
Acrylic	X		
Urethane	X		

Nitric Acid, 70%

	Maximum Temperature °F (°C)
Phenolics	X
Epoxy	X
Furans	X
Vinyl ester	X
Epoxy polyamide	X
Coal tar epoxy	X
Coal tar	X
Urethanes	X
Neoprene	X
Polysulfides	X
Hypalon	X
Plastisols	70 (23)
PFA	450 (232)
FEP	400 (204)
PTFE	450 (232)
ETFE	80 (27)
ECTFE	150 (66)
Fluoroelastomers	190 (88)
PVDF	120 (49)
Isophthalic polyesters	X
Bisphenol A fumurate	X
Hydrogenated polyester	
Halogenated polyester	80 (27)
Methyl-appended silicone	X

Nitric Acid, Conc.

	Maximum Temperature °F (°C)
Phenolics	X
Epoxy	X
Furans	X
Vinyl ester	X
Epoxy polyamide	X
Coal tar epoxy	X
Coal tar	X
Urethanes	X
Neoprene	X
Polysulfides	X
Hypalon	X
Plastisols	60 (16)
PFA	450 (232)
FEP	400 (204)
PTFE	450 (232)
ETFE	210 (99)
ECTFE	250 (121)
Fluoroelastomers	90 (32)
PVDF	210 (99)
Isophthalic polyesters	X
Bisphenol A fumurate	X
Hydrogenated polyester	
Halogenated polyester	90 (32)
Methyl-appended silicone	X

Nitrobenzene

	Maximum Temperature °F (°C)
Phenolics	80 (27)
Epoxy	X
Furans	260 (127)
Vinyl ester	100 (38)
Epoxy polyamide	X
Coal tar epoxy	X
Coal tar	X
Urethanes	X
Neoprene	X
Polysulfides	X
Hypalon	X
Plastisols	X
PFA	200 (93)
FEP	400 (204)
PTFE	400 (204)
ETFE	300 (149)
ECTFE	140 (60)
Fluoroelastomers	X
PVDF	140 (60)
Isophthalic polyesters	X
Bisphenol A fumurate	X
Hydrogenated polyester	
Halogenated polyester	X
Methyl-appended silicone	X

Oil, Vegetable

	Maximum Temperature °F (°C)
Phenolics	
Epoxy	90 (32)
Furans	260 (127)
Vinyl ester	180 (82)
Epoxy polyamide	100 (38)
Coal tar epoxy	100 (38)
Coal tar	
Urethanes	
Neoprene	240 (116)
Polysulfides	X
Hypalon	
Plastisols	140 (60)
PFA	200 (93)
FEP	400 (204)
PTFE	400 (204)
ETFE	290 (143)
ECTFE	300 (149)
Fluoroelastomers	200 (93)
PVDF	220 (104)
Isophthalic polyesters	150 (66)
Bisphenol A fumurate	220 (104)
Hydrogenated polyester	
Halogenated polyester	220 (104)
Methyl-appended silicone	

Oxalic Acid, 10%

	Maximum Temperature °F (°C)
Phenolics	200 (93)
Epoxy	100 (38)
Furans	200 (93)
Vinyl ester	200 (93)
Epoxy polyamide	100 (38)
Coal tar epoxy	100 (38)
Coal tar	
Urethanes	
Neoprene	200 (93)
Polysulfides	X
Hypalon	200 (93)
Plastisols	140 (60)
PFA	
FEP	400 (204)
PTFE	400 (204)
ETFE	200 (93)
ECTFE	140 (60)
Fluoroelastomers	400 (204)
PVDF	140 (60)
Isophthalic polyesters	160 (71)
Bisphenol A fumurate	200 (93)
Hydrogenated polyester	200 (93)
Halogenated polyester	200 (93)
Methyl-appended silicone	

Mineral Oil

	Maximum Temperature °F (°C)
Phenolics	160 (71)
Epoxy	210 (99)
Furans	
Vinyl ester	250 (121)
Epoxy polyamide	100 (38)
Coal tar epoxy	100 (38)
Coal tar	
Urethanes	90 (32)
Neoprene	200 (93)
Polysulfides	80 (27)
Hypalon	200 (93)
Plastisols	140 (60)
PFA	200 (93)
FEP	400 (204)
PTFE	400 (204)
ETFE	300 (149)
ECTFE	300 (149)
Fluoroelastomers	400 (204)
PVDF	250 (121)
Isophthalic polyesters	200 (93)
Bisphenol A fumurate	200 (93)
Hydrogenated polyester	
Halogenated polyester	90 (32)
Silicone	300 (149)

Motor Oil

	Maximum Temperature °F (°C)
Phenolics	160 (71)
Epoxy	110 (43)
Furans	
Vinyl ester	250 (121)
Epoxy polyamide	
Coal tar epoxy	110 (43)
Coal tar	
Urethanes	
Neoprene	
Polysulfides	80 (27)
Hypalon	
Plastisols	140 (60)
PFA	200 (93)
FEP	400 (204)
PTFE	400 (204)
ETFE	
ECTFE	300 (149)
Fluoroelastomers	190 (88)
PVDF	250 (121)
Isophthalic polyesters	160 (71)
Bisphenol A fumurate	
Hydrogenated polyester	
Halogenated polyester	
Methyl-appended silicone	

Oxalic Acid, Sat.

	Maximum Temperature °F (°C)
Phenolics (dry)	110 (43)
Epoxy	100 (38)
Furans	200 (93)
Vinyl ester	200 (93)
Epoxy polyamide	100 (38)
Coal tar epoxy	100 (38)
Coal tar	
Urethanes	
Neoprene	X
Polysulfides	X
Hypalon	X
Plastisols	140 (60)
PFA	
FEP	400 (204)
PTFE	400 (204)
ETFE	200 (93)
ECTFE	140 (60)
Fluoroelastomers	400 (204)
PVDF	120 (49)
Isophthalic polyesters	160 (71)
Bisphenol A fumurate	200 (93)
Hydrogenated polyester	200 (93)
Halogenated polyester	200 (93)
Methyl-appended silicone	

Perchloric Acid, 10%

	Maximum Temperature °F (°C)
Phenolics	
Epoxy	X
Furans	X
Vinyl ester	140 (60)
Epoxy polyamide	X
Coal tar epoxy	X
Coal tar	X
Urethanes	X
Neoprene	X
Polysulfides	
Hypalon	90 (32)
Plastisols	X
PFA	200 (93)
FEP	400 (204)
PTFE	400 (204)
ETFE	200 (93)
ECTFE	140 (60)
Fluoroelastomers	400 (204)
PVDF	200 (93)
Isophthalic polyesters	X
Bisphenol A fumurate	X
Hydrogenated polyester	X
Halogenated polyester	90 (32)
Methyl-appended silicone	X

Perchloric Acid, 70%

	Maximum Temperature °F (°C)
Phenolics (dry)	
Epoxy	X
Furans	200 (93)
Vinyl ester	X
Epoxy polyamide	X
Coal tar epoxy	X
Coal tar	X
Urethanes	X
Neoprene	X
Polysulfides	
Hypalon	90 (32)
Plastisols	X
PFA	200 (93)
FEP	400 (204)
PTFE	400 (204)
ETFE	140 (60)
ECTFE	140 (60)
Fluoroelastomers	400 (204)
PVDF	100 (38)
Isophthalic polyesters	X
Bisphenol A fumurate	X
Hydrogenated polyester	X
Halogenated polyester	90 (32)
Methyl-appended silicone	X

Phenol

	Maximum Temperature °F (°C)
Phenolics	X
Epoxy	X
Furans	X
Vinyl ester	X
Epoxy polyamide	X
Coal tar epoxy	X
Coal tar	X
Urethanes	X
Neoprene	X
Polysulfides	X
Hypalon	X
Plastisols	X
PFA	450 (232)
FEP	400 (204)
PTFE	450 (232)
ETFE	210 (99)
ECTFE	150 (66)
Fluoroelastomers	210 (99)
PVDF	200 (93)
Isophthalic polyesters	X
Bisphenol A fumurate	X
Hydrogenated polyester	X
Halogenated polyester, 5%	90 (32)
Methyl-appended silicone	X

Phosphoric Acid, 50–80%

	Maximum Temperature °F (°C)
Phenolics (dry)	X
Epoxy	110 (43)
Furans, 50%	212 (100)
Vinyl ester	210 (99)
Epoxy polyamide	X
Coal tar epoxy	X
Coal tar	X
Urethanes	
Neoprene	150 (66)
Polysulfides	X
Hypalon	200 (93)
Plastisols	140 (60)
PFA	450 (232)
FEP	400 (204)
PTFE	450 (232)
ETFE	270 (132)
ECTFE	250 (121)
Fluoroelastomers	300 (149)
PVDF	220 (104)
Isophthalic polyesters	180 (82)
Bisphenol A fumurate	220 (104)
Hydrogenated polyester	210 (99)
Halogenated polyester	250 (121)
Methyl-appended silicone	X

Phthalic Acid

	Maximum Temperature °F (°C)
Phenolics (dry)	100 (38)
Epoxy	X
Furans	200 (93)
Vinyl ester	200 (93)
Epoxy polyamide	
Coal tar epoxy	
Coal tar	
Urethanes	
Neoprene	200 (93)
Polysulfides	
Hypalon	140 (60)
Plastisols	X
PFA	
FEP	400 (204)
PTFE	400 (204)
ETFE	200 (93)
ECTFE	200 (93)
Fluoroelastomers	90 (32)
PVDF	200 (93)
Isophthalic polyesters	160 (71)
Bisphenol A fumurate	200 (93)
Hydrogenated polyester	200 (93)
Halogenated polyester	80 (27)
Methyl-appended silicone	

Potassium Acetate

	Maximum Temperature °F (°C)
Phenolics	
Epoxy	160 (71)
Furans	200 (93)
Vinyl ester	200 (93)
Epoxy polyamide	100 (38)
Coal tar epoxy	100 (38)
Coal tar	
Urethanes	
Neoprene	
Polysulfides	
Hypalon	
Plastisols	
PFA	250 (121)
FEP	200 (93)
PTFE	400 (204)
ETFE	
ECTFE	
Fluoroelastomers	80 (27)
PVDF	200 (93)
Isophthalic polyesters	160 (71)
Bisphenol A fumurate	200 (93)
Hydrogenated polyester	
Halogenated polyester	200 (93)
Methyl-appended silicone	X

Potassium Bromide, 30%

	Maximum Temperature °F (°C)
Phenolics, 10%	160 (71)
Epoxy	200 (93)
Furans	200 (93)
Vinyl ester	200 (93)
Epoxy polyamide	100 (38)
Coal tar epoxy	100 (38)
Coal tar	
Urethanes	90 (32)
Neoprene	160 (71)
Polysulfides	
Hypalon	240 (116)
Plastisols	140 (60)
PFA	200 (93)
FEP	400 (204)
PTFE	400 (204)
ETFE	300 (149)
ECTFE	300 (149)
Fluoroelastomers	190 (88)
PVDF	200 (93)
Isophthalic polyesters	160 (71)
Bisphenol A fumurate	200 (93)
Hydrogenated polyester	
Halogenated polyester	200 (93)
Methyl-appended silicone	

Potassium Carbonate, 50%

	Maximum Temperature °F (°C)
Phenolics	
Epoxy	200 (93)
Furans	240 (116)
Vinyl ester	120 (49)
Epoxy polyamide, 25%	100 (38)
Coal tar epoxy, 25%	100 (38)
Coal tar	
Urethanes	
Neoprene	200 (93)
Polysulfides	
Hypalon	200 (93)
Plastisols	140 (60)
PFA	200 (93)
FEP	400 (204)
PTFE	400 (204)
ETFE	280 (138)
ECTFE	280 (138)
Fluoroelastomers	180 (82)
PVDF	200 (93)
Isophthalic polyesters	X
Bisphenol A fumurate, 10%	180 (82)
Hydrogenated polyester	X
Halogenated polyester	110 (43)
Methyl-appended silicone	

Potassium Chloride, 30%

	Maximum Temperature °F (°C)
Phenolics	
Epoxy	200 (93)
Furans	260 (127)
Vinyl ester	200 (93)
Epoxy polyamide	100 (38)
Coal tar epoxy	100 (38)
Coal tar	
Urethanes	110 (43)
Neoprene	160 (71)
Polysulfides	
Hypalon	240 (116)
Plastisols	140 (60)
PFA	200 (93)
FEP	400 (204)
PTFE	400 (204)
ETFE	280 (138)
ECTFE	280 (138)
Fluoroelastomers	400 (204)
PVDF	260 (127)
Isophthalic polyesters	160 (71)
Bisphenol A fumurate	200 (93)
Hydrogenated polyester	190 (88)
Halogenated polyester	190 (88)
Silicone	400 (204)

Potassium Cyanide, 30%

	Maximum Temperature °F (°C)
Phenolics	
Epoxy	200 (93)
Furans	260 (127)
Vinyl ester	X
Epoxy polyamide	140 (60)
Coal tar epoxy	140 (60)
Coal tar	
Urethanes	90 (32)
Neoprene	200 (93)
Polysulfides	
Hypalon	200 (93)
Plastisols	140 (60)
PFA	200 (93)
FEP	400 (204)
PTFE	400 (204)
ETFE	300 (149)
ECTFE	300 (149)
Fluoroelastomers	400 (204)
PVDF	240 (116)
Isophthalic polyesters	X
Bisphenol A fumurate	200 (93)
Hydrogenated polyester	200 (93)
Halogenated polyester	140 (60)
Silicone	400 (204)

Potassium Hydroxide, 50%

	Maximum Temperature °F (°C)
Phenolics	160 (71)
Epoxy	100 (38)
Furans	200 (93)
Vinyl ester	X
Epoxy polyamide	100 (38)
Coal tar epoxy	100 (38)
Coal tar	
Urethanes	90 (32)
Neoprene	200 (93)
Polysulfides	80 (27)
Hypalon	200 (93)
Plastisols	140 (60)
PFA	200 (93)
FEP	400 (204)
PTFE	400 (204)
ETFE	200 (93)
ECTFE	140 (60)
Fluoroelastomers	X
PVDF	200 (93)
Isophthalic polyesters	X
Bisphenol A fumurate	160 (71)
Hydrogenated polyester	X
Halogenated polyester	X
Methyl-appended silicone	

Potassium Hydroxide, 90%

	Maximum Temperature °F (°C)
Phenolics	
Epoxy	100 (38)
Furans	200 (93)
Vinyl ester	X
Epoxy polyamide	X
Coal tar epoxy	X
Coal tar	X
Urethanes	90 (32)
Neoprene	200 (93)
Polysulfides	
Hypalon	200 (93)
Plastisols	140 (60)
PFA	200 (93)
FEP	400 (204)
PTFE	400 (204)
ETFE	
ECTFE	140 (60)
Fluoroelastomers	X
PVDF	200 (93)
Isophthalic polyesters	X
Bisphenol A fumurate	
Hydrogenated polyester	X
Halogenated polyester	X
Methyl-appended silicone	X

Potassium Nitrate, 80%

	Maximum Temperature °F (°C)
Phenolics	200 (93)
Epoxy	200 (93)
Furans	260 (127)
Vinyl ester	200 (93)
Epoxy polyamide	100 (38)
Coal tar epoxy	100 (38)
Coal tar	
Urethanes	90 (32)
Neoprene	200 (93)
Polysulfides	
Hypalon	240 (116)
Plastisols	140 (60)
PFA	200 (93)
FEP	400 (204)
PTFE	400 (204)
ETFE	280 (138)
ECTFE	280 (138)
Fluoroelastomers	400 (204)
PVDF	260 (127)
Isophthalic polyesters	180 (82)
Bisphenol A fumurate	200 (93)
Hydrogenated polyester	180 (82)
Halogenated polyester	180 (82)
Methyl-appended silicone	400 (204)

Potassium Permanganate, 10%

	Maximum Temperature °F (°C)
Phenolics	80 (27)
Epoxy	140 (60)
Furans	260 (127)
Vinyl ester	200 (93)
Epoxy polyamide	100 (38)
Coal tar epoxy	100 (38)
Coal tar	
Urethanes	100 (38)
Neoprene	100 (38)
Polysulfides	
Hypalon	240 (116)
Plastisols	140 (60)
PFA	200 (93)
FEP	400 (204)
PTFE	460 (238)
ETFE	280 (138)
ECTFE	280 (138)
Fluoroelastomers	160 (71)
PVDF	260 (127)
Isophthalic polyesters	X
Bisphenol A fumurate	200 (93)
Hydrogenated polyester	210 (99)
Halogenated polyester	150 (66)
Methyl-appended silicone	

Potassium Permanganate, 20%

	Maximum Temperature °F (°C)
Phenolics	90 (32)
Epoxy	140 (60)
Furans	160 (71)
Vinyl ester	
Epoxy polyamide	100 (38)
Coal tar epoxy	100 (38)
Coal tar	
Urethanes	
Neoprene	100 (38)
Polysulfides	
Hypalon	240 (116)
Plastisols	90 (32)
PFA	200 (93)
FEP	400 (204)
PTFE	400 (204)
ETFE	280 (138)
ECTFE	280 (138)
Fluoroelastomers	160 (71)
PVDF	260 (127)
Isophthalic polyesters	100 (38)
Bisphenol A fumurate	200 (93)
Hydrogenated polyester	
Halogenated polyester	140 (60)
Methyl-appended silicone	

Potassium Sulfate, 10%

	Maximum Temperature °F (°C)
Phenolics	
Epoxy	200 (93)
Furans	240 (116)
Vinyl ester	200 (93)
Epoxy polyamide	100 (38)
Coal tar epoxy	100 (38)
Coal tar	
Urethanes	90 (32)
Neoprene	200 (93)
Polysulfides	90 (32)
Hypalon	240 (116)
Plastisols	140 (60)
PFA	200 (93)
FEP	400 (204)
PTFE	400 (204)
ETFE	280 (138)
ECTFE	280 (138)
Fluoroelastomers	400 (204)
PVDF	260 (127)
Isophthalic polyesters	100 (38)
Bisphenol A fumurate	200 (93)
Hydrogenated polyester	200 (93)
Halogenated polyester	190 (88)
Methyl-appended silicone	

Propylene Glycol

	Maximum Temperature °F (°C)
Phenolics	160 (71)
Epoxy	100 (38)
Furans	240 (116)
Vinyl ester	200 (93)
Epoxy polyamide	100 (38)
Coal tar epoxy	100 (38)
Coal tar	
Urethanes	
Neoprene	90 (32)
Polysulfides	
Hypalon	
Plastisols	X
PFA	400 (204)
FEP	400 (204)
PTFE	400 (204)
ETFE	
ECTFE	
Fluoroelastomers	300 (149)
PVDF	240 (116)
Isophthalic polyesters	180 (82)
Bisphenol A fumurate	200 (93)
Hydrogenated polyester	200 (93)
Halogenated polyester	100 (38)
Methyl-appended silicone	

Pyridine

	Maximum Temperature °F (°C)
Phenolics	X
Epoxy	X
Furans	X
Vinyl ester	X
Epoxy polyamide	X
Coal tar epoxy	X
Coal tar	X
Urethanes	
Neoprene	X
Polysulfides	
Hypalon	X
Plastisols	X
PFA	200 (93)
FEP	400 (204)
PTFE	460 (238)
ETFE	140 (60)
ECTFE	X
Fluoroelastomers	X
PVDF	X
Isophthalic polyesters	X
Bisphenol A fumurate	X
Hydrogenated polyester	X
Halogenated polyester	X
Methyl-appended silicone	X

Salicylic Acid

	Maximum Temperature °F (°C)
Phenolics	
Epoxy	200 (93)
Furans	240 (116)
Vinyl ester	140 (60)
Epoxy polyamide	100 (38)
Coal tar epoxy	100 (38)
Coal tar	
Urethanes	
Neoprene	X
Polysulfides	
Hypalon	X
Plastisols	X
PFA	200 (93)
FEP	400 (204)
PTFE	400 (204)
ETFE	240 (116)
ECTFE	240 (116)
Fluoroelastomers	280 (138)
PVDF	200 (93)
Isophthalic polyesters	100 (38)
Bisphenol A fumurate	140 (60)
Hydrogenated polyester	
Halogenated polyester	120 (49)
Methyl-appended silicone	

Sodium Acetate

	Maximum Temperature °F (°C)
Phenolics	
Epoxy	200 (93)
Furans	260 (127)
Vinyl ester	200 (93)
Epoxy polyamide	100 (38)
Coal tar epoxy	100 (38)
Coal tar	
Urethanes	
Neoprene	200 (93)
Polysulfides	
Hypalon	X
Plastisols	140 (60)
PFA	200 (93)
FEP	400 (204)
PTFE	400 (204)
ETFE	280 (138)
ECTFE	280 (138)
Fluoroelastomers	X
PVDF	260 (127)
Isophthalic polyesters	180 (82)
Bisphenol A fumurate	180 (82)
Hydrogenated polyester	200 (93)
Halogenated polyester	200 (93)
Methyl-appended silicone	X

Sodium Bicarbonate, 20%

	Maximum Temperature °F (°C)
Phenolics	X
Epoxy	200 (93)
Furans	240 (116)
Vinyl ester	200 (93)
Epoxy polyamide	100 (38)
Coal tar epoxy	100 (38)
Coal tar	
Urethanes	
Neoprene	200 (93)
Polysulfides	
Hypalon	240 (116)
Plastisols	140 (60)
PFA	200 (93)
FEP	400 (204)
PTFE	400 (204)
ETFE	280 (138)
ECTFE	280 (138)
Fluoroelastomers	400 (204)
PVDF	260 (127)
Isophthalic polyesters, 10%	180 (82)
Bisphenol A fumurate	160 (71)
Hydrogenated polyester	
Halogenated polyester	140 (60)
Silicone	400 (204)

Sodium Bisulfate

	Maximum Temperature °F (°C)
Phenolics	260 (127)
Epoxy	200 (93)
Furans	240 (116)
Vinyl ester	200 (93)
Epoxy polyamide	100 (38)
Coal tar epoxy	100 (38)
Coal tar	
Urethanes	
Neoprene	200 (93)
Polysulfides	
Hypalon	100 (38)
Plastisols	140 (60)
PFA	200 (93)
FEP	400 (204)
PTFE	400 (204)
ETFE	280 (138)
ECTFE	280 (138)
Fluoroelastomers	180 (82)
PVDF	260 (127)
Isophthalic polyesters	180 (82)
Bisphenol A fumurate	200 (93)
Hydrogenated polyester	
Halogenated polyester	200 (93)
Methyl-appended silicone	

Sodium Carbonate

	Maximum Temperature °F (°C)
Phenolics	160 (71)
Epoxy	200 (38)
Furans, 50%	240 (116)
Vinyl ester	180 (82)
Epoxy polyamide	100 (38)
Coal tar epoxy	100 (38)
Coal tar	
Urethanes	
Neoprene	200 (93)
Polysulfides	X
Hypalon	240 (116)
Plastisols	140 (60)
PFA	200 (93)
FEP	400 (204)
PTFE	400 (204)
ETFE	280 (138)
ECTFE	280 (138)
Fluoroelastomers	180 (82)
PVDF	260 (127)
Isophthalic polyesters, 20%	90 (32)
Bisphenol A fumurate	160 (71)
Hydrogenated polyester, 10%	100 (38)
Halogenated polyester, 10%	180 (82)
Methyl-appended silicone	300 (149)

Sodium Chlorate

	Maximum Temperature °F (°C)
Phenolics, 50%	160 (71)
Epoxy	100 (38)
Furans	160 (71)
Vinyl ester	220 (104)
Epoxy polyamide	
Coal tar epoxy, 50%	100 (38)
Coal tar	
Urethanes	
Neoprene	80 (27)
Polysulfides	
Hypalon	240 (116)
Plastisols	140 (60)
PFA	200 (93)
FEP	400 (204)
PTFE	400 (204)
ETFE	280 (138)
ECTFE	280 (138)
Fluoroelastomers	180 (82)
PVDF	260 (127)
Isophthalic polyesters	X
Bisphenol A fumurate	200 (93)
Hydrogenated polyester	200 (93)
Halogenated polyester, 48%	200 (93)
Methyl-appended silicone	

Sodium Chloride

	Maximum Temperature °F (°C)
Phenolics	160 (71)
Epoxy	200 (93)
Furans	240 (116)
Vinyl ester	180 (82)
Epoxy polyamide	110 (43)
Coal tar epoxy	110 (43)
Coal tar	X
Urethanes	80 (27)
Neoprene	200 (93)
Polysulfides	80 (27)
Hypalon	240 (116)
Plastisols	140 (60)
PFA	200 (93)
FEP	400 (204)
PTFE	400 (204)
ETFE	280 (138)
ECTFE	280 (138)
Fluoroelastomers	400 (204)
PVDF	260 (127)
Isophthalic polyesters	200 (93)
Bisphenol A fumurate	200 (93)
Hydrogenated polyester	200 (93)
Halogenated polyester	210 (99)
Methyl-appended silicone, 10%	400 (204)

Sodium Cyanide

	Maximum Temperature °F (°C)
Phenolics	
Epoxy	200 (93)
Furans	240 (116)
Vinyl ester	200 (93)
Epoxy polyamide	100 (38)
Coal tar epoxy	100 (38)
Coal tar	
Urethanes	
Neoprene	180 (82)
Polysulfides	
Hypalon	240 (116)
Plastisols	140 (60)
PFA	200 (93)
FEP	400 (204)
PTFE	400 (204)
ETFE	280 (138)
ECTFE	280 (138)
Fluoroelastomers	400 (204)
PVDF	260 (127)
Isophthalic polyesters	150 (66)
Bisphenol A fumurate	160 (71)
Hydrogenated polyester	
Halogenated polyester, 50%	140 (60)
Silicone	140 (60)

Sodium Hydroxide, 10%

	Maximum Temperature °F (°C)
Phenolics	300 (149)
Epoxy	190 (88)
Furans	X
Vinyl ester	170 (77)
Epoxy polyamide	100 (38)
Coal tar epoxy	100 (38)
Coal tar	
Urethanes	90 (32)
Neoprene	230 (110)
Polysulfides	X
Hypalon	200 (93)
Plastisols	140 (60)
PFA	450 (232)
FEP	400 (204)
PTFE	450 (232)
ETFE	230 (110)
ECTFE	300 (149)
Fluoroelastomers	X
PVDF	230 (110)
Isophthalic polyesters	X
Bisphenol A fumurate	130 (54)
Hydrogenated polyester	100 (38)
Halogenated polyester	110 (43)
Methyl-appended silicone	90 (27)

Sodium Hydroxide, 50%

	Maximum Temperature °F (°C)
Phenolics	X
Epoxy	200 (93)
Furans	X
Vinyl ester	220 (104)
Epoxy polyamide	100 (38)
Coal tar epoxy	100 (38)
Coal tar	
Urethanes	90 (32)
Neoprene	230 (110)
Polysulfides	X
Hypalon	200 (93)
Plastisols	140 (60)
PFA	450 (232)
FEP	400 (204)
PTFE	450 (232)
ETFE	230 (110)
ECTFE	250 (121)
Fluoroelastomers	X
PVDF	220 (104)
Isophthalic polyesters	X
Bisphenol A fumurate	220 (104)
Hydrogenated polyester	X
Halogenated polyester	X
Methyl-appended silicone	90 (27)

Sodium Hypochlorite, 20%

	Maximum Temperature °F (°C)
Phenolics	X
Epoxy	X
Furans	X
Vinyl ester	180 (82)
Epoxy polyamide	X
Coal tar epoxy	X
Coal tar	
Urethanes	X
Neoprene	X
Polysulfides	X
Hypalon	200 (93)
Plastisols	140 (60)
PFA	450 (232)
FEP	400 (204)
PTFE	450 (232)
ETFE	300 (149)
ECTFE	300 (149)
Fluoroelastomers	400 (204)
PVDF	280 (138)
Isophthalic polyesters	X
Bisphenol A fumurate	X
Hydrogenated polyester	160 (71)
Halogenated polyester	X
Methyl-appended silicone	X

Sodium Hypochlorite, Conc.

	Maximum Temperature °F (°C)
Phenolics	X
Epoxy	X
Furans	X
Vinyl ester	100 (38)
Epoxy polyamide	
Coal tar epoxy	X
Coal tar	
Urethanes	
Neoprene	X
Polysulfides	X
Hypalon	X
Plastisols	140 (60)
PFA	450 (232)
FEP	400 (204)
PTFE	450 (232)
ETFE	300 (149)
ECTFE	300 (149)
Fluoroelastomers	400 (204)
PVDF	280 (138)
Isophthalic polyesters	X
Bisphenol A fumurate	X
Hydrogenated polyester	
Halogenated polyester	X
Methyl-appended silicone	X

Sodium Nitrate

	Maximum Temperature °F (°C)
Phenolics	80 (27)
Epoxy	200 (93)
Furans	160 (71)
Vinyl ester	200 (93)
Epoxy polyamide	
Coal tar epoxy	80 (27)
Coal tar	
Urethanes	90 (32)
Neoprene	200 (93)
Polysulfides	
Hypalon	140 (60)
Plastisols	140 (60)
PFA	200 (93)
FEP	400 (204)
PTFE	400 (204)
ETFE	300 (149)
ECTFE	300 (149)
Fluoroelastomers	X
PVDF	280 (138)
Isophthalic polyesters	180 (82)
Bisphenol A fumurate	220 (104)
Hydrogenated polyester	210 (99)
Halogenated polyester	250 (121)
Methyl-appended silicone	X

Sodium Peroxide, 10%

	Maximum Temperature °F (°C)
Phenolics	
Epoxy	90 (32)
Furans	
Vinyl ester	160 (71)
Epoxy polyamide	X
Coal tar epoxy	X
Coal tar	X
Urethanes	
Neoprene	200 (93)
Polysulfides	
Hypalon	250 (121)
Plastisols	140 (60)
PFA	200 (93)
FEP	400 (204)
PTFE	400 (204)
ETFE	300 (149)
ECTFE	300 (149)
Fluoroelastomers	400 (204)
PVDF	260 (127)
Isophthalic polyesters	X
Bisphenol A fumurate	220 (104)
Hydrogenated polyester	
Halogenated polyester	X
Methyl-appended silicone	X

Stearic Acid

	Maximum Temperature °F (°C)
Phenolics	210 (99)
Epoxy	220 (104)
Furans	260 (127)
Vinyl ester	220 (104)
Epoxy polyamide	X
Coal tar epoxy	X
Coal tar	X
Urethanes	90 (32)
Neoprene	200 (93)
Polysulfides	
Hypalon	140 (60)
Plastisols	140 (60)
PFA	200 (93)
FEP	400 (204)
PTFE	400 (204)
ETFE	300 (149)
ECTFE	300 (149)
Fluoroelastomers	100 (38)
PVDF	280 (138)
Isophthalic polyesters	180 (82)
Bisphenol A fumurate	200 (93)
Hydrogenated polyester	210 (99)
Halogenated polyester	250 (121)
Methyl-appended silicone	X

Styrene

	Maximum Temperature °F (°C)
Phenolics	
Epoxy	100 (38)
Furans	360 (182)
Vinyl ester	100 (38)
Epoxy polyamide	
Coal tar epoxy	X
Coal tar	X
Urethanes	
Neoprene	X
Polysulfides	
Hypalon	X
Plastisols	X
PFA	
FEP	100 (38)
PTFE	350 (177)
ETFE	210 (99)
ECTFE	
Fluoroelastomers	300 (149)
PVDF	190 (88)
Isophthalic polyesters	X
Bisphenol A fumurate	100 (38)
Hydrogenated polyester	100 (38)
Halogenated polyester	X
Methyl-appended silicone	X

Sulfur Dioxide, Wet

	Maximum Temperature °F (°C)
Phenolics	300 (149)
Epoxy	150 (66)
Furans	260 (127)
Vinyl ester	210 (99)
Epoxy polyamide	100 (38)
Coal tar epoxy	100 (38)
Coal tar	
Urethanes	
Neoprene	X
Polysulfides	
Hypalon	X
Plastisols	X
PFA	200 (93)
FEP	400 (204)
PTFE	400 (204)
ETFE	230 (110)
ECTFE	150 (66)
Fluoroelastomers	X
PVDF	210 (99)
Isophthalic polyesters	90 (32)
Bisphenol A fumurate	220 (104)
Hydrogenated polyester	210 (99)
Halogenated polyester	250 (121)
Methyl-appended silicone	

Sulfur Trioxide

	Maximum Temperature °F (°C)
Phenolics	300 (149)
Epoxy	
Furans	X
Vinyl ester	210 (99)
Epoxy polyamide	X
Coal tar epoxy	X
Coal tar	X
Urethanes	
Neoprene	X
Polysulfides	
Hypalon	X
Plastisols	140 (60)
PFA	200 (93)
FEP	400 (204)
PTFE	400 (204)
ETFE	80 (27)
ECTFE	80 (27)
Fluoroelastomers	190 (88)
PVDF	X
Isophthalic polyesters	90 (32)
Bisphenol A fumurate	250 (121)
Hydrogenated polyester	
Halogenated polyester	120 (49)
Methyl-appended silicone	X

Sulfuric Acid, 10%

	Maximum Temperature °F (°C)
Phenolics	250 (121)
Epoxy	140 (60)
Furans	160 (71)
Vinyl ester	200 (93)
Epoxy polyamide	X
Coal tar epoxy	X
Coal tar	X
Urethanes	X
Neoprene	150 (66)
Polysulfides	X
Hypalon	200 (93)
Plastisols	140 (60)
PFA	450 (232)
FEP	400 (204)
PTFE	450 (232)
ETFE	300 (149)
ECTFE	250 (121)
Fluoroelastomers	350 (177)
PVDF	250 (121)
Isophthalic polyesters	180 (71)
Bisphenol A fumurate	220 (104)
Hydrogenated polyester	210 (99)
Halogenated polyester	260 (127)
Methyl-appended silicone	X

Sulfuric Acid, 50%

	Maximum Temperature °F (°C)
Phenolics	250 (121)
Epoxy	X
Furans	260 (127)
Vinyl ester	210 (99)
Epoxy polyamide	X
Coal tar epoxy	X
Coal tar	X
Urethanes	X
Neoprene	100 (38)
Polysulfides	X
Hypalon	200 (93)
Plastisols	140 (60)
PFA	450 (232)
FEP	400 (204)
PTFE	450 (232)
ETFE	300 (149)
ECTFE	250 (121)
Fluoroelastomers	350 (177)
PVDF	220 (104)
Isophthalic polyesters	150 (66)
Bisphenol A fumurate	220 (104)
Hydrogenated polyester	210 (99)
Halogenated polyester	200 (93)
Methyl-appended silicone	X

Sulfuric Acid, 70%

	Maximum Temperature °F (°C)
Phenolics	210 (93)
Epoxy	X
Furans	
Vinyl ester	180 (82)
Epoxy polyamide	X
Coal tar epoxy	X
Coal tar	X
Urethanes	X
Neoprene	X
Polysulfides	X
Hypalon	160 (71)
Plastisols	140 (60)
PFA	450 (232)
FEP	400 (204)
PTFE	450 (232)
ETFE	300 (149)
ECTFE	250 (121)
Fluoroelastomers	350 (177)
PVDF	220 (104)
Isophthalic polyesters	X
Bisphenol A fumurate	160 (71)
Hydrogenated polyester	90 (32)
Halogenated polyester	190 (88)
Methyl-appended silicone	X

Sulfuric Acid, 90%

	Maximum Temperature °F (°C)
Phenolics	X
Epoxy	X
Furans	X
Vinyl ester	X
Epoxy polyamide	X
Coal tar epoxy	X
Coal tar	X
Urethanes	X
Neoprene	X
Polysulfides	X
Hypalon	X
Plastisols	X
PFA	450 (232)
FEP	400 (204)
PTFE	450 (232)
ETFE	300 (149)
ECTFE	150 (66)
Fluoroelastomers	350 (177)
PVDF	210 (99)
Isophthalic polyesters	X
Bisphenol A fumurate	X
Hydrogenated polyester	X
Halogenated polyester	X
Methyl-appended silicone	X

Sulfuric Acid, 98%

	Maximum Temperature °F (°C)
Phenolics	X
Epoxy	X
Furans	X
Vinyl ester	X
Epoxy polyamide	X
Coal tar epoxy	X
Coal tar	X
Urethanes	X
Neoprene	X
Polysulfides	X
Hypalon	X
Plastisols	X
PFA	450 (232)
FEP	400 (204)
PTFE	450 (232)
ETFE	300 (149)
ECTFE	150 (66)
Fluoroelastomers	350 (177)
PVDF	140 (60)
Isophthalic polyesters	X
Bisphenol A fumurate	X
Hydrogenated polyester	X
Halogenated polyester	X
Methyl-appended silicone	X

Sulfuric Acid, 100%

	Maximum Temperature °F (°C)
Phenolics	X
Epoxy	X
Furans	X
Vinyl ester	X
Epoxy polyamide	X
Coal tar epoxy	X
Coal tar	X
Urethanes	X
Neoprene	X
Polysulfides	X
Hypalon	X
Plastisols	X
PFA	450 (232)
FEP	400 (204)
PTFE	450 (232)
ETFE	300 (149)
ECTFE	80 (27)
Fluoroelastomers	180 (82)
PVDF	X
Isophthalic polyesters	X
Bisphenol A fumurate	X
Hydrogenated polyester	X
Halogenated polyester, 50%	X
Methyl-appended silicone	X

Sulfurous Acid

	Maximum Temperature °F (°C)
Phenolics	X
Epoxy, 20%	240 (116)
Furans	160 (71)
Vinyl ester, 10%	120 (49)
Epoxy polyamide	110 (43)
Coal tar epoxy	100 (38)
Coal tar	
Urethanes	
Neoprene	100 (38)
Polysulfides	
Hypalon	160 (71)
Plastisols	140 (60)
PFA	450 (232)
FEP	400 (204)
PTFE	450 (232)
ETFE	210 (99)
ECTFE	250 (121)
Fluoroelastomers	400 (204)
PVDF	220 (104)
Isophthalic polyesters	X
Bisphenol A fumurate	110 (43)
Hydrogenated polyester, 25%	210 (99)
Halogenated polyester, 10%	80 (27)
Methyl-appended silicone	X

Thionyl Chloride

	Maximum Temperature °F (°C)
Phenolics	200 (93)
Epoxy	X
Furans	X
Vinyl ester	X
Epoxy polyamide	
Coal tar epoxy	
Coal tar	
Urethanes	
Neoprene	X
Polysulfides	
Hypalon	
Plastisols	X
PFA[a]	450 (232)
FEP[a]	400 (204)
PTFE[a]	450 (232)
ETFE	210 (99)
ECTFE	150 (66)
Fluoroelastomers	X
PVDF	X
Isophthalic polyesters	X
Bisphenol A fumurate	X
Hydrogenated polyester	
Halogenated polyester	X
Methyl-appended silicone	

[a] Corrodent will permeate.

Toluene

	Maximum Temperature °F (°C)
Phenolics	200 (93)
Epoxy	X
Furans	212 (100)
Vinyl ester	120 (49)
Epoxy polyamide	X
Coal tar epoxy	X
Coal tar	X
Urethanes	X
Neoprene	X
Polysulfides	X
Hypalon	X
Plastisols	X
PFA	200 (93)
FEP	400 (204)
PTFE	400 (204)
ETFE	250 (121)
ECTFE	140 (60)
Fluoroelastomers	400 (204)
PVDF	210 (99)
Isophthalic polyesters	100 (38)
Bisphenol A fumurate	X
Hydrogenated polyester	80 (27)
Halogenated polyester	100 (38)
Methyl-appended silicone	X

Trichloroethylene

	Maximum Temperature °F (°C)
Phenolics	160 (71)
Epoxy	X
Furans	160 (71)
Vinyl ester	X
Epoxy polyamide	X
Coal tar epoxy	X
Coal tar	X
Urethanes	X
Neoprene	X
Polysulfides	X
Hypalon	X
Plastisols	X
PFA	200 (93)
FEP	400 (204)
PTFE	400 (204)
ETFE	270 (132)
ECTFE	300 (149)
Fluoroelastomers	400 (204)
PVDF	260 (127)
Isophthalic polyesters	X
Bisphenol A fumurate	X
Hydrogenated polyester	
Halogenated polyester	120 (49)
Methyl-appended silicone	X

Turpentine

	Maximum Temperature °F (°C)
Phenolics	110 (43)
Epoxy	150 (66)
Furans	
Vinyl ester	150 (66)
Epoxy polyamide	X
Coal tar epoxy	X
Coal tar	X
Urethanes	X
Neoprene	X
Polysulfides	80 (27)
Hypalon	X
Plastisols	X
PFA	200 (93)
FEP	400 (204)
PTFE	400 (204)
ETFE	270 (132)
ECTFE	300 (149)
Fluoroelastomers	400 (204)
PVDF	280 (138)
Isophthalic polyesters	80 (27)
Bisphenol A fumurate	80 (27)
Hydrogenated polyester	
Halogenated polyester	120 (49)
Methyl-appended silicone	X

Water, Salt

	Maximum Temperature °F (°C)
Phenolics	160 (71)
Epoxy, 10%	210 (99)
Furans	
Vinyl ester	160 (71)
Epoxy polyamide	130 (54)
Coal tar epoxy	130 (54)
Coal tar	100 (38)
Urethanes	X
Neoprene	210 (99)
Polysulfides	80 (27)
Hypalon	250 (121)
Plastisols	140 (60)
PFA	200 (93)
FEP	400 (204)
PTFE	400 (204)
ETFE	250 (121)
ECTFE	300 (149)
Fluoroelastomers	190 (88)
PVDF	280 (138)
Isophthalic polyesters	160 (71)
Bisphenol A fumurate	180 (82)
Hydrogenated polyester	210 (99)
Halogenated polyester	
Methyl-appended silicone	210 (99)

White Liquor

	Maximum Temperature °F (°C)
Phenolics	160 (71)
Epoxy	200 (93)
Furans	140 (60)
Vinyl ester	180 (82)
Epoxy polyamide	150 (66)
Coal tar epoxy	100 (38)
Coal tar	X
Urethanes	X
Neoprene	140 (60)
Polysulfides	
Hypalon	
Plastisols	140 (60)
PFA	
FEP	400 (204)
PTFE	400 (204)
ETFE	
ECTFE	250 (121)
Fluoroelastomers	190 (88)
PVDF	200 (93)
Isophthalic polyesters	X
Bisphenol A fumurate	180 (82)
Hydrogenated polyester	
Halogenated polyester	X
Methyl-appended silicone	

Xylene

	Maximum Temperature °F (°C)
Phenolics	150 (66)
Epoxy	X
Furans	260 (127)
Vinyl ester	140 (60)
Epoxy polyamide	X
Coal tar epoxy	X
Coal tar	X
Urethanes	X
Neoprene	X
Polysulfides	80 (27)
Hypalon	X
Plastisols	X
PFA	200 (93)
FEP	400 (204)
PTFE	400 (204)
ETFE	250 (121)
ECTFE	150 (66)
Fluoroelastomers	400 (204)
PVDF	210 (99)
Isophthalic polyesters	X
Bisphenol A fumurate	90 (32)
Hydrogenated polyester	90 (32)
Halogenated polyester	150 (66)
Methyl-appended silicone	X

Reference

1. P.A. Schweitzer. 2004. *Corrosion Resistance Tables*, Vols. 1–4, 5th ed., New York: Marcel Dekker.

6

Sheet Linings

Designers of tanks and process vessels are faced with the problem of choosing the most reliable material of construction at a reasonable cost. When handling corrosive materials, a choice must often be made between using an expensive metallic material of construction or using a low-cost material from which to fabricate the shell and then installing a corrosion-resistant lining. Carbon steel has been and still is the material predominantly selected; although, there has been a tendency over the past few years to use a fiberglass-reinforced plastic shell. This latter choice has the advantage of providing atmospheric corrosion protection of the shell exterior.

For many years, vessels have been successfully lined with various rubber formulations, both natural and synthetic. Many such vessels have given over twenty years of reliable service.

With the development of newer synthetic elastomeric and plastic materials, the variety of available lining materials has greatly increased.

As with any material, the corrosion resistance, allowable operating values, and cost varies with each. Care must be taken when selecting the lining material that it is compatible with the corrodent being handled at the operating temperatures and pressures required.

6.1 A Shell Design

For a lining to perform satisfactorily, the vessel shell must meet certain design configurations. Although these details may vary slightly depending on the specific lining material to be used, there are certain basic principles that apply in all cases:

1. Vessel must be of butt-welded construction.
2. All internal welds must be ground flush.
3. All weld spatter must be removed.

4. All sharp corners must be ground to a minimum of 1/8 in. radius.

5. All outlets to be of flanged or pad type. Certain lining materials require that nozzles be not less than 2 in. (55 mm) in diameter.

6. No protrusions are permitted inside of the vessel.

Once the lining has been installed, there should be no welding permitted on the exterior of the vessel.

After the fabrication has been completed, the interior surface of the vessel must be prepared to accept the lining. This is a very critical step. Unless the surface is properly prepared, proper bonding of the lining to the shell will not be achieved. The basic requirement is that the surface be absolutely clean. To ensure proper bonding, all surfaces to be lined should be abrasive-blasted to white metal in accordance with SSPC specification Tp5-63 or NACE specification 1. A white metal surface condition is defined as being one where all rust, scales, paint, and the like have been removed and the surface has a uniform gray–white appearance. Streaks or stains of rust or other contaminants are not allowed. A near-white blast-cleaned finish equal to SSPC SP-10 is allowed on occasion. This is a more economical finish. In any case, it is essential that the finish be as the lining contractor has specified. Some lining contractors will fabricate the vessel as well as prepare the surface. When the total responsibility is placed on the lining contractor, the problem is simplified, and usually, a better quality product will be the result.

When a vessel shell is fabricated from a reinforced thermosetting plastic (RTP), several advantages are realized. The RTPs generally have a wider range of corrosion resistance, but relatively low allowable operating temperatures. When a fluoropolymer-type lining is applied to an RTP shell, the temperature to which the backup RTP is exposed, has been reduced in addition to preventing the RTP from becoming exposed to the chemicals in the process system. An upper temperature limit for using RTP dual laminates is 350°F (177°C).

The dual-laminate construction lessens the problem of permeation through liners. If there is permeation, it is believed to pass through the RTP structure at a rate equal to or greater than through the fluoropolymer itself, resulting in no potential for collection of permeate at the thermoplastic-to-thermoset interface. If delamination does not occur, permeation is not a problem.

6.2 Considerations in Liner Selection

Before a lining material is selected, careful consideration should be given to several broad categories, specifically materials being handled, operating conditions, and conditions external to the vessel.

The following information must be known about the materials being handled:

1. What are the primary chemicals being handled and at what concentrations?
2. Are there any secondary chemicals, and if so, at what concentration?
3. Are there any trace impurities or chemicals?
4. Are there any solids present, and if so, what are their particle sizes and concentrations?
5. If a vessel, will there be agitation and to what degree? If a pipeline, what are the flow rates (maximum and minimum)?
6. What are the fluid purity requirements?

The answers to the above questions will narrow the selection to those materials that are compatible. This next set of questions will narrow the selection further by eliminating materials that do not have the required physical and/or mechanical properties required.

1. What is the normal operating temperature and temperature range?
2. What peak temperatures can be reached during shutdown, startup, process upset, etc.?
3. Will any mixing areas exist where exothermic or heat-of-mixing temperatures can develop?
4. What is the normal operating pressure?
5. What vacuum conditions and range are possible during operation, startup, shutdown, or upset conditions?
6. Will there be temperature cycling?
7. What cleaning methods will be used?

Finally, consideration should be given the conditions external to the vessel or pipe.

1. What are the ambient temperature conditions?
2. What is the maximum surface temperature during operation?
3. What are the insulation requirements?
4. What is the nature of the external environment? This can dictate finish requirements and/or affect the selection of the shell material.
5. What are the external heating requirements?
6. Is grounding necessary?

With the answers to these questions, an appropriate selection of liner and shell can be made.

6.3 Design Considerations

In addition to selecting a lining material that is resistant to the corrodent being handled, there are three other factors to be considered in the design: permeation, absorption, and environmental stress cracking. Permeation and absorption can cause

1. Bond failure and blistering, resulting from the accumulation of fluids at the bond when the substrate is less permeable than the liner or from corrosion/reaction products if the substrate is attacked by the permeant.
2. Failure of the substrate from corrosive attack.
3. Loss of contents through the substrate and liner as a result of the eventual failure of the substrate. In unbonded linings, it is important that the space between the liner and substrate be vented to the atmosphere, not only to allow minute quantities of permeant vapor to escape but also to prevent expansion of entrapped air from collapsing the liner.

6.3.1 Permeation

All materials are somewhat permeable to chemical molecules, but plastic materials tend to be an order of magnitude greater in their permeability rates than metals. Polymers can be permeated by gases, vapors, or liquids. Permeation is strictly a physical phenomenon; there is no chemical attack on the polymer. It is a molecular migration either through microvoids in the polymer (if the structure is more or less porous) or between polymer molecules.

Permeation is a function of two variables: one relating to diffusion between molecular chains and the other to the solubility of the permeant in the polymer. The driving forces of diffusion are the concentration gradients in liquids and the partial pressure gradient for gases. Solubility is a function of the affinity of the permeant for the polymer.

Material passing through cracks and voids is not related to permeation. These are two distinct happenings. They are not related in any way.

Permeation is affected by the following factors:

1. Temperature and pressure
2. Permeant concentration
3. Thickness of the polymer

An increase in temperature will increase the permeation rate because the solubility of the permeant in the polymer will increase, and as the temperature rises, the polymer chain movement is stimulated, permitting more permeants to diffuse among the chain more easily. For many gases, the permeant rates increase linearly with the partial pressure gradient, and the same effect is experienced with concentration gradients of liquids. If the permeant is highly soluble in the polymer, the permeability increase may not be linear.

The thickness of the polymer affects the permeation. An increase in thickness will generally decrease permeation by the square of the thickness. However, there are disadvantages to this approach. First, as the lining thickness is increased, thermal stresses on the bond are increased, resulting in bond failure. Temperature changes and large differences in coefficients of thermal expansion are the most common causes of bond failure. The thickness and modulus of elasticity of the lining material are two of the factors that influence these stresses. In addition, as the thickness of the sheet lining material increases, it becomes more difficult to form, and heat may have to be supplied. Also, the thicker sheets are much more difficult to weld. A third factor is cost. As the thickness of the material increases, not only does the material cost more but the labor costs also increase because of the greater difficulty of working with the material. If polymers such as fluorinated ethylene propylene (FEP), polytetrafluoroethylene (PTFE), or polyvinylidene fluoride (PVDF) are being used, the cost may become prohibitive.

The density of the polymer in addition to its thickness will also have an effect on the permeation rate. The higher the specific gravity of the sheet, the fewer the voids that will be present through which permeation can take place. A comparison of the specific gravity between two different polymers will not give an indication of the relative permeation rates. However, a comparison between two liners of the same polymer will provide the difference in the relative permeation rates. The liner having the greater density will have the lower permeation rate.

Other chemical and physiochemical properties affecting permeation are

1. Ease of condensation of the permeant. Chemicals that readily condense will permeate at higher rates.
2. The higher the intermolecular chain forces (e.g., van der Waal's hydrogen bonding) of the polymer, the lower the permeation rate.
3. The higher the level of crystallinity in the polymer, the lower the permeation rate.
4. The greater the degrees of crosslinking within the polymer, the lower the permeation rate.
5. Chemical similarity between the polymer and the permeant. When the polymer and permeant both have similar functional groups, the permeant rate will increase.
6. The smaller the molecule of the permeant, the greater the permeation rate.

The magnitude of any of the effects will be a function of the combination of polymer and permeant in actual service.

6.3.2 Absorption

Polymers have the potential to absorb varying amounts of corrodents that they come into contact with, particularly organic liquids. This can result in swelling, cracking, and penetration to the substrate. Swelling can cause softening of the polymer, introduce high stresses, and precipitate failure of the bond. If the polymer has a high absorption rate, permeation will probably take place. An approximation of the expected permeation and/or absorption of a polymer can be based on the absorption of water. This data is usually available. Table 6.1 provides the water absorption rates for the more common polymers used for linings.

The failure because of absorption can best be understood by considering the "steam cycle" test described in the ASTM standards for lined pipe. A section of lined pipe is subjected to thermal and pressure fluctuations. This is repeated for 100 cycles. The steam creates a temperature and pressure gradient through the liner, causing absorption of a small quantity of steam that condenses to water within the inner wall. Upon pressure release, or on reintroduction of steam, the entrapped water can expand to vapor causing an original micropore. The repeated pressure and thermal cycling enlarges the micropores, ultimately producing visible water-filled blisters within the liner.

In an actual process, the polymer may absorb process fluids, and repeated temperature or pressure cycling can cause blisters. Eventually, the corrodent may find its way to the substrate.

Related effects can occur when process chemicals are absorbed that may later react, decompose, or solidify within the structure of the plastic. Prolonged retention of the chemicals may lead to their decomposition within

TABLE 6.1

Water Absorption Rates for Common Polymers

Polymer	Water Absorption 24 h at 73°F/23°C (%)
PVC	0.05
CPVC	0.03
PP (Homo)	0.02
PP (Co)	0.03
EHMW PE	<0.01
ECTFE	<0.1
PVDF	<0.04
PEA	<0.03
ETFE	0.029
PTFE	<0.01
FEP	<0.01

the polymer. Although unusual, it is possible for absorbed monomers to polymerized.

Several steps can be taken to reduce absorption. For example, thermal insulation of the substrate will reduce the temperature gradient across the vessel, thereby, preventing condensation and subsequent expansion of the absorbed fluids. This also reduces the rate and magnitude of temperature changes, keeping blisters to a minimum. The use of operating procedures or devices that limit the ratio of process pressure reductions or temperature increases will provide additional protection.

6.3.3 Environmental Stress Cracking

Stress cracks develop when a tough polymer is stressed for an extended period of time under loads that are small relative to the polymer's yield point. Cracking will occur with little elongation of the material. The higher the molecular weight of the polymer, the less likelihood of environmental stress cracking, other things being equal. Molecular weight is a function of length of individual chains that make up the polymer. Longer chain polymers tend to crystallize less than polymers of lower molecular weight or shorter chains, and they also have a greater load-bearing capacity.

Crystallinity is an important factor affecting stress corrosion cracking. The less the crystallization that takes place, the less the likelihood of stress cracking. Unfortunately, the lower the crystallinity, the greater the likelihood of permeation.

Resistance to stress cracking can be reduced by the absorption of substances that chemically resemble the polymer and will plasticize it. In addition, the mechanical strength will also be reduced. Halogenated chemicals, particularly those consisting of small molecules containing fluorine or chlorine, are especially likely to be similar to the fluoropolymers and should be tested for their effect.

The presence of contaminants in a fluid may act as an accelerator. For example, polypropylene can safely handle sulfuric or hydrochloric acids, but iron or copper contamination in concentrated. sulfuric or hydrochloric acids can result in the stress cracking of polypropylene.

6.4 Causes of Lining Failure

Linings, if properly selected, installed, and maintained, and if the vessel has been properly designed, fabricated, and prepared to accept the lining, promise many useful years of service. However, on occasion, there have been lining failures that can be attributed to one or more of the following causes.

6.4.1 Liner Selection

Selection is the first step. Essential to this step is a careful analysis of the materials to be handled, their concentrations, and operating conditions as outlined in the beginning of this chapter. Consideration must also be given to the physical and mechanical properties of the liner to ensure that it meets the specified operating conditions. If there is any doubt, corrosion testing should be undertaken to guarantee the resistance of the liner material.

6.4.2 Inadequate Surface Preparation

Surface preparation is extremely important. All specifications for surface preparation must be followed. If not done properly, poor bonding can result and/or mechanical damage to the liner is possible.

6.4.3 Thermal Stresses

If not properly designed for, thermal stresses produced by thermal cycling can eventually result in bond failure.

6.4.4 Permeation

Certain lining materials are subject to permeation when in contact with specific corrodents. When the possibility of permeation exists, an alternate lining material should be selected. Permeation can result in debonding resulting from corrosion products or fluids accumulating at the interface between the liner and substrate. In addition, corrosion of the substrate can result leading to leakage problems and eventual failure of the substrate.

6.4.5 Absorption

As with permeation, adsorption of the corrodent by the liner material can result in swelling of the liner, cracking, and eventual penetration to the substrate. This can lead to high stresses and debonding.

6.4.6 Welding Flaws

It is essential that qualified personnel perform the welding and that only qualified, experienced contractors be used to install linings. A welding flaw is a common cause of lining failure.

6.4.7 Debonding

Debonding can also occur as a result of the use of the wrong bonding agent. Care should be taken that the proper bonding agent is employed for the specific lining being used.

6.4.8 Operation

Lined vessels should be properly identified when installed with the allowable operating characteristics of the liner posted to avoid damage to the liner during clean-up or repair operations. Most failures from this cause result while vessels are being cleaned or repaired. If live steam is used to clean the vessel, allowable operating temperatures may be exceeded. If the vessel is solvent-cleaned, then chemical attack may occur.

6.5 Elastomeric Linings

Elastomers, sometimes referred to as rubbers, have given many years of service in providing protection to steel vessels. Each of these materials can be compounded to improve certain, of its properties. Because of this, it is necessary that a complete specification for a lining using these materials include specific properties that are required for the application. These include resilience, hysteresis, static or dynamic sheer and compression modulus, flex fatigue and cracking, creep resistance to oils and chemicals, permeability, and brittle point, all in the temperature range to be encountered in service. This will permit a competent manufacturer to propose the proper lining material for the application.

Elastomeric linings are sheet-applied and bonded to the steel substrate. Choice of bonding material to be used is dependend on the specific elastomer to be installed. Repair of these linings is relatively simple. Many older vessels with numerous repair patches are still satisfactorily operating.

The most common elastomers used for lining applications along with their operating temperature range are shown in Table 6.2.

Elastomeric materials can fail as the result of chemical action and/or mechanical damage. Chemical deterioration occurs as the result of a chemical reaction between an elastomer and the medium or by the absorption of the medium into the elastomer. This attack results in the swelling of the elastomer and a reduction in its tensile strength.

The degree of deterioration is a function of the temperature and the concentration of the corrodent. In general, the higher the temperature and the higher the concentration of the corrodent, the greater will be the chemical attack. Elastomers, unlike metals, absorb varying quantities of the material they are in contact with, especially organic liquids. This can result in swelling, cracking, and penetration to the substrate of an elastomeric lined vessel. Swelling can cause softening of the elastomer, and in a lined vessel, introduce high stresses and failure of the bond. If an elastomeric lining has high absorption, permeation will probably result. Some elastomers, such as the fluorocarbons, are easily permeated but have very little absorption. An approximation of the expected permeation and/or absorption of an elastomer can be based on the absorption of water. This data is usually available.

TABLE 6.2

Elastomers Used as Liners

	Temperature Range			
	°F		°C	
Elastomer	Minimum	Maximum	Minimum	Maximum
Natural rubber, NR	−59	175	−50	80
Butyl rubber, IIR	−30	300	−34	149
Chlorobutyl rubber, CIIR	−30	300	−34	149
Neoprene, CR	−13	203	−25	95
Hypalon, CSM	−20	250	−30	121
Urethane rubber, AU	−65	250	−54	121
EPDM rubber	−65	300	−54	149
Nitrile rubber, NBR, Buna-N	−40	250	−40	121
Polyester elastomer, PE	−40	302	−40	150
Perfluoroelastomers, FPM	−58	600	−50	316
Fluoroelastomers, FKM	−10	400	−18	204

6.5.1 Natural Rubber

The maximum temperature for continuous use of natural rubber (NR) is 175°F (80°C). The degree of curing that NR is subjected to will determine whether it is classified as soft, semihard, or hard. Soft rubber is the form primarily used for lining material, although, some hard linings are produced.

Natural rubber provides excellent resistance to most inorganic salt solutions, alkalies, and nonoxidizing acids. Hydrochloric acid will react with soft rubber to form rubber hydrochloride and, therefore, it is not recommended that NR be used in contact with this acid. Strong oxidizing media such as nitric acid, concentrated sulfuric acid, permanganates, dichromates, chlorine dioxide, and sodium hypochlorite will severely attack rubber. Mineral and vegetable oils, gasoline, benzene, toluene, and chlorinated hydrocarbons also affect rubber. Cold water tends to preserve NR. Natural rubber offers good resistance to radiation and alcohols.

6.5.1.1 Soft Natural Rubber Linings

These linings have the advantage of:

Ease of application, cure, and repair unaffected by mechanical stresses or rapid temperature changes, making it ideal for outdoor tanks.

Good chemical resistance within temperature limitation. See Table 6.3.

Excellent to superior physical properties. Maximum tensile elongation, abrasion, and tear resistance.

Low cost.

TABLE 6.3

Compatibility of Soft Natural Rubber with Selected Corrodents

Chemical	Maximum Temperature	
	°F	°C
Acetaldehyde	X	X
Acetamide	X	X
Acetic acid, 10%	150	66
Acetic acid, 50%	X	X
Acetic acid, 80%	X	X
Acetic acid, glacial	X	X
Acetic anhydride	X	X
Acetone	140	60
Acetyl chloride	X	X
Alum	140	60
Aluminum chloride, aqueous	140	60
Aluminum chloride, dry	160	71
Aluminum fluoride	X	X
Aluminum nitrate	X	X
Aluminum sulfate	140	60
Ammonium carbonate	140	60
Ammonium chloride, 10%	140	60
Ammonium chloride, 50%	140	60
Ammonium chloride, sat.	140	60
Ammonium fluoride, 10%	X	X
Ammonium fluoride, 25%	X	X
Ammonium hydroxide, 25%	140	60
Ammonium hydroxide sat.	140	60
Ammonium nitrate	140	60
Ammonium phosphate	140	60
Ammonium sulfate, 10–40%	140	60
Ammonium sulfide	140	60
Amyl acetate	X	X
Amyl alcohol	140	60
Amyl chloride	X	X
Aniline	X	X
Aqua regia, 3:1	X	X
Barium carbonate	140	60
Barium chloride	140	60
Barium hydroxide	140	60
Barium sulfate	140	60
Barium sulfide	140	60
Benzaldehyde	X	X
Benzene	X	X
Benzene sulfonic acid, 10%	X	X
Benzoic acid	140	60
Benzly alcohol	X	X
Benzyl chloride	X	X
Borax	140	60
Boric acid	140	60
Butyl acetate	X	X
Butyl alcohol	140	60

(continued)

TABLE 6.3 *Continued*

Chemical	Maximum Temperature	
	°F	°C
Butyric acid	X	X
Calcium bisulfite	140	60
Calcium carbonate	140	60
Calcium chlorate	140	60
Calcium chloride	140	60
Calcium hydroxide, 10%	140	60
Calcium hydroxide, sat.	140	60
Calcium hypochlorite	X	X
Calcium nitrate	X	X
Calcium oxide	140	60
Calcium sulfate	140	60
Carbon bisulfide	X	X
Carbon disulfide	X	X
Carbon monoxide	X	X
Carbon tetrachloride	X	X
Carbonic acid	140	60
Cellosolve	X	X
Chloracetic acid, 50% water	X	X
Chloracetic acid	X	X
Chlorine gas, dry	X	X
Chlorine gas, wet	X	X
Chlorine, liquid	X	X
Chlorobenzene	X	X
Chloroform	X	X
Chlorosulfonic acid	X	X
Chromic acid, 10%	X	X
Chromic acid, 50%	X	X
Citric acid, 15%	140	60
Citric acid, conc.	X	X
Copper carbonate	X	X
Copper chloride	X	X
Copper cyanide	140	60
Copper sulfate	140	60
Cresol	X	X
Cupric chloride, 5%	X	X
Cupric chloride, 50%	X	X
Cyclohexane	X	X
Dichloroethane (ethylene dichloride)	X	X
Ethylene glycol	140	60
Ferric chloride	140	60
Ferric chloride, 50% in water	140	60
Ferric nitrate, 10–50%	X	X
Ferrous chloride	140	60
Ferrous nitrate	X	X
Fluorine gas, dry	X	X
Hydrobromic acid, dil.	140	60
Hydrobromic acid, 20%	140	60

(continued)

TABLE 6.3 *Continued*

Chemical	Maximum Temperature	
	°F	°C
Hydrobromic acid, 50%	140	60
Hydrochloric acid, 20%	X	X
Hydrochloric acid, 38%	140	60
Hydrofluoric acid, 30%	X	X
Hydrofluoric acid, 70%	X	X
Hydrofluoric acid, 100%	X	X
Lactic acid, 25%	X	X
Lactic acid, conc.	X	X
Magnesium chloride	140	60
Malic acid	X	X
Methyl chloride	X	X
Methyl ethyl ketone	X	X
Methyl isobutyl ketone	X	X
Muriatic acid	140	60
Nitric acid, 5%	X	X
Nitric acid, 20%	X	X
Nitric acid, 70%	X	X
Nitric acid, anhydrous	X	X
Nitrous acid, conc.	X	X
Phenol	X	X
Phosphoric acid, 50–80%	140	60
Potassium bromide, 30%	140	60
Sodium carbonate	140	60
Sodium chloride	140	60
Sodium hydroxide, 10%	140	60
Sodium hydroxide, 50%	X	X
Sodium hydroxide, conc.	X	X
Sodium hypochlorite, 20%	X	X
Sodium hypochlorite, conc.	X	X
Sodium sulfide, to 50%	140	60
Stannic chloride	140	60
Stannous chloride	140	60
Sulfuric acid, 10%	140	60
Sulfuric acid, 50%	X	X
Sulfuric acid, 70%	X	X
Sulfuric acid, 90%	X	X
Sulfuric acid, 98%	X	X
Sulfuric acid, 100%	X	X
Sulfuric acid fuming	X	X
Sulfurous acid	X	X
Zinc chloride	140	60

The chemicals listed are in the pure state or in a saturated solution unless otherwise indicated. Compatibility is shown to the maximum allowable temperature for which data is available. Incompatibility is shown by an X. A blank space indicates that data is unavailable.

Source: From P.A. Schweitzer. 2004. *Corrosion Resistance Tables*, Vols. 1–4, 5th ed., New York: Marcel Dekker.

Limitations of these linings include:

Not oil or flame resistant
Not ozone, sunlight, or weather resistant
Temperature limited to 140°F/60°C
Cannot be used with dilute hydrochloric acid (5–10%) or spent acids

Typical applications for soft natural rubber linings include chemical storage tanks to handle 37% phosphoric acid, hydrofluorosilicic acid, alum, chlorides, and sulfates of ammonia, cadmium, iron, phosphorus, and sodium or inorganic salts in general. This material is suitable to line trailer tanks for transporting the above chemicals. Soft natural rubber is also used to line pipes and fittings.

6.5.1.2 Multiply Natural Rubber Linings

Multiply natural rubber linings consist of three layers of rubber: soft natural rubber/hard natural rubber/soft natural rubber. The advantages of these linings are:

Better permeation resistance than soft natural rubber
Heat resistance to 160°T/71°C
Ease of application, cure, and repair
Good flexibility, reducing the danger of cracking in cold weather, rapid temperature changes, or mechanical stresses
Moderate cost

The limitations of these linings are:

Not oil or flame resistant
Cannot be used in transport trucks

Refer to Table 6.4 for the compatibility of multiply linings with selected corrodents. Applications for multiply linings include those for soft natural rubber linings with the exceptions noted above.

6.5.1.3 Semihard Natural Rubber Linings

Semihard natural rubber linings provide:

Ease of application, cure, and repair
Excellent chemical and permeation resistance. Refer to Table 6.5.
Heat resistance to 180°F/82°C
Moderate cost

TABLE 6.4

Compatibility of Multiple Ply (Soft/Hard/Soft) Natural Rubber with Selected Corrodents

Chemical	Maximum Temperature	
	°F	°C
Acetaldehyde	X	X
Acetamide	X	X
Acetic acid, 10%	X	X
Acetic acid, 50%	X	X
Acetic acid, 80%	X	X
Acetic acid, glacial	X	X
Acetic anhydride	X	X
Acetone	140	60
Alum	160	71
Aluminum chloride, aqueous	160	71
Aluminum chloride, dry	160	71
Aluminum fluoride	X	X
Aluminum nitrate	X	X
Aluminum oxychloride		
Aluminum sulfate	160	71
Ammonium carbonate	160	71
Ammonium chloride, 10%	160	71
Ammonium chloride, 50%	160	71
Ammonium chloride, sat.	160	71
Ammonium fluoride, 10%	X	X
Ammonium fluoride, 25%	X	X
Ammonium hydroxide, 25%	100	38
Ammonium hydroxide, sat.	100	38
Ammonium nitrate	160	71
Ammonium persulfate		
Ammonium phosphate	160	71
Ammonium sulfate, 10–40%	160	71
Ammonium sulfide	160	71
Amyl acetate	X	X
Amyl alcohol	100	38
Amyl chloride	X	X
Aniline	X	X
Aqua regia, 3:1	X	X
Barium carbonate	160	71
Barium chloride	160	71
Barium hydroxide	160	71
Barium sulfate	160	71
Barium sulfide	160	71
Benzaldehyde	X	X
Benzene	X	X
Benzene sulfonic acid, 10%	X	X
Benzoic acid	160	71
Benzyl alcohol	X	X
Benzyl chloride	X	X
Borax	160	71
Boric acid	140	60

(continued)

TABLE 6.4 *Continued*

Chemical	Maximum Temperature	
	°F	°C
Butyl acetate	X	X
Butyl alcohol	160	71
Butyric acid	X	X
Calcium bisulfite	160	71
Calcium carbonate	160	71
Calcium chlorate	140	60
Calcium chloride	140	60
Calcium hydroxide, 10%	160	71
Calcium hydroxide, sat.	160	71
Calcium hypochlorite	X	X
Calcium nitrate	X	X
Calcium oxide	160	71
Calcium sulfate	160	71
Carbon bisulfide	X	X
Carbon disulfide	X	X
Carbon monoxide	X	X
Carbon tetrachloride	X	X
Carbonic acid	160	71
Cellosolve	X	X
Chloracetic acid, 50% water	X	X
Chloracetic acid	X	X
Chlorine gas, dry	X	X
Chlorine gas, wet	X	X
Chlorine, liquid	X	X
Chlorobenzene	X	X
Chloroform	X	X
Chlorosulfonic acid	X	X
Chromic acid, 10%	X	X
Chromic acid, 50%	X	X
Citric acid, 15%	X	X
Citric acid, conc.	X	X
Copper carbonate	X	X
Copper chloride	X	X
Copper cyanide	160	71
Copper sulfate	160	71
Cresol	X	X
Cupric chloride, 5%	X	X
Cupric chloride, 50%	X	X
Cyclohexane	X	X
Dichloroethane (ethylene dichloride)	X	X
Ethylene glycol	160	71
Ferric chloride	160	71
Ferric chloride, 50% in water	160	71
Ferric nitrate, 10–50%	X	X
Ferrous chloride	140	60
Ferrous nitrate	X	X
Fluorine gas, dry	X	X

(continued)

TABLE 6.4 *Continued*

Chemical	Maximum Temperature	
	°F	°C
Hydrobromic acid, dil.	160	71
Hydrobromic acid, 20%	160	71
Hydrobromic acid, 50%	160	71
Hydrochloric acid, 20%	X	X
Hydrochloric acid, 38%	160	71
Hydrofluoric acid, 30%	X	X
Hydrofluoric acid, 70%	X	X
Hydrofluoric acid, 100%	X	X
Lactic acid, 25%	X	X
Lactic acid, conc.	X	X
Magnesium chloride	160	71
Malic acid	100	38
Methyl chloride	X	X
Methyl ethyl ketone	X	X
Methyl isobutyl ketone	X	X
Muriatic acid	140	60
Nitric acid, 5%	X	X
Nitric acid, 20%	X	X
Nitric acid, 70%	X	X
Nitric acid, anhydrous	X	X
Nitrous acid, conc.	X	X
Phenol	X	X
Phosphoric acid, 50–80%	160	71
Potassium bromide, 30%	160	71
Sodium carbonate	160	71
Sodium chloride	160	71
Sodium hydroxide, 10%	160	71
Sodium hydroxide, 50%	X	X
Sodium hydroxide, conc.	X	X
Sodium hypochlorite, 20%	X	X
Sodium hypochlorite, conc.	X	X
Sodium sulfide, to 50%	160	71
Stannic chloride	160	71
Stannous chloride	160	71
Sulfuric acid, 10%	160	71
Sulfuric acid, 50%	X	X
Sulfuric acid, 70%	X	X
Sulfuric acid, 90%	X	X
Sulfuric acid, 98%	X	X
Sulfuric acid, 100%	X	X
Sulfuric acid, fuming	X	X
Sulfurous acid	X	X
Zinc chloride	160	71

The chemicals listed are in the pure state or in a saturated solution unless otherwise indicated. Compatibility is shown to the maximum allowable temperature for which data is available. Incompatibility is shown by an X. A blank space indicates that data is unavailable.

Source: From P.A. Schweitzer. 2004. *Corrosion Resistance Tables*, Vols. 1–4, 5th ed., New York: Marcel Dekker.

TABLE 6.5

Compatibility of Semihard Natural Rubber with Selected Corrodents

Chemical	Maximum Temperature	
	°F	°C
Acetamide	X	X
Acetic acid, glacial	X	X
Acetic anhydride	X	X
Acetone	X	X
Alum	180	82
Aluminum chloride, aqueous	180	82
Aluminum fluoride	X	X
Aluminum nitrate	100	38
Aluminum sulfate	180	82
Ammonium carbonate	180	82
Ammonium chloride, 10%	180	82
Ammonium chloride, 50%	180	82
Ammonium chloride, sat.	180	82
Ammonium fluoride, 10%	X	X
Ammonium fluoride, 25%	X	X
Ammonium nitrate	180	82
Ammonium phosphate	180	82
Ammonium sulfate, 10–40%	180	82
Ammonium sulfide	180	82
Amyl alcohol	180	82
Aniline	X	X
Barium carbonate	180	82
Barium chloride	180	82
Barium hydroxide	180	82
Barium sulfate	180	82
Benzene	X	X
Benzoic acid	180	82
Borax	180	82
Boric acid	180	82
Butyl acetate	100	38
Butyl alcohol	180	82
Butyric acid	100	38
Calcium bisulfite	180	82
Calcium carbonate	180	82
Calcium chloride	180	82
Calcium hydroxide, 10%	180	82
Calcium hydroxide, sat.	180	82
Calcium hypochlorite	X	X
Chloracetic acid, 50% water	100	38
Chloracetic acid	100	38
Chromic acid, 10%	X	X
Chromic acid, 50%	X	X
Citric acid, 15%	100	38
Citric acid, conc.	100	38
Copper carbonate	180	82
Copper chloride	X	X
Copper cyanide	180	82

(continued)

TABLE 6.5 *Continued*

Chemical	Maximum Temperature	
	°F	°C
Copper sulfate	180	82
Cupric chloride, 5%	X	X
Cupric chloride, 50%	X	X
Ethylene glycol	180	82
Ferric chloride	180	82
Ferric chloride, 50% in water	180	82
Ferric nitrate, 10–50%	100	38
Ferrous chloride	180	82
Ferrous nitrate	100	38
Hydrobromic acid, dil.	180	82
Hydrobromic acid, 20%	180	82
Hydrobromic acid, 50%	180	82
Hydrochloric acid, 20%	180	82
Hydrochloric acid, 38%	180	82
Hydrofluoric acid, 70%	X	X
Hydrofluoric acid, 100%	X	X
Lactic acid, 25%	100	38
Lactic acid, conc.	100	38
Magnesium chloride	180	82
Malic acid	100	38
Methyl chloride	100	38
Nitric acid, 5%	100	38
Nitric acid, 20%	X	X
Nitric acid, 70%	X	X
Nitric acid, anhydrous	X	X
Nitrous acid, conc.	100	38
Phenol	X	X
Phosphoric acid, 50–80%	180	82
Potassium bromide, 30%	180	82
Sodium carbonate	180	82
Sodium chloride	180	82
Sodium hydroxide, 10%	180	82
Sodium hydroxide, 50%	100	38
Sodium hydroxide, conc.	100	38
Sodium hypochlorite, 20%	X	X
Sodium hypochlorite, conc.	X	X
Sodium sulfide, to 50%	180	82
Stannic chloride	180	82
Stannous chloride	180	82
Sulfuric acid, 10%	180	82
Sulfuric acid, 50%	100	38
Sulfuric acid, 70%	X	X
Sulfuric acid, 90%	X	X
Sulfuric acid, 98%	X	X
Sulfuric acid, 100%	X	X
Sulfuric acid, fuming	X	X

(continued)

TABLE 6.5 *Continued*

Chemical	Maximum Temperature	
	°F	°C
Sulfurous acid	150	66
Zinc chloride	180	82

The chemicals listed are in the pure state or in a saturated solution unless otherwise indicated. Compatibility is shown to the maximum allowable temperature for which data is available. Incompatibility is shown by an X. A blank space indicates that data is unavailable.

Source: From P.A. Schweitzer. 2004. *Corrosion Resistance Tables*, Vols. 1–4, 5th ed., New York: Marcel Dekker.

These linings are limited by their inability to resist oil or flames and the fact that they are subject to damage by cold weather, sudden extreme temperature changes, or mechanical stresses. In addition, they are not abrasion resistant. Applications include chemical process tanks, pumps, fans, pipe, and fittings.

6.5.1.4 Hard Natural Rubber Linings

Hard natural rubber linings have the advantage of:

> Ease of application, cure, and repair
> Better chemical, heat, and permeation resistance than soft natural rubber. Refer to Table 6.6.
> Heat resistance to 200°F/93°C
> Moderate cost

Limitations of these linings are:

> Not oil or flame resistant
> Subject to damage by cold weather, exposure, sudden extreme temperature changes, or mechanical stresses
> Not abrasion resistant
> Not to be used in transport vehicles

Typical applications include chemical storage tanks, pickling tanks, pipe and fittings, plating tanks, and various accessories as agitators, pumps, etc.

6.5.2 Isoprene

Natural rubber is chemically a natural *cis*-polyisoprene. Isoprene (IR) is a synthetic *cis*-polyisoprene. The corrosion resistance of IR is the same as that

TABLE 6.6

Compatibility of Hard Natural Rubber with Selected Corrodents

	Maximum Temperature	
Chemical	°F	°C
Acetamide	X	X
Acetic acid, 10%	200	93
Acetic acid, 50%	200	93
Acetic acid, 80%	150	66
Acetic acid, glacial	100	38
Acetic anhydride	100	38
Acetone	X	X
Acrylonitrile	90	32
Adipic acid	80	27
Allyl alcohol	X	X
Alum	200	93
Aluminum chloride, aqueous	200	93
Aluminum fluoride	X	X
Aluminum hydroxide	200	93
Aluminum nitrate	190	88
Aluminum sulfate	200	93
Ammonium bifluoride	200	93
Ammonium carbonate	200	93
Ammonium chloride, 10%	200	93
Ammonium chloride, 50%	200	93
Ammonium chloride, sat.	200	93
Ammonium fluoride, 10%	X	X
Ammonium fluoride, 25%	X	X
Ammonium hydroxide, 25%	X	X
Ammonium hydroxide, sat.	X	X
Ammonium nitrate	150	66
Ammonium persulfate	200	93
Ammonium phosphate	200	93
Ammonium sulfate, 10–40%	200	93
Ammonium sulfide	200	93
Amyl alcohol	200	93
Aniline	X	X
Aqua regia, 3:1	X	X
Barium carbonate	200	93
Barium chloride	200	93
Barium hydroxide	X	X
Barium sulfate	200	93
Barium sulfide	200	93
Benzaldehyde	X	X
Benzene	X	X
Benzoic acid	200	93
Borax	200	93
Boric acid	200	93
Butyl acetate	X	X
Butyl alcohol	160	71
Butyric acid	150	66
Calcium bisulfite	200	93

(continued)

TABLE 6.6 *Continued*

Chemical	Maximum Temperature	
	°F	°C
Calcium carbonate	200	93
Calcium chloride	200	93
Calcium hydroxide, 10%	200	93
Calcium hydroxide, sat.	200	93
Calcium hypochlorite	X	X
Calcium nitrate	200	93
Calcium oxide	200	93
Calcium sulfate	200	93
Carbon bisulfide	X	X
Carbon disulfide	X	X
Carbon monoxide	X	X
Carbon tetrachloride	X	X
Carbonic acid	200	93
Chloracetic acid, 50% water	120	49
Chloracetic acid	120	49
Chlorine gas, dry	X	X
Chlorine gas, wet	190	88
Chlorine, liquid	X	X
Chlorobenzene	X	X
Chloroform	X	X
Chlorosulfonic acid	X	X
Chromic acid, 10%	X	X
Chromic acid, 50%	X	X
Citric acid, 15%	150	66
Citric acid, conc.	150	66
Copper carbonate	200	93
Copper chloride	100	38
Copper cyanide	200	93
Copper sulfate	200	93
Ethylene glycol	200	93
Ferric chloride	200	93
Ferric chloride, 50% in water	200	93
Ferric nitrate, 10–50%	150	66
Ferrous chloride	200	93
Ferrous nitrate	150	66
Hydrobromic acid, dil.	200	93
Hydrobromic acid, 20%	200	93
Hydrobromic acid, 50%	200	93
Hydrochloric acid, 20%	200	93
Hydrochloric acid, 38%	200	93
Hydrocyanic acid, 10%	200	93
Hydrofluoric acid, 30%	X	X
Hydrofluoric acid, 70%	X	X
Hydrofluoric acid, 100%	X	X
Hypochlorous acid	150	66
Lactic acid, 25%	150	66
Lactic acid, conc.	150	66

(continued)

TABLE 6.6 *Continued*

Chemical	Maximum Temperature	
	°F	°C
Magnesium chloride	200	93
Malic acid	150	66
Methyl ethyl ketone	X	X
Muriatic acid	200	93
Nitric acid, 5%	150	66
Nitric acid, 20%	X	X
Nitric acid, 70%	X	X
Nitric acid, anhydrous	X	X
Nitrous acid, conc.	150	66
Phenol	X	X
Phosphoric acid, 50–80%	200	93
Potassium bromide, 30%	200	93
Sodium carbonate	200	93
Sodium chloride	200	93
Sodium hydroxide, 10%	200	93
Sodium hydroxide, 50%	150	66
Sodium hydroxide, conc.	150	66
Sodium hypochlorite, 20%	X	X
Sodium hypochlorite, conc.	X	X
Sodium sulfide, to 50%	200	93
Stannic chloride	200	93
Stannous chloride	200	93
Sulfuric acid, 10%	200	93
Sulfuric acid, 70%	X	X
Sulfurous acid	200	93
Zinc chloride	200	93

The chemicals listed are in the pure state or in a saturated solution unless otherwise indicated. Compatibility is shown to the maximum allowable temperature for which data is available. Incompatibility is shown by an X. A blank space indicates that data is unavailable.

Source: From P.A. Schweitzer. 2004. *Corrosion Resistance Tables*, Vols. 1–4, 5th ed., New York: Marcel Dekker.

of NR. The major difference is that isoprene has no odor and, therefore, can be used in the handling of certain food products.

6.5.3 Neoprene

The maximum temperature that neoprene (CR) can be used under is 180–200°F (82–93°C). Excellent service is experienced in contact with aliphatic compounds(methyl and ethyl alcohols, ethylene glycols, etc.), aliphatic hydrocarbons, and most freon refrigerants. However, the outstanding properties of neoprene is its resistance to attack from solvents, waxes, fats, oils, greases, and many other petroleum-based products. Dilute mineral acids, inorganic salt solutions, and alkalies can also be handled successfully.

Chlorinated and aromatic hydrocarbons, organic esters, aromatic hydroxy compounds, and certain ketones have an adverse effect on neoprene and, consequently, only limited serviceability can be expected with them. Highly oxidizing acid and salt solutions also cause surface deterioration and loss of strength. Included in this category are nitric acid and concentrated sulfuric acid. Table 6.7 provides the compatibility of neoprene with selected corrodents. Reference [1] provides a more detailed listing.

The advantages of neoprene include its good chemical resistance, heat resistance to 200°F/93°C, good oil resistance, better resistance to ozone, sunlight, and weather than natural rubber, its excellent abrasion resistance, and the fact it is unaffected toy cold weather, rapid temperature changes, or mechanical stresses.

On the negative side is the fact that it is more expensive than natural rubber and is more difficult to apply.

Typical applications include caustic storage and transportation tanks, sulfuric acid pickle tanks, chemical process and storage tanks, and miscelleneous process equipment accessories.

6.5.4 Butyl Rubber

The maximum temperature that butyl rubber (IIR) can be exposed to on a continuous basis is 250–300°F (120–148°C). Butyl rubber is very nonpolar. It has exceptional resistance to dilute mineral acids, alkalies, phosphate ester oils, acetone, ethylene, ethylene glycol, and water. Resistance to concentrated acids, except nitric and sulfuric, is good. It is attacked by petroleum oils, gasoline, and most solvents (except oxygenated solvents) but is resistant to swelling by vegetable and animal oils.

Refer to Table 6.8 for the compatibility of butyl rubber with selected corrodents. Reference [1] provides a more detailed listing. Butyl rubber linings nave the following advantages:

Unaffected by cold water or rapid temperature changes
Heat resistant to 200°F/93°C
Resistant to ozone, sunlight, and aging
Possessed of good chemical and gaseous permeation resistance
Resistant to water and oxygenated solvents
Resistant to sliding abrasion
Able to be used for hydrochloric acid

Limitations of butyl rubber are:

Not flame or oil resistant
More costly and difficult to apply than natural rubber

TABLE 6.7

Compatibility of Neoprene with Selected Corrodents

Chemical	Maximum Temperature	
	°F	°C
Acetaldehyde	200	93
Acetamide	200	93
Acetic acid, 10%	160	71
Acetic acid, 50%	160	71
Acetic acid, 80%	160	71
Acetic acid, glacial	X	X
Acetic anhydride	X	X
Acetone	X	X
Acetyl chloride	X	X
Acrylic acid	X	X
Acrylonitrile	140	60
Adipic acid	160	71
Allyl alcohol	120	49
Allyl chloride	X	X
Alum	200	93
Aluminum acetate		
Aluminum chloride, aqueous	150	66
Aluminum chloride, dry		
Aluminum fluoride	200	93
Aluminum hydroxide	180	82
Aluminum nitrate	200	93
Aluminum oxychloride		
Aluminum sulfate	200	93
Ammonia gas	140	60
Ammonium bifluoride	X	X
Ammonium carbonate	200	93
Ammonium chloride, 10%	150	66
Ammonium chloride, 50%	150	66
Ammonium chloride, sat.	150	66
Ammonium fluoride, 10%	200	93
Ammonium fluoride, 25%	200	93
Ammonium hydroxide, 25%	200	93
Ammonium hydroxide, sat.	200	93
Ammonium nitrate	200	93
Ammonium persulfate	200	93
Ammonium phosphate	150	66
Ammonium sulfate, 10–40%	150	66
Ammonium sulfide	160	71
Ammonium sulfite		
Amyl acetate	X	X
Amyl alcohol	200	93
Amyl chloride	X	X
Aniline	X	X
Antimony trichloride	140	60
Aqua regia, 3:1	X	X
Barium carbonate	150	66
Barium chloride	150	66
Barium hydroxide	230	110

(continued)

TABLE 6.7 *Continued*

Chemical	Maximum Temperature	
	°F	°C
Barium sulfate	200	93
Barium sulfide	200	93
Benzaldehyde	X	X
Benzene	X	X
Benzene sulfonic acid, 10%	100	38
Benzoic acid	150	66
Benzyl alcohol	X	X
Benzyl chloride	X	X
Borax	200	93
Boric acid	150	66
Bromine gas, dry	X	X
Bromine gas, moist	X	X
Bromine liquid	X	X
Butadiene	140	60
Butyl acetate	60	16
Butyl alcohol	200	93
n-Butylamine		
Butyl phthalate		
Butyric acid	X	X
Calcium bisulfite	X	X
Calcium carbonate	200	93
Calcium carbonate	200	93
Calcium chlorate	200	93
Calcium chloride	150	66
Calcium hydroxide, 10%	230	110
Calcium hydroxide, sat.	230	110
Calcium hypochlorite	X	X
Calcium nitrate	150	66
Calcium oxide	200	93
Calcium sulfate, water	150	66
Caprylic acid		
Carbon bisulfide	X	X
Carbon dioxide, dry	200	93
Carbon dioxide, wet	200	93
Carbon disulfide	X	X
Carbon monoxide	X	X
Carbon tetrachloride	X	X
Carbonic acid	150	66
Cellosolve	X	X
Chloracetic acid, 50% water	X	X
Chloracetic acid	X	X
Chlorine gas, dry	X	X
Chlorine gas, wet	X	X
Chlorine, liquid	X	X
Chlorobenzene	X	X
Chloroform	X	X
Chlorosulfonic acid	X	X
Chromic acid, 10%	140	60

(continued)

TABLE 6.7 *Continued*

Chemical	Maximum Temperature	
	°F	°C
Chromic acid, 50%	100	38
Chromyl chloride		
Citric acid, 15%	150	66
Citric acid, conc.	150	66
Copper acetate	160	71
Copper carbonate		
Copper chloride	200	93
Copper cyanide	160	71
Copper sulfate	200	93
Cresol	X	X
Cupric chloride, 5%	200	93
Cupric chloride, 50%	160	71
Cyclohexane	X	X
Cyclohexanol	X	X
Dichloroacetic acid	X	X
Dichloroethane (ethylene dichloride)	X	X
Ethylene glycol	100	38
Ferric chloride	160	71
Ferric chloride, 50% in water	160	71
Ferric nitrate, 10–50%	200	93
Ferrous chloride	90	32
Ferrous nitrate	200	93
Fluorine gas, dry	X	X
Fluorine gas, moist	X	X
Hydrobromic acid, dil.	X	X
Hydrobromic acid, 20%	X	X
Hydrobromic acid, 50%	X	X
Hydrochloric acid, 20%	X	X
Hydrochloric acid, 38%	X	X
Hydrocyanic acid, 10%	X	X
Hydrofluoric acid, 30%	X	X
Hydrofluoric acid, 70%	X	X
Hydrofluoric acid, 100%	X	X
Hypochlorous acid	X	X
Iodine solution, 10%	80	27
Ketones, general	X	X
Lactic acid, 25%	140	60
Lactic acid, conc.	90	32
Magnesium chloride	200	93
Malic acid		
Manganese chloride	200	93
Methyl chloride	X	X
Methyl ethyl ketone	X	X
Methyl isobutyl ketone	X	X
Muriatic acid	X	X
Nitric acid, 5%	X	X
Nitric acid, 20%	X	X
Nitric acid, 70%	X	X

(continued)

TABLE 6.7 *Continued*

Chemical	Maximum Temperature	
	°F	°C
Nitric acid, anhydrous	X	X
Nitrous acid, conc.	X	X
Oleum	X	X
Perchloric acid, 10%		
Perchloric acid, 70%	X	X
Phenol	X	X
Phosphoric acid, 50–80%	150	66
Picric acid	200	93
Potassium bromide, 30%	160	71
Salicylic acid		
Silver bromide, 10%		
Sodium carbonate	200	93
Sodium chloride	200	93
Sodium hydroxide, 10%	230	110
Sodium hydroxide, 50%	230	110
Sodium hydroxide, conc.	230	110
Sodium hypochlorite, 20%	X	X
Sodium hypochlorite, conc.	X	X
Sodium sulfide, to 50%	200	93
Stannic chloride	200	93
Stannous chloride	X	X
Sulfuric acid, 10%	150	66
Sulfuric acid, 50%	100	38
Sulfuric acid, 70%	X	X
Sulfuric acid, 90%	X	X
Sulfuric acid, 98%	X	X
Sulfuric acid, 100%	X	X
Sulfuric acid, fuming	X	X
Sulfurous acid	100	38
Thionyl chloride	X	X
Toluene	X	X
Trichloroacetic acid	X	X
White liquor	140	60
Zinc chloride	160	71

The chemicals listed are in the pure state or in a saturated solution unless otherwise indicated. Compatibility is shown to the maximum allowable temperature for which data is available. Incompatibility is shown by an X. A blank space indicates that data is unavailable.

Source: From P.A. Schweitzer. 2004. *Corrosion Resistance Tables*, Vols. 1–4, 5th ed., New York: Marcel Dekker.

Applications include linings for hydrochloric acid storage and processing vessels and mixed acid wastes. Butyl rubber is also used to line pipe and fittings.

6.5.5 Chlorosulfonated Polyethylene (Hypalon)

The maximum temperature that Hypalon can be exposed to on a continuous basis is 250°F (121°C). Hypalon is highly resistant to attack by hydrocarbon

TABLE 6.8

Compatibility of Butyl Rubber with Selected Corrodents

Chemical	Maximum Temperature	
	°F	°C
Acetaldehyde	80	27
Acetic acid, 10%	150	66
Acetic acid, 50%	110	43
Acetic acid, 80%	110	43
Acetic acid, glacial	X	X
Acetic anhydride	X	X
Acetone	100	38
Acrylonitrile	X	X
Adipic acid	X	X
Allyl alcohol	190	88
Allyl chloride	X	X
Alum	200	93
Aluminum acetate	200	93
Aluminum chloride, aqueous	200	93
Aluminum chloride, dry	200	93
Aluminum fluoride	180	82
Aluminum hydroxide	100	38
Aluminum nitrate	100	38
Aluminum sulfate	200	93
Ammonium bifluoride	X	X
Ammonium carbonate	190	88
Ammonium chloride, 10%	200	93
Ammonium chloride, 50%	200	93
Ammonium chloride, sat.	200	93
Ammonium fluoride, 10%	150	66
Ammonium fluoride, 25%	150	66
Ammonium hydroxide, 25%	190	88
Ammonium hydroxide, sat.	190	88
Ammonium nitrate	200	93
Ammonium persulfate	190	88
Ammonium phosphate	150	66
Ammonium sulfate, 10–40%	150	66
Amyl acetate	X	X
Amyl alcohol	150	66
Aniline	150	66
Antimony trichloride	150	66
Barium chloride	150	66
Barium hydroxide	190	88
Barium sulfide	190	88
Benzaldehyde	90	32
Benzene	X	X
Benzene sulfonic acid, 10%	90	32
Benzoic acid	150	66
Benzyl alcohol	190	88
Benzyl chloride	X	X
Borax	190	88
Boric acid	150	66

(continued)

TABLE 6.8 *Continued*

Chemical	Maximum Temperature	
	°F	°C
Butyl acetate	X	X
Butyl alcohol	140	60
Butyric acid	X	X
Calcium bisulfite	120	49
Calcium carbonate	150	66
Calcium chlorate	190	88
Calcium chloride	190	88
Calcium hydroxide, 10%	190	88
Calcium hydroxide, sat.	190	88
Calcium hypochlorite	X	X
Calcium nitrate	190	88
Calcium sulfate	100	38
Carbon dioxide, dry	190	88
Carbon dioxide, wet	190	88
Carbon disulfide	190	88
Carbon monoxide	X	X
Carbon tetrachloride	90	32
Carbonic acid	150	66
Cellosolve	150	66
Chloracetic acid, 50% water	150	66
Chloracetic acid	100	38
Chlorine gas, dry	X	X
Chlorine, liquid	X	X
Chlorobenzene	X	X
Chloroform	X	X
Chlorosulfonic acid	X	X
Chromic acid, 10%	X	X
Chromic acid, 50%	X	X
Citric acid, 15%	190	88
Citric acid, conc.	190	88
Copper chloride	150	66
Copper sulfate	190	88
Cresol	X	X
Cupric chloride, 5%	150	66
Cupric chloride, 50%	150	66
Cyclohexane	X	X
Dichloroethane (ethylene dichloride)	X	X
Ethylene glycol	200	93
Ferric chloride	175	79
Ferric chloride, 50% in water	160	71
Ferric nitrate, 10–50%	190	88
Ferrous chloride	175	79
Ferrous nitrate	190	88
Fluorine gas, dry	X	X
Hydrobromic acid, dil.	125	52
Hydrobromic acid, 20%	125	52

(continued)

TABLE 6.8 *Continued*

Chemical	Maximum Temperature	
	°F	°C
Hydrobromic acid, 50%	125	52
Hydrochloric acid, 20%	125	52
Hydrochloric acid, 38%	125	52
Hydrocyanic acid, 10%	140	60
Hydrofluoric acid, 30%	150	66
Hydrofluoric acid, 70%	150	66
Hydrofluoric acid, 100%	150	66
Hypochlorous acid	X	X
Lactic acid, 25%	125	52
Lactic acid, conc.	125	52
Magnesium chloride	200	93
Malic acid	X	X
Methyl chloride	90	32
Methyl ethyl ketone	100	38
Methyl isobutyl ketone	80	27
Muriatic acid	X	X
Nitric acid, 5%	200	93
Nitric acid, 20%	150	66
Nitric acid, 70%	X	X
Nitric acid, anhydrous	X	X
Nitrous acid, conc.	125	52
Oleum	X	X
Perchloric acid, 10%	150	66
Phenol	150	66
Phosphoric acid, 50–80%	150	66
Salicylic acid	80	27
Sodium chloride	200	93
Sodium hydroxide, 10%	150	66
Sodium hydroxide, 50%	150	66
Sodium hydroxide, conc.	150	66
Sodium hypochlorite, 20%	X	X
Sodium hypochlorite, conc.	X	X
Sodium sulfide, 50%	150	66
Stannic chloride	150	66
Stannous chloride	150	66
Sulfuric acid, 10%	200	93
Sulfuric acid, 50%	150	66
Sulfuric acid, 70%	X	X
Sulfuric acid, 90%	X	X
Sulfuric acid, 98%	X	X
Sulfuric acid, 100%	X	X
Sulfuric acid, fuming	X	X
Sulfurous acid	200	93
Thionyl chloride	X	X
Toluene	X	X

(continued)

TABLE 6.8 *Continued*

| | Maximum Temperature | |
Chemical	°F	°C
Trichloroacetic acid	X	X
Zinc chloride	200	93

The chemicals listed are in the pure state or in a saturated solution unless otherwise indicated. Compatibility is shown to the maximum allowable temperature for which data is available. Incompatibility is shown by an X. A blank space indicates that data is unavailable.

Source: From P.A. Schweitzer. 2004. *Corrosion Resistance Tables*, Vols. 1–4, 5th ed., New York: Marcel Dekker.

oils and fuels, even at elevated temperatures. It is also resistant to such oxidizing chemicals as sodium hypochlorite, sodium peroxide, ferric chloride, and sulfuric, chromic, and hydrofluoric acids. Concentrated hydrochloric acid (37%) can be handled below 158°F (70°C). Below this temperature, Hypalon can handle all concentrations without adverse effect. Nitric acid up to 60% concentration at room temperature can also be handled without adverse effect. Hypalon is also resistant to salt solutions, alcohols, and both weak and concentrated alkalies.

Aliphatic, aromatic, and chlorinated hydrocarbons, aldehydes, and ketones will attack Hypalon.

Hypalon is one of the most weather resistant elastomers available. Oxidation takes place at a very low rate. Sunlight and ultraviolet light have little, if any, adverse effect on its physical properties. It is inherently resistant to ozone attack without the need for the addition of special antioxidants or antiozonates to the formulation.

Refer to Table 6.9 for the compatibility of Hypalon with selected corrodents. Reference [1] provides a more detailed listing.

Hypalon's advantages include some oil resistance, excellent resistance to ozone, sunlight, weathering, and chemicals as well as resistance to permeation and abrasion. It is unaffected by cold weather or rapid temperature changes. Hypalon is difficult to apply, is more costly than natural rubber, is not resistant to oxygenated solvents, and requires a pressure cure.

Applications include linings of tanks and tank cars containing chromic and high concentrations of sulfuric and nitric acids and other oxidizing chemicals. It is also used for pond liners.

6.5.6 Urethane Rubbers

The urethane rubbers are produced from a number of polyurethane polymers. Urethane rubbers operate through the range of 50–212°F (10–100°C). Standard compounds exhibit good low temperature impact resistance and low brittle points. At very low temperatures, the material

TABLE 6.9

Compatibility of Hypalon with Selected Corrodents

Chemical	Maximum Temperature	
	°F	°C
Acetaldehyde	60	16
Acetamide	X	X
Acetic acid, 10%	200	93
Acetic acid, 50%	200	93
Acetic acid, 80%	200	93
Acetic acid, glacial	X	X
Acetic anhydride	200	93
Acetone	X	X
Acetyl chloride	X	X
Acrylonitrile	140	60
Adipic acid	140	60
Allyl alcohol	200	93
Aluminum fluoride	200	93
Aluminum hydroxide	200	93
Aluminum nitrate	200	93
Aluminum sulfate	180	82
Ammonia carbonate	140	60
Ammonia gas	90	32
Ammonium chloride, 10%	190	88
Ammonium chloride, 50%	190	88
Ammonium chloride, sat.	190	88
Ammonium fluoride, 10%	200	93
Ammonium hydroxide, 25%	200	93
Ammonium hydroxide, sat.	200	93
Ammonium nitrate	200	93
Ammonium persulfate	80	27
Ammonium phosphate	140	60
Ammonium sulfate, 10–40%	200	93
Ammonium sulfide	200	93
Amyl acetate	60	16
Amyl alcohol	200	93
Amyl chloride	X	X
Aniline	140	60
Antimony trichloride	140	60
Barium carbonate	200	93
Barium chloride	200	93
Barium hydroxide	200	93
Barium sulfate	200	93
Barium sulfide	200	93
Benzaldehyde	X	X
Benzene	X	X
Benzene sulfonic acid, 10%	X	X
Benzoic acid	200	93
Benzyl alcohol	140	60
Benzyl chloride	X	X
Borax	200	93
Boric acid	200	93

(continued)

TABLE 6.9 *Continued*

Chemical	Maximum Temperature	
	°F	°C
Bromine gas, dry	60	16
Bromine gas, moist	60	16
Bromine liquid	60	16
Butadiene	X	X
Butyl acetate	60	16
Butyl alcohol	200	93
Butyric acid	X	X
Calcium bisulfite	200	93
Calcium carbonate	90	32
Calcium chlorate	90	32
Calcium chloride	200	93
Calcium hydroxide, 10%	200	93
Calcium hydroxide, sat.	200	93
Calcium hypochlorite	200	93
Calcium nitrate	100	38
Calcium oxide	200	93
Calcium sulfate	200	93
Caprylic acid	X	X
Carbon dioxide, dry	200	93
Carbon dioxide, wet	200	93
Carbon disulfide	200	93
Carbon monoxide	X	X
Carbon tetrachloride	200	93
Carbonic acid	X	X
Chloracetic acid	X	X
Chlorine gas, dry	X	X
Chlorine gas, wet	90	32
Chlorobenzene	X	X
Chloroform	X	X
Chlorosulfonic acid	X	X
Chromic acid, 10%	150	66
Chromic acid, 50%	150	66
Chromyl chloride		
Citric acid, 15%	200	93
Citric acid, conc.	200	93
Copper acetate	X	X
Copper chloride	200	93
Copper cyanide	200	93
Copper sulfate	200	93
Cresol	X	X
Cupric chloride, 5%	200	93
Cupric chloride, 50%	200	93
Cyclohexane	X	X
Cyclohexanol	X	X
Dichloroethane (ethylene dichloride)	X	X
Ethylene glycol	200	93
Ferric chloride	200	93

(continued)

TABLE 6.9 *Continued*

Chemical	Maximum Temperature	
	°F	°C
Ferric chloride, 50% in water	200	93
Ferric nitrate, 10–50%	200	93
Ferrous chloride	200	93
Fluorine gas, dry	140	60
Hydrobromic acid, dil.	90	32
Hydrobromic acid, 20%	100	38
Hydrobromic acid, 50%	100	38
Hydrochloric acid, 20%	160	71
Hydrochloric acid, 38%	140	60
Hydrocyanic acid, 10%	90	32
Hydrofluoric acid, 30%	90	32
Hydrofluoric acid, 70%	90	32
Hydrofluoric acid, 100%	90	32
Hypochlorous acid	X	X
Ketones, general	X	X
Lactic acid, 25%	140	60
Lactic acid, conc.	80	27
Magnesium chloride	200	93
Manganese chloride	180	82
Methyl chloride	X	X
Methyl ethyl ketone	X	X
Methyl isobutyl ketone	X	X
Muriatic acid	140	60
Nitric acid, 5%	100	38
Nitric acid, 20%	100	38
Nitric acid, 70%	X	X
Nitric acid, anhydrous	X	X
Oleum	X	X
Perchloric acid, 10%	100	38
Perchloric acid, 70%	90	32
Phenol	X	X
Phosphoric acid, 50–80%	200	93
Picric acid	80	27
Pottasium bromide, 30%	200	93
Sodium carbonate	200	93
Sodium chloride	200	93
Sodium hydroxide, 10%	200	93
Sodium hydroxide, 50%	200	93
Sodium hydroxide, conc.	200	93
Sodium hypochlorite, 20%	200	93
Sodium hypochlorite, conc.		
Sodium sulfide, to 50%	200	93
Stannic chloride	90	32
Stannous chloride	200	93
Sulfuric acid, 10%	200	93
Sulfuric acid, 50%	200	93
Sulfuric acid, 70%	160	71

(continued)

TABLE 6.9 *Continued*

Chemical	Maximum Temperature	
	°F	°C
Sulfuric acid, 90%	X	X
Sulfuric acid, 98%	X	X
Sulfuric acid, 100%	X	X
Sulfurous acid	160	71
Toluene	X	X
Zinc chloride	200	93

The chemicals listed are in the pure state or in a saturated solution unless otherwise indicated. Compatibility is shown to the maximum allowable temperature for which data is available. Incompatibility is shown by an X.

Source: From P.A. Schweitzer. 2004. *Corrosion Resistance Tables*, Vols. 1–4, 5th ed., New York: Marcel Dekker.

remains flexible and has good resistance to thermal shock. The standard compositions can be used at temperatures as low as −80°F (−18°C).

Urethane rubbers have little or no resistance to burning. Some slight improvement can be made in this area by special compounding, but the material will still ignite in an actual fire situation.

The urethane rubbers are resistant to most mineral and vegetable oils, greases, and fuels, and to aliphatic and chlorinated hydrocarbons.

Aromatic hydrocarbons, polar solvents, esters, and ketones will attack urethane. Alcohols will soften and swell the urethane rubbers. These rubbers have limited service in weak acid solutions and cannot be used with concentrated acids. Nor are they resistant to steam or caustic, but they are resistant to the swelling and deteriorating effects of being immersed in water.

The urethane rubbers exhibit excellent resistance to ozone attack and have good resistance to weathering. However, extended exposure to ultraviolet light will reduce their physical properties and will cause the rubber to darken. Table 6.10 lists the compatibility of urethane rubbers with selected corrodents, and Table 6.11 lists the compatibility of polyether urethane rubbers with selected corrodents.

Urethane is used to line tanks and piping. It is also applied as a coating to prevent seepage of the corrodent into the concrete and attack of the reinforcing steel. Protection from abrasion and corrosion is also supplied.

6.5.7 Polyester Elastomer

In general, the fluid resistance of these polyester rubbers increases with increasing hardness. Since these rubbers contain no plasticizers, they are not susceptible to the solvent extraction or heat volatilization of such additives. Many fluids and chemicals will extract plasticizers from elastomers causing a significant increase in stiffness (modulus) and volume shrinkage.

TABLE 6.10

Compatibility of Urethane Rubbers with Selected Corrodents

Chemical	Maximum Temperature	
	°F	°C
Acetamide	X	X
Acetic acid, 10%	X	X
Acetic acid, 50%	X	X
Acetic acid, 80%	X	X
Acetic acid, glacial	X	X
Acetone	X	X
Ammonia gas		
Ammonium chloride, 10%	90	32
Ammonium chloride, 28%	90	32
Ammonium chloride, 50%	90	32
Ammonium chloride, sat.	90	32
Ammonium hydroxide, 10%	90	32
Ammonium hydroxide, 25%	90	32
Ammonium hydroxide, sat.	80	27
Ammonium sulfate, 10–40%		
Amyl alcohol	X	X
Aniline	X	X
Benzene	X	X
Benzoic acid	X	X
Bromine water, dil.	X	X
Bromine water, sat.	X	X
Butane	100	38
Butyl acetate	X	X
Butyl alcohol	X	X
Calcium chloride, dil.	X	X
Calcium chloride, sat.	X	X
Calcium hydroxide, 10%	X	X
Calcium hydroxide, 20%	90	32
Calcium hydroxide, 30%	90	32
Calcium hydroxide, sat.	90	32
Cane sugar liquors		
Carbon bisulfide		
Carbon tetrachloride	X	X
Carbonic acid	90	32
Castor oil	90	32
Cellosolve	X	X
Chlorine water, sat.	X	X
Chlorobenzene	X	X
Chloroform	X	X
Chromic acid, 10%	X	X
Chromic acid, 30%	X	X
Chromic acid, 40%	X	X
Chromic acid, 50%	X	X
Citric acid, 5%	X	X
Citric acid, 10%	X	X
Citric acid, 15%	X	X
Citric acid, conc.		

(continued)

TABLE 6.10 *Continued*

	Maximum Temperature	
Chemical	°F	°C
Copper sulfate	90	32
Corn oil		
Cottonseed oil	90	32
Cresol	X	X
Dextrose	X	X
Dibutyl phthalate	X	X
Ethers, general	X	X
Ethyl acetate	X	X
Ethyl alcohol	X	X
Ethylene chloride		
Ethylene glycol	90	32
Fomaldehyde, dil.	X	X
Formaldehyde, 37%	X	X
Formaldehyde, 50%	X	X
Glycerine	90	32
Hydrochloric acid, dil.	X	X
Hydrochloric acid, 20%	X	X
Hydrochloric acid, 35%	X	X
Hydrochloric acid, 38%	X	X
Hydrochloric acid, 50%	X	X
Isopropyl alcohol	X	X
Jet fuel, JP 4	X	X
Jet fuel, JP 5	X	X
Kerosene	90	32
Lard oil	90	32
Linseed oil	90	32
Magnesium hydroxide	90	32
Mercury	90	32
Methyl alcohol	90	32
Methyl ethyl ketone	X	X
Methyl isobutyl ketone	X	X
Mineral oil	90	32
Muriatic acid	X	X
Nitric acid	X	X
Oils and fats	80	27
Phenol	X	X
Potassium bromide, 30%	90	32
Potassium chloride, 30%	90	32
Potassium hydroxide, to 50%	90	32
Potassium sulfate, 10%	90	32
Silver nitrate	90	32
Soaps	X	X
Sodium carbonate	X	X
Sodium chloride	80	27
Sodium hydroxide, to 50%	90	32
Sodium hypochlorite, all conc.	X	X
Sulfuric acid, all conc.	X	X

(continued)

TABLE 6.10 *Continued*

Chemical	Maximum Temperature	
	°F	°C
Toluene	X	X
Trichloroethylene	X	X
Trisodium phosphate	90	32
Turpentine	X	X
Water, salt	X	X
Water, sea	X	X
Whiskey	X	X
Xylene	X	X

The chemicals listed are in the pure state or in a saturated solution unless otherwise indicated. Compatibility is shown to the maximum allowable temperature for which data is available. Incompatibility is shown by an X. A blank space indicates that data is unavailable.

Source: From P.A. Schweitzer. 2004. *Corrosion Resistance Tables*, Vols. 1–4, 5th ed., New York: Marcel Dekker.

Overall, PE elastomers are resistant to the same class of chemicals and fluids that the polyurethanes are and can be satisfactorily used at higher temperatures in the same fluids.

Polyester elastomers have excellent resistance to nonpolar materials such as oils and hydraulic fluids, even at elevated temperatures. At room temperature, PE elastomers are resistant to most polar fluids such as acids, bases, amines, and glycols. Resistance is very poor at temperatures of 158°F/70°C or above. These rubbers should not be in applications requiring continuous exposure to polar fluids at elevated temperatures.

The compatibility of PE elastomers with selected corrodents is shown in Table 6.12. PE elastomer are used to line tanks, ponds, swimming pools, and drums.

6.5.8 Chlorobutyl Rubber

Chlorobutyl rubber (CIIR) is chlorinated isobutylene–isoprene. CIIR exhibits the same general corrosion resistance as natural rubber but can be used at higher temperatures. Unlike butyl rubber, CIIR cannot be used with hydrochloric acid. Refer to Table 6.13 for the compatibility of chlorobutyl rubber with selected corrodents and Reference [1] for additional listings.

Chlorobutyl rubber has the following advantages:

Heat resistance to 200°F/93°C

Unaffected by cold weather or rapid temperature changes

Good chemical and permeation resistance

Good resistance to ozone, sunlight, and weathering

Easier to apply than butyl rubber

TABLE 6.11

Compatibility of Polyether Urethane Rubbbers with Selected Corrodents

	Maximum Tempture	
Chemical	°F	°C
Acetamide	X	X
Acetic acid, 10%	X	X
Acetic acid, 50%	X	X
Acetic acid, 80%	X	X
Acetic acid, glacial	X	X
Acetone	X	X
Acetyl chloride	X	X
Ammonium chloride, 10%	130	54
Ammonium chloride, 28%	130	54
Ammonium chloride, 50%	130	54
Ammonium chloride, sat.	130	54
Ammonium hydroxide, 10%	110	43
Ammonium hydroxide, 25%	110	43
Ammonium hydroxide, sat.	100	38
Ammonium sulfate, 10–40%	120	49
Amyl alcohol	X	X
Aniline	X	X
Benzene	X	X
Benzoic acid	X	X
Benzyl alcohol	X	X
Borax	90	32
Butane	80	27
Butyl acetate	X	X
Butyl alcohol	X	X
Calcium chloride, dil.	130	54
Calcium chloride, sat.	130	54
Calcium hydroxide, 10%	130	54
Calcium hydroxide, 20%	130	54
Calcium hydroxide, 30%	130	54
Calcium hydroxide, sat.	130	54
Calcium hypochlorite, 5%	130	54
Carbon bisulfide	X	X
Carbon tetrachloride	X	X
Carbonic acid	80	27
Castor oil	130	54
Caustic potash	130	54
Chloroacetic acid	X	X
Chlorobenzene	X	X
Chloroform	X	X
Chromic acid, 10%	X	X
Chromic acid, 30%	X	X
Chromic acid, 40%	X	X
Chromic acid, 50%	X	X
Citric acid, 5%	90	32
Citric acid, 10%	90	32
Citric acid, 15%	90	32
Cottonseed oil	130	38

(continued)

TABLE 6.11 *Continued*

Chemical	Maximum Tempture	
	°F	°C
Cresylic acid	X	X
Cyclohexane	80	27
Dioxane	X	X
Diphenyl	X	X
Ethyl acetate	X	X
Ethyl benzene	X	X
Ethyl chloride	X	X
Ethylene oxide	X	X
Ferric chloride, 75%	100	38
Fluorosilicic acid	100	38
Formic acid	X	X
Freon F-11	X	X
Freon F-12	X	X
Glycerine	130	38
Hydrofluoric acid, dil.	X	X
Hydrochloric acid, 20%	X	X
Hydrochloric acid, 35%	X	X
Hydrochloric acid, 38%	X	X
Hydrochloric acid, 50%	X	X
Hydrofluoric acid, 30%	X	X
Hydrofluoric acid, 40%	X	X
Hydrofluoric acid, 50%	X	X
Hydrofluoric acid, 70%	X	X
Hydrofluoric acid, 100%	X	X
Isobutyl alcohol	X	X
Isopropyl acetate	X	X
Lactic acid, all conc.	X	X
Methyl ethyl ketone	X	X
Methyl isobutyl ketone	X	X
Monochlorobenzene	X	X
Muriatic acid	X	X
Nitrobenzene	X	X
Olive oil	120	49
Phenol	X	X
Potassium acetate	X	X
Potassium chloride, 30%	130	54
Potassium hydroxide	130	54
Potassium sulfate, 10%	130	54
Propane	X	X
Propyl acetate	X	X
Propyl alcohol	X	X
Sodium chloride	130	54
Sodium hydroxide, 30%	140	60
Sodium hypochlorite	X	X
Sodium peroxide	X	X
Sodium phosphate, acid	130	54
Sodium phosphate, alk.	130	54

(continued)

TABLE 6.11   *Continued*

Chemical	Maximum Tempture	
	°F	°C
Sodium phosphate, neut.	130	54
Toluene	X	X
Trichloroethylene	X	X
Water, demineralized	130	54
Water, distilled	130	54
Water, salt	130	54
Water, sea	130	54
Xylene	X	X

The chemicals listed are in the pure state or in a saturated solution unless otherwise indicated. Compatibility is shown to the maximum allowable temperature for which data is available. Incompatibility is shown by an X. A blank space indicates that data is unavailable.

Source: From P.A. Schweitzer. 2004. *Corrosion Resistance Tables*, Vols. 1–4, 5th ed., New York: Marcel Dekker.

Limitations of CIIR include its inability to be used with hydrochloric acid or oils. It is more expensive than natural rubber.

6.5.9   Ethylene Propylene Rubber

Ethylene Propylene Rubber (EPDM) has exceptional heat resistance, being able to operate at temperatures of 300–350°F/148–176°C, while also finding applications as low as −70°F/−56°C. As with other elastomeric materials, it can be compounded to improve specific properties. EPDM remains flexible at low temperatures. Standard compounds have brittle points of −90°F/−68°C or below.

EPDM resists attack from oxygenated solvents such as acetone, methyl ethyl ketone, ethyl acetate, weak acids, and alkalies, detergents, phosphate esters, alcohols, and glycols. It exhibits exceptional resistance to hot water and high pressure steam. The elastomer, being hydrocarbon-based, is not resistant to hydrocarbon solvents or oils, chlorinated hydrocarbons, or turpentine.

Ethylene propylene terpolymer rubbers are, in general, resistant to most of the same corrodents as EPDM but do not have as broad a resistance to mineral acids and some organics.

Ethylene propylene rubbers are particularly resistant to sun, weather, and ozone attack. Ozone resistance is inherent in the polymer, and for all practical purposes, it can be considered immune to ozone attack.

The compatibility of EPDM rubber with selected corrodents will be found in Table 6.14 with additional listings in Reference [1].

TABLE 6.12

Compatibility of Polyester Elastomers with Selected Corrodents

	Maximum Temperature	
Chemical	°F	°C
Acetic acid, 10%	80	27
Acetic acid, 50%	80	27
Acetic acid, 80%	80	27
Acetic acid, glacial	100	38
Acetic acid, vapor	90	32
Ammonium chloride, 10%	90	32
Ammonium chloride, 28%	90	32
Ammonium chloride, 50%	90	32
Ammonium chloride, sat.	90	32
Ammonium sulfate, 10–40%	80	27
Amyl acetate	80	27
Aniline	X	X
Beer	80	27
Benzene	80	27
Borax	80	27
Boric acid	80	27
Bromine, liquid	X	X
Butane	80	27
Butyl acetate	80	27
Calcium chloride, dil.	80	27
Calcium chloride, sat.	80	27
Calcium hydroxide, 10%	80	27
Calcium hydroxide, 20%	80	27
Calcium hydroxide, 30%	80	27
Calcium hydroxide, sat.	80	27
Calcium hypochlorite, 5%	80	27
Carbon bisulfide	80	27
Carbon tetrachloride	X	X
Castor oil	80	27
Chlorine gas, dry	X	X
Chlorine gas, wet	X	X
Chlorobenzene	X	X
Chloroform	X	X
Chlorosulfonic acid	X	X
Citric acid, 5%	80	27
Citric acid, 10%	80	27
Citric acid, 15%	80	27
Citric acid, conc.	80	27
Copper sulfate	80	27
Cottonseed oil	80	27
Cupric chloride, to 50%	80	27
Cyclohexane	80	27
Dibutyl phthalate	80	27
Dioctyl phthalate	80	27
Ethyl acetate	80	27
Ethyl alcohol	80	27
Ethylene chloride	X	X

(continued)

TABLE 6.12 *Continued*

Chemical	Maximum Temperature	
	°F	°C
Ethylene glycol	80	27
Formaldehyde, dil.	80	27
Formaldehyde, 37%	80	27
Formic acid, 10–85%	80	27
Glycerine	80	27
Hydrochloric acid, dil.	80	27
Hydrochloric acid, 20%	X	X
Hydrochloric acid, 35%	X	X
Hydrochloric acid, 38%	X	X
Hydrochloric acid, 50%	X	X
Hydrofluoric acid, dil.	X	X
Hydrofluoric acid, 30%	X	X
Hydrofluoric acid, 40%	X	X
Hydrofluoric acid, 50%	X	X
Hydrofluoric acid, 70%	X	X
Hydrofluoric acid, 100%	X	X
Hydrogen	80	27
Hydrogen sulfide, dry	80	27
Lactic acid	80	27
Methyl ethyl ketone	80	27
Methylene chloride	X	X
Mineral oil	80	27
Muriatic acid	X	X
Nitric acid	X	X
Nitrobenzene	X	X
Ozone	80	27
Phenol	X	X
Potassium dichromate, 30%	80	27
Potassium hydroxide	80	27
Pyridine	80	27
Soap solutions	80	27
Sodium chloride	80	27
Sodium hydroxide, all conc.	80	27
Stannous chloride, 15%	80	27
Stearic acid	80	27
Sulfuric acid, all conc.	X	X
Toluene	80	27
Water, demineralized	160	71
Water, distilled	160	71
Water, salt	160	71
Water, sea	160	71
Xylene	80	27
Zinc chloride	80	27

The chemicals listed are in the pure state or in a saturated solution unless otherwise indicated. Compatibility is shown to the maximum allowable temperature for which data is available. Incompatibility is shown by an X. A blank space indicates that data is unavailable.

Source: From P.A. Schweitzer. 2004. *Corrosion Resistance Tables*, Vols. 1–4, 5th ed., New York: Marcel Dekker.

TABLE 6.13

Compatibility of Cholorobutyl Rubber with Selected Corrodents

Chemical	Maximum Temperature	
	°F	°C
Acetic acid, 10%	150	60
Acetic acid, 50%	150	60
Acetic acid, 80%	150	60
Acetic acid, glacial	X	X
Acetic anhydride	X	X
Acetone	100	38
Alum	200	93
Aluminum chloride, aqueous	200	93
Aluminum nitrate	190	88
Aluminum sulfate	200	93
Ammonium carbonate	200	93
Ammonium chloride, 10%	200	93
Ammonium chloride, 50%	200	93
Ammonium chloride, sat.	200	93
Ammonium nitrate	200	93
Ammonium phosphate	150	66
Ammonium sulfate, 10–40%	150	66
Amyl alcohol	150	66
Aniline	150	66
Antimony trichloride	150	66
Barium chloride	150	66
Benzoic acid	150	66
Boric acid	150	66
Calcium chloride	160	71
Calcium nitrate	160	71
Calcium sulfate	160	71
Carbon monoxide	100	38
Carbonic acid	150	66
Chloracetic acid	100	38
Chromic acid, 10%	X	X
Chromic acid, 50%	X	X
Citric acid, 15%	90	32
Copper chloride	150	66
Copper cyanide	160	71
Copper sulfate	160	71
Cupric chloride, 5%	150	66
Cupric chloride, 50%	150	66
Ethylene glycol	200	93
Ferric chloride	175	79
Ferric chloride, 50% in water	100	38
Ferric nitrate, 10–50%	160	71
Ferrous chloride	175	79
Hydrobromic acid, dil.	125	52
Hydrobromic acid, 20%	125	52
Hydrobromic acid, 50%	125	52
Hydrochloric acid, 20%	X	X
Hydrochloric acid, 38%	X	X

(continued)

TABLE 6.13 *Continued*

Chemical	Maximum Temperature	
	°F	°C
Hydrofluoric acid, 70%	X	X
Hydrofluoric acid, 100%	X	X
Lactic acid, 25%	125	52
Lactic acid, conc.	125	52
Magnesium chloride	200	93
Nitric acid, 5%	200	93
Nitric acid, 20%	150	66
Nitric acid, 70%	X	X
Nitric acid, anhydrous	X	X
Nitrous acid, conc.	125	52
Phenol	150	66
Phosphoric acid, 50–80%	150	66
Sodium chloride	200	93
Sodium hydroxide, 10%	150	66
Sodium sulfide, to 50%	150	66
Sulfuric acid, 10%	200	93
Sulfuric acid, 70%	X	X
Sulfuric acid, 90%	X	X
Sulfuric acid, 98%	X	X
Sulfuric acid, 100%	X	X
Sulfuric acid, fuming	X	X
Sulfurous acid	200	93
Zinc chloride	200	93

The chemicals listed are in the pure state or in a saturated solution unless otherwise indicated. Compatibility is shown to the maximum allowable temperature for which data is available. Incompatibility is shown by an X. A blank space indicates that data is unavailable.

Source: From P.A. Schweitzer. 2004. *Corrosion Resistance Tables*, Vols. 1–4, 5th ed., New York: Marcel Dekker.

The advantages of EPDM as a lining material are many:

It has a high heat resistance.

It is unaffected by cold or rapid temperature changes.

It has excellent resistance to sunlight, ozone, and weather.

It has good abrasion and chemical resistance.

On the negative side is the fact that it is neither oil nor flame resistant, and it is difficult to apply.

EPDM has been used to line sodium hypochlorite storage and make-up tanks and has been used as a lining for other bleach solutions. Pumps have also been lined with EPDM.

The compatibility of ethylene propylene terpolymer (EPT) rubbers with selected corrodents can be found in Reference [1].

TABLE 6.14

Compatibility of EDPM Rubber with Selected Corrodents

Chemical	Maximum Temperature	
	°F	°C
Acetaldehyde	200	93
Acetamide	200	93
Acetic acid, 10%	140	60
Acetic acid, 50%	140	60
Acetic acid, 80%	140	60
Acetic acid, glacial	140	60
Acetic anhydride	X	X
Acetone	200	93
Acetyl chloride	X	X
Acrylonitrile	140	60
Adipic acid	200	93
Allyl alcohol	200	93
Allyl chloride	X	X
Alum	200	93
Aluminum fluoride	190	88
Aluminum hydroxide	200	93
Aluminum nitrate	200	93
Aluminum sulfate	190	88
Ammonia gas	200	93
Ammonium bifluoride	200	93
Ammonium carbonate	200	93
Ammonium chloride, 10%	200	93
Ammonium chloride, 50%	200	93
Ammonium chloride, sat.	200	93
Ammonium fluoride, 10%	200	93
Ammonium fluoride, 25%	200	93
Ammonium hydroxide, 25%	100	38
Ammonium hydroxide, sat.	100	38
Ammonium nitrate	200	93
Ammonium persulfate	200	93
Ammonium phosphate	200	93
Ammonium sulfate, 10–40%	200	93
Ammonium sulfide	200	93
Amyl acetate	200	93
Amyl alcohol	200	93
Amyl chloride	X	X
Aniline	140	60
Antimony trichloride	200	93
Aqua regia, 3:1	X	X
Barium carbonate	200	93
Barium chloride	200	93
Barium hydroxide	200	93
Barium sulfate	200	93
Barium sulfide	140	60
Benzaldehyde	150	66
Benzene	X	X
Benzene sulfonic acid, 10%	X	X

(continued)

TABLE 6.14 *Continued*

Chemical	Maximum Temperature	
	°F	°C
Benzoic acid	X	X
Benzyl alcohol	X	X
Benzyl chloride	X	X
Borax	200	93
Boric acid	190	88
Bromine gas, dry	X	X
Bromine gas, moist	X	X
Bromine liquid	X	X
Butadiene	X	X
Butyl acetate	140	60
Butyl alcohol	200	93
Butyric acid	140	60
Calcium bisulfite	X	X
Calcium carbonate	200	93
Calcium chlorate	140	60
Calcium chloride	200	93
Calcium hydroxide, 10%	200	93
Calcium hydroxide, sat.	200	93
Calcium hypochlorite	200	93
Calcium nitrate	200	93
Calcium oxide	200	93
Calcium sulfate	200	93
Carbon bisulfide	X	X
Carbon dioxide, dry	200	93
Carbon dioxide, wet	200	93
Carbon disulfide	200	93
Carbon monoxide	X	X
Carbon tetrachloride	200	93
Carbonic acid	X	X
Cellosolve	200	93
Chloracetic acid	160	71
Chlorine gas, dry	X	X
Chlorine gas, wet	X	X
Chlorine, liquid	X	X
Chlorobenzene	X	X
Chloroform	X	X
Chlorosulfonic acid	X	X
Chromic acid, 50%	X	X
Citric acid, 15%	200	93
Citric acid, conc.	200	93
Copper acetate	100	38
Copper carbonate	200	93
Copper chloride	200	93
Copper cyanide	200	93
Copper sulfate	200	93
Cresol	X	X
Cupric chloride, 5%	200	93

(continued)

TABLE 6.14 *Continued*

Chemical	Maximum Temperature	
	°F	°C
Cupric chloride, 50%	200	93
Cyclohexane	X	X
Cyclohexanol	X	X
Dichloroethane (ethylene dichloride)	X	X
Ethylene glycol	200	93
Ferric chloride	200	93
Ferric chloride, 50% in water	200	93
Ferric nitrate, 10–50%	200	93
Ferrous chloride	200	93
Ferrous nitrate	200	93
Fluorine gas, moist	60	16
Hydrobromic acid, dil.	90	32
Hydrobromic acid, 20%	140	60
Hydrobromic acid, 50%	140	60
Hydrochloric acid, 20%	100	38
Hydrochloric acid, 38%	90	32
Hydrocyanic acid, 10%	200	93
Hydrofluoric acid, 30%	60	16
Hydrofluoric acid, 70%	X	X
Hydrofluoric acid, 100%	X	X
Hypochlorous acid	200	93
Iodine solution, 10%	140	60
Ketones, general	X	X
Lactic acid, 25%	140	60
Lactic acid, conc.		
Magnesium chloride	200	93
Malic acid	X	X
Methyl chloride	X	X
Methyl ethyl ketone	80	27
Methyl isobutyl ketone	60	16
Nitric acid, 5%	60	16
Nitric acid, 20%	60	16
Nitric acid, 70%	X	X
Nitric acid, anhydrous	X	X
Oleum	X	X
Perchloric acid, 10%	140	60
Phosphoric acid, 50–80%	140	60
Picric acid	200	93
Potassium bromide, 30%	200	93
Salicylic acid	200	93
Sodium carbonate	200	
Sodium chloride	140	60
Sodium hydroxide, 10%	200	93
Sodium hydroxide, 50%	180	82
Sodium hydroxide, conc.	180	82
Sodium hypochlorite, 20%	200	93
Sodium hypochlorite, conc.	200	93

(continued)

TABLE 6.14 _Continued_

	Maximum Temperature	
Chemical	°F	°C
Sodium sulfide, to 50%	200	93
Stannic chloride	200	93
Stannous chloride	200	93
Sulfuric acid, 10%	150	66
Sulfuric acid, 50%	150	66
Sulfuric acid, 70%	140	60
Sulfuric acid, 90%	X	X
Sulfuric acid, 98%	X	X
Sulfuric acid, 100%	X	X
Sulfuric acid, fuming	X	X
Toluene	X	X
Trichloroacetic acid	80	27
White liquor	200	93
Zinc chloride	200	93

The chemicals listed are in the pure state or in a saturated solution unless otherwise indicated. Compatibility is shown to the maximum allowable temperature for which data is available. Incompatibility is shown by an X. A blank space indicates that data is unavailable.

Source: From P.A. Schweitzer. 2004. _Corrosion Resistance Tables_, Vols. 1–4, 5th ed., New York: Marcel Dekker.

6.5.10   Nitrile Rubber (NBR, Buna-N)

Nitrile rubber is a copolymer of butadiene and acrylonitrile. The main advantages of the nitrile rubbers are their low cost, good oil and abrasion resistance, and good low temperature and swell characteristics.

The nitrile rubbers exhibit good resistance to solvents, oil, water, and hydraulic fluids. A very slight swelling occurs in the presence of aliphatic hydrocarbons, fatty acids, alcohols, and glycols. The deterioration of physical properties as a result of this swelling is small, making NBR suitable for gas and oil resistant applications.

The use of highly polar solvents such as acetone and methyl ethyl ketone, chlorinated hydrocarbons, ethers and esters should be avoided because these materials will attack the nitrile rubbers. Table 6.15 shows the compatibility of the nitrile rubbers with selected corrodents.

Nitrile rubbers are unaffected by cold weather or rapid temperature change and provide excellent resistance to aliphatic hydrocarbons. They also have good abrasion resistance. However, this material is more costly than natural rubber and is more difficult to apply. Buna-N is used to line tanks and piping.

6.5.11   Fluoroelastomers

Fluoroelastomers (FKM) are fluorine containing hydrocarbon polymers with a saturated structure obtained by polymerizing fluorinated monomers such

TABLE 6.15

Compatibility of Nitrile Rubber with Selected Corrodents

Chemical	Maximum Temperature	
	°F	°C
Acetaldehyde	X	X
Acetamide	180	82
Acetic acid, 10%	X	X
Acetic acid, 50%	X	X
Acetic acid, 80%	X	X
Acetic acid, glacial	X	X
Acetic anhydride	X	X
Acetone	X	X
Acetyl chloride	X	X
Acrylic acid	X	X
Acrylonitrile	X	X
Adipic acid	180	82
Allyl alcohol	180	82
Allyl chloride	X	X
Alum	150	66
Aluminum chloride, aqueous	150	66
Aluminum hydroxide	180	82
Aluminum nitrate	190	88
Aluminum sulfate	200	93
Ammonia gas	190	88
Ammonium carbonate	X	X
Ammonium nitrate	150	66
Ammonium phosphate	150	66
Amyl alcohol	150	66
Aniline	X	X
Barium chloride	125	52
Benzene	150	66
Benzoic acid	150	66
Boric acid	150	66
Calcium hypochlorite	X	X
Carbonic acid	100	38
Ethylene glycol	100	38
Ferric chloride	150	66
Ferric nitrate, 10–50%	150	66
Hydrofluoric acid, 70%	X	X
Hydrofluoric acid, 100%	X	X
Nitric acid, 20%	X	X
Nitric acid, 70%	X	X
Nitric acid, anhydrous	X	X
Phenol	X	X
Phosphoric acid, 50–80%	150	66
Sodium carbonate	125	52
Sodium chloride	200	93
Sodium hydroxide, 10%	150	66
Sodium hypochlorite, 20%	X	X
Sodium hypochlorite, conc.	X	X
Stannic chloride	150	66
Sulfuric acid, 10%	150	66

(continued)

TABLE 6.15 *Continued*

Chemical	Maximum Temperature	
	°F	°C
Sulfuric acid, 50%	150	66
Sulfuric acid, 70%	X	X
Sulfuric acid, 90%	X	X
Sulfuric acid, 98%	X	X
Sulfuric acid, 100%	X	X
Sulfuric acid, fuming	X	X
Zinc chloride	150	66

The chemicals listed are in the pure state or in a saturated solution unless otherwise indicated. Compatibility is shown to the maximum allowable temperature for which data is available. Incompatibility is shown by an X. A blank space indicates that data is unavailable.

Source: From P.A. Schweitzer. 2004. *Corrosion Resistance Tables*, Vols. 1–4, 5th ed., New York: Marcel Dekker.

as vinylidene fluoride, hexafluoroprene, and tetrafluoroethylene. The result is a high-performance rubber with exceptional resistance to oils and chemicals at elevated temperatures.

As with other rubbers, fluoropolymers are capable of being compounded with various additives to enhance specific properties for particular applications.

Fluoropolymers are manufactured under various tradenames by different manufacturers. Three typical materials are listed below:

Trade Name	Manufacturer
Viton	DuPont
Technoflon	Ausimont
Fluorel	3M

The temperature resistance of the fluoroelastomers is exceptionally good over a wide temperature range. At high temperatures, their mechanical properties are retained better than those of any other elastomer. Compounds remain usefully elastic indefinitely when exposed to aging up to 400°F/204°C. Continuous service limits are generally considered to be as follows:

3000 h at 450°F/232°C
1000 h at 500°F/260°C
240 h at 550°F/268°C
48 h at 600°F/313°C

These compounds are suitable for continuous use at 400°F/205°C depending upon the corrodent.

On the low temperature side, these rubbers are generally serviceable in dynamic applications down to $-10°F/-23°C$. Flexibility at low temperatures is a function of the material thickness. The thinner the cross section, the less stiff the material is at every temperature. The brittle point at a thickness of 0.075 in. (1.9 mm) is in the neighborhood of $-50°F/-45°C$.

Being halogen-containing polymers, these elastomers are more resistant to burning than exclusively hydrocarbon rubbers. Normally compounded material will burn when directly exposed to flame, but it will stop burning when the flame is removed. However, it must be remembered that under an actual fire condition, the fluoroelastomers will burn. During combustion, fluorinated products such as hydrochloric acid can be given off. Special compounding can improve the flame resistance. One such formulation has been developed for the space program that will ignite under conditions of the NASA test that specifies 100% oxygen at 6.2 psi absolute.

Fluoroelastomers have been approved by the U.S. Food and Drug Administration for use in repeated contact with food products.

The fluoropolymers provide excellent resistance to oils, fuels, lubricants, most mineral acids, many aliphatic and aromatic hydrocarbons (carbon tetrachloride, benzene, toluene, and xylene) that act as solvents for other rubbers, gasoline, naphtha, chlorinated solvents, and pesticides. Special formulations can be produced to obtain resistance to hot mineral acids.

These elastomers are not suitable for use with low molecular weight esters and ethers, certain amines, or hot anhydrous hydra-fluoric or chlorosulfonic acids. Table 6.16 provides the compatibility of fluoroelastomers with selected corrodents. Reference [1] provides additional information.

The main applications of the fluoroelastomers are in those applications requiring resistance to high operating temperatures together with high chemical resistance to aggressive media and to those characterized by severe operating conditions that no other elastomer can withstand. Linings and coatings are such applications. Their chemical stability solves the problem of chemical corrosion by making it possible to use them for such purposes as:

A protective lining for power station stacks operated with high sulfur fuels

Tank linings for the chemical industry

The advantages of the fluoroelastomers are in their high degree of resistance to aggressive media. Their cost is less than the perfluoroelastomers but greater than other elastomeric materials used for the lining of vessels.

TABLE 6.16

Compatibility of Fluoroelastomers with Selected Corrodents

Chemical	Maximum Temperature	
	°F	°C
Acetaldehyde	X	X
Acetamide	210	199
Acetic acid, 10%	190	88
Acetic acid, 50%	180	82
Acetic acid, 80%	180	82
Acetic acid, glacial	X	X
Acetic anhydride	X	X
Acetone	X	X
Acetyl chloride	400	204
Acrylic acid	X	X
Acrylonitrile	X	X
Adipic acid	190	82
Allyl alcohol	190	88
Allyl chloride	100	38
Alum	190	88
Aluminum acetate	180	82
Aluminum chloride, aqueous	400	204
Aluminum fluoride	400	204
Aluminum hydroxide	190	88
Aluminum nitrate	400	204
Aluminum oxychloride	X	X
Aluminum sulfate	390	199
Ammonia gas	X	X
Ammonium bifluoride	140	60
Ammonium carbonate	190	88
Ammonium chloride, 10%	400	204
Ammonium chloride, 50%	300	149
Ammonium chloride, sat.	300	149
Ammonium fluoride, 10%	140	60
Ammonium fluoride, 25%	140	60
Ammonium hydroxide, 25%	190	88
Ammonium hydroxide, sat.	190	88
Ammonium nitrate	X	X
Ammonium persulfate	140	60
Ammonium phosphate	180	82
Ammonium sulfate, 10–40%	180	82
Ammonium sulfide	X	X
Amyl acetate	X	X
Amyl alcohol	200	93
Amyl chloride	190	88
Aniline	230	110
Antimony trichloride	190	88
Aqua regia, 3:1	190	88
Barium carbonate	250	121
Barium chloride	400	204
Barium hydroxide	400	204
Barium sulfate	400	204

(continued)

TABLE 6.16 *Continued*

Chemical	Maximum Temperature	
	°F	°C
Barium sulfide	400	204
Benzaldehyde	X	X
Benzene	400	204
Benzene sulfonic acid, 10%	190	88
Benzoic acid	400	204
Benzyl alcohol	400	204
Benzyl chloride	400	204
Borax	190	88
Boric acid	400	204
Bromine gas, dry, 25%	180	82
Bromine gas, moist, 25%	180	82
Bromine liquid	350	177
Butadiene	400	204
Butyl acetate	X	X
Butyl alcohol	400	204
Butyl phthalate	80	27
n-Butylamine	X	X
Butyric acid	120	49
Calcium bisulfide	400	204
Calcium bisulfite	400	204
Calcium carbonate	190	88
Calcium chlorate	190	88
Calcium chloride	300	149
Calcium hydroxide, 10%	300	149
Calcium hydroxide, sat.	400	204
Calcium hypochlorite	400	204
Calcium nitrate	400	204
Calcium sulfate	200	93
Carbon bisulfide	400	204
Carbon dioxide, dry	80	27
Carbon dioxide, wet	X	X
Carbon disulfide	400	204
Carbon monoxide	400	204
Carbon tetrachloride	350	177
Carbonic acid	400	204
Cellosolve	X	X
Chloracetic acid	X	X
Chloracetic acid, 50% water	X	X
Chlorine gas, dry	190	88
Chlorine gas, wet	190	88
Chlorine, liquid	190	88
Chlorobenzene	400	204
Chloroform	400	204
Chlorosulfonic acid	X	X
Chromic acid, 10%	350	177
Chromic acid, 50%	350	177
Citric acid, 15%	300	149

(continued)

TABLE 6.16 *Continued*

Chemical	Maximum Temperature	
	°F	°C
Citric acid, conc.	400	204
Copper acetate	X	X
Copper carbonate	190	88
Copper chloride	400	204
Copper cyanide	400	204
Copper sulfate	400	204
Cresol	X	X
Cupric chloride, 5%	180	82
Cupric chloride, 50%	180	82
Cyclohexane	400	204
Cyclohexanol	400	204
Dichloroethane (ethylene dichloride)	190	88
Ethylene glycol	400	204
Ferric chloride	400	204
Ferric chloride, 50% in water	400	204
Ferric nitrate, 10–50%	400	204
Ferrous chloride	180	82
Ferrous nitrate	210	99
Fluorine gas, dry	X	X
Fluorine gas, moist	X	X
Hydrobromic acid, dil.	400	204
Hydrobromic acid, 20%	400	204
Hydrobromic acid, 50%	400	204
Hydrochloric acid, 20%	350	177
Hydrochloric acid, 38%	350	177
Hydrocyanic acid, 10%	400	204
Hydrofluoric acid, 30%	210	99
Hydrofluoric acid, 70%	350	177
Hydrofluoric acid, 100%	X	X
Hypochlorous acid	400	204
Iodine solution, 10%	190	88
Ketones, general	X	X
Lactic acid, 25%	300	149
Lactic acid, conc.	400	204
Magnesium chloride	390	199
Malic acid	390	199
Manganese chloride	180	82
Methyl chloride	190	88
Methyl ethyl ketone	X	X
Methyl isobutyl ketone	X	X
Muriatic acid	350	149
Nitric acid, 5%	400	204
Nitric acid, 20%	400	204
Nitric acid, 70%	190	88
Nitric acid, anhydrous	190	88
Nitrous acid, conc.	90	32
Oleum	190	88

(continued)

TABLE 6.16 *Continued*

Chemical	Maximum Temperature	
	°F	°C
Perchloric acid, 10%	400	204
Perchloric acid, 70%	400	204
Phenol	210	99
Phosphoric acid, 50–80%	300	149
Picric acid	400	204
Potassium bromide, 30%	190	88
Salicylic acid	300	149
Sodium carbonate	190	88
Sodium chloride	400	204
Sodium hydroxide, 10%	X	X
Sodium hydroxide, 50%	X	X
Sodium hydroxide, conc.	X	X
Sodium hypochlorite, 20%	400	204
Sodium hypochlorite, conc.	400	204
Sodium sulfide, to 50%	190	88
Stannic chloride	400	204
Stannous chloride	400	204
Sulfuric acid, 10%	350	149
Sulfuric acid, 50%	350	149
Sulfuric acid, 70%	350	149
Sulfuric acid, 90%	350	149
Sulfuric acid, 98%	350	149
Sulfuric acid, 100%	180	82
Sulfuric acid, turning	200	93
Sulfurous acid	400	204
Thionyl chloride	X	X
Toluene	400	204
Trichloroacetic acid	190	88
White liquor	190	88
Zinc chloride	400	204

The chemicals listed are in the pure state or in a saturated solution unless otherwise indicated. Compatibility is shown to the maximum allowable temperature for which data is available. Incompatibility is shown by an X. A blank space indicates that data is unavailable.

Source: From P.A. Schweitzer. 2004. *Corrosion Resistance Tables*, Vols. 1–4, 5th ed., New York: Marcel Dekker.

6.6 Thermoplastic Linings

Thermoplastic linings are used more extensively than any other type of linings. The most common thermoplasts used are

Polyvinyl chloride (PVC)

Chlorinated polyvinyl chloride (CPVC)

Polyvinylidene fluoride (PVDF)

Polypropylene (PP)
Polyethylene (PE)
Tetrafluoroethylene (PTFE)
Fluorinated ethylene propylene (FEP)
Perfluoralkoxy (PEA)
Ethylene-tetrafluoroethylene (ETFE)
Ethylene-chlorotrifluorethylene (ECTFE)

These materials are capable of providing a wide range of corrosion resistance. Table 6.17 lists the general area of corrosion resistance for each of the thermoplasts, but it is only a general guide. The resistance of a lining material to a specific corrodent should be checked.

These materials are used to line vessels as well as pipes and fittings. When vessels are lined, the linings, with the exception of plasticized PVC, are fabricated from sheet stock that must be cut, shaped, and joined. Joining is usually accomplished by hot-gas welding. Several problems exist when a thermoplastic lining is bonded to a metal shell, primarily the large differences in the coefficient of thermal expansion and the difficulty of adhesion. If the vessel is to be used under ambient temperature, such as a storage vessel, the problems of thermal expansion differences are eliminated. However, the problem of adhesion is still present. Because of this, many linings are installed as loose linings in the vessel.

Techniques have been developed to overcome the problem of adhesion that make use of an intermediate bond. One approach is to heat the thermoplastic sheet then impress it into one surface a fiber cloth or nonwoven web. This provides half of the bond. Bonding to the metal surface is accomplished by the use of an epoxy adhesive that will bond to both the fiber and the metal.

Welding techniques for joining the thermoplastic sheets are very critical. Each weld must be continuous, leak-tight, and mechanically strong. It is

TABLE 6.17

General Corrosion Resistance of Thermoplastic Lining Materials

Material	Strong Acids	Strong Bases	Chlorinated Solvents	Esters and Ketones	Strong Oxidants
PVC (type 1)	F	F	P	P	P
CPVC	F	F	F	P	P
PVDF	E	P	E	P	E
PP	G	E	P	F	P
PE	F	G	P	P	E
PTFE	E	E	E	E	E
FEP	E	E	E	E	E
PFA	E	E	E	E	E
ETFE	E	E	E	E	E
ECTFE	E	E	E	E	E

E, excellent; G, good; F, fair; P, poor.

important that only qualified welders be used for this operation. Poor welding is a common cause of lining failure.

6.6.1 Polyvinyl Chloride

The maximum allowable temperature for continuous operation of type 1 polyvinyl chloride (PVC) is 140°F (60°C). Plasticized PVC can be bonded directly to a metal substrate. Unplasticized PVC cannot be bonded directly. It is bonded to a plasticized material forming a dual laminate that is then bonded to a metal substrate.

Plasticized PVC does not have the same range of corrosion resistance as the unplasticized PVC (type 1), Unplasticized PVC is resistant to attack by most acids and strong alkalies as well as gasoline, kerosene, aliphatic alcohols, and hydrocarbons. It is subject to attack by aromatics, chlorinated organic com-pounds, and lacquer solvents.

In addition to handling highly corrosive and abrasive chemicals, these linings have found many applications in marine environments. Table 6.18 lists the compatibility of type 1 PVC with selected corrodents and Table 6.19 does the same for type 2 PVC. Additional data can be found in Reference [1].

6.6.2 Chlorinated Polyvinyl Chloride

Chlorinated polyvinyl chloride (CPVC) can be continuously operated at a maximum temperature of 200°F (93°C). CPVC is very similar in properties to PVC except that it has a higher allowable operating temperature and a somewhat better resistance to chlorinated solvents. It is extremely difficult to hot-weld the joints in a sheet lining. This factor should be considered when selecting this material.

CPVC is resistant to most acids, alkalies, salts, halogens, and many corrosive waters. In general, it should not be used to handle most polar organic materials, including chlorinated or aromatic hydrocarbons, esters, and ketones.

Table 6.20 lists the compatibility of CPVC with selected corrodents. Reference [1] provides a more detailed listing.

6.6.3 Polyvinylidene Fluoride

Polyvinylidene fluoride (PVDF) may be continuously operated at a maximum temperature of 275°F (135°C). It is one of the most popular lining materials because of its range of corrosion resistance and high allowable operating temperature. Unless a dual laminate is used, linings of PVDF will be loose.

PVDF is resistant to most acids, bases, and organic solvents. It also has the ability to handle wet or dry chlorine, bromine, and other halogens.

It is not resistant to strong alkalies, fuming acids, polar solvents, amines, ketones, and esters. When used with strong alkalies, it stress-cracks.

TABLE 6.18

Compatibility of Type 1 PVC with Selected Corrodents

Chemical	Maximum Temperature	
	°F	°C
Acetaldehyde	X	X
Acetamide	X	X
Acetic acid, 10%	140	60
Acetic acid, 50%	140	60
Acetic acid, 80%	140	60
Acetic acid, glacial	130	54
Acetic anhydride	X	X
Acetone	X	X
Acetyl chloride	X	X
Acrylic acid	X	X
Acrylonitrile	X	X
Adipic acid	140	60
Allyl alcohol	90	32
Allyl chloride	X	X
Alum	140	60
Aluminum acetate	100	38
Aluminum chloride, aqueous	140	60
Aluminum chloride, dry	140	60
Aluminum fluoride	140	60
Aluminum hydroxide	140	60
Aluminum nitrate	140	60
Aluminum oxychloride	140	60
Aluminum sulfate	140	60
Ammonia gas	140	60
Ammonium bifluoride	90	32
Ammonium carbonate	140	60
Ammonium chloride, 10%	140	60
Ammonium chloride, 50%	140	60
Ammonium chloride, sat.	140	60
Ammonium fluoride, 10%	140	60
Ammonium fluoride, 25%	140	60
Ammonium hydroxide, 25%	140	60
Ammonium hydroxide, sat.	140	60
Ammonium nitrate	140	60
Ammonium persulfate	140	60
Ammonium phosphate	140	60
Ammonium sulfate, 10–40%	140	60
Ammonium sulfide	140	60
Ammonium sulfite	120	49
Amyl acetate	X	X
Amyl alcohol	140	60
Amyl chloride	X	X
Aniline	X	X
Antimony trichloride	140	60
Aqua regia, 3:1	X	X
Barium carbonate	140	60
Barium chloride	140	60

(continued)

TABLE 6.18 *Continued*

Chemical	Maximum Temperature	
	°F	°C
Barium hydroxide	140	60
Barium sulfate	140	60
Barium sulfide	140	60
Benzaldehyde	X	X
Benzene	X	X
Benzene sulfonic acid, 10%	140	60
Benzoic acid	140	60
Benzyl alcohol	X	X
Benzyl chloride	60	16
Borax	140	60
Boric acid	140	60
Bromine gas, dry	X	X
Bromine gas, moist	X	X
Bromine liquid	X	X
Butadiene	140	60
Butyl acetate	X	X
Butyl alcohol	X	X
Butyl phthalate	80	27
n-Butylamine	X	X
Butyric acid	60	16
Calcium bisulfide	140	60
Calcium bisulfite	140	60
Calcium carbonate	140	60
Calcium chlorate	140	60
Calcium chloride	140	60
Calcium hydroxide, 10%	140	60
Calcium hydroxide, sat.	140	60
Calcium hypochlorite	140	60
Calcium nitrate	140	60
Calcium oxide	140	60
Calcium sulfate	140	60
Caprylic acid	120	49
Carbon bisulfide	X	X
Carbon dioxide, dry	140	60
Carbon dioxide, wet	140	60
Carbon disulfide	X	X
Carbon monoxide	140	60
Carbon tetrachloride	X	X
Carbonic acid	140	60
Cellosolve	X	X
Chloracetic acid	140	60
Chloracetic acid, 50% water	140	60
Chlorine gas, dry	140	60
Chlorine gas, wet	X	X
Chlorine, liquid	X	X
Chlorobenzene	X	X
Chloroform	X	X

(continued)

TABLE 6.18 *Continued*

Chemical	Maximum Temperature	
	°F	°C
Chlorosulfonic acid	60	16
Chromic acid, 10%	140	60
Chromic acid, 50%	140	60
Chromyl chloride	120	49
Citric acid, 15%	140	60
Citric acid, conc.	140	60
Copper acetate	80	27
Copper carbonate	140	60
Copper chloride	140	60
Copper cyanide	140	60
Copper sulfate	140	60
Cresol	130	54
Cupric chloride, 5%	140	60
Cupric chloride, 50%	140	60
Cyclohexane	80	27
Cyclohexanol	X	X
Dichloroacetic acid	120	49
Dichloroethane (ethylene dichloride)	X	X
Ethylene glycol	140	60
Ferric chloride	140	60
Ferric chloride, 50% in water	140	60
Ferric nitrate, 10–50%	140	60
Ferrous chloride	140	60
Ferrous nitrate	140	60
Fluorine gas, dry	X	X
Fluorine gas, moist	X	X
Hydrobromic acid, dil.	140	60
Hydrobromic acid, 20%	140	60
Hydrobromic acid, 50%	140	60
Hydrochloric acid, 20%	140	60
Hydrochloric acid, 38%	140	60
Hydrocyanic acid, 10%	140	60
Hydrofluoric acid, 100%	X	X
Hydrofluoric acid, 30%	120	49
Hydrofluoric acid, 70%	68	20
Hypochlorous acid	140	60
Iodine solution, 10%	100	38
Ketones, general	X	X
Lactic acid, 25%	140	60
Lactic acid, conc.	80	27
Magnesium chloride	140	60
Malic acid	140	60
Manganese chloride	90	32
Methyl chloride	X	X
Methyl ethyl ketone	X	X
Methyl isobutyl ketone	X	X
Muriatic acid	140	60

(continued)

TABLE 6.18 *Continued*

Chemical	Maximum Temperature	
	°F	°C
Nitric acid, 5%	140	60
Nitric acid 20%	140	60
Nitric acid, 70%	140	60
Nitric acid, anhydrous	X	X
Nitrous acid, conc.	60	16
Oleum	X	X
Pherchloric acid, 10%	140	60
Perchloric acid, 70%	60	16
Phenol	X	X
Phosphoric acid, 50–80%	140	60
Picric acid	X	X
Potassium bromide, 30%	140	60
Salicylic acid	140	60
Silver bromide, 10%	140	60
Sodium carbonate	140	60
Sodium chloride	140	60
Sodium hydroxide, 10%	140	60
Sodium hydroxide, 50%	140	60
Sodium hydroxide, conc.	140	60
Sodium hypochlorite, 20%	140	60
Sodium hypochlorite, conc.	140	60
Sodium sulfide, to 50%	140	60
Stannic chloride	140	60
Stannous chloride	140	60
Sulfuric acid, 10%	140	60
Sulfuric acid, 50%	140	60
Sulfuric acid, 70%	140	60
Sulfuric acid, 90%	140	60
Sulfuric acid, 98%	X	X
Sulfuric acid, 100%	X	X
Sulfuric acid, fuming	X	X
Sulfurous acid	140	60
Thionyl chloride	X	X
Toluene	X	X
Trichloroacetic acid	90	32
White liquor	140	60
Zinc chloride	140	60

The chemicals listed are in the pure state or in a saturated solution unless otherwise indicated. Compatibility is shown to the maximum allowable temperature for which data is available. Incompatibility is shown by an X. A blank space indicates that data is unavailable.

Source: From P.A. Schweitzer. 2004. *Corrosion Resistance Tables*, Vols. 1–4, 5th ed., New York: Marcel Dekker.

TABLE 6.19

Compatibility of Type 2 PVC with Selected Corrodents

Chemical	Maximum Temperature	
	°F	°C
Acetaldehyde	X	X
Acetamide	X	X
Acetic acid, 10%	100	38
Acetic acid, 50%	90	32
Acetic acid, 80%	X	X
Acetic acid, glacial	X	X
Acetic anhydride	X	X
Acetone	X	X
Acetyl chloride	X	X
Acrylic acid	X	X
Acrylonitrile	X	X
Adipic acid	140	60
Allyl alcohol	90	32
Allyl chloride	X	X
Alum	140	60
Aluminum acetate	100	38
Aluminum chloride, aqueous	140	60
Aluminum fluoride	140	60
Aluminum hydroxide	140	60
Aluminum nitrate	140	60
Aluminum oxychloride	140	60
Aluminum sulfate	140	60
Ammonia gas	140	60
Ammonium bifluoride	90	32
Ammonium carbonate	140	60
Ammonium chloride, 10%	140	60
Ammonium chloride, 50%	140	60
Ammonium chloride, sat.	140	60
Ammonium fluoride, 10%	90	32
Ammonium fluoride, 25%	90	32
Ammonium hydroxide, 25%	140	60
Ammonium hydroxide, sat.	140	60
Ammonium nitrate	140	60
Ammonium persulfate	140	60
Ammonium phosphate	140	60
Ammonium sulfate, 10–40%	140	60
Ammonium sulfide	140	60
Amyl acetate	X	X
Amyl alcohol	X	X
Amyl chloride	X	X
Aniline	X	X
Antimony trichloride	140	60
Aqua regia, 3:1	X	X
Barium carbonate	140	60
Barium chloride	140	60
Barium hydroxide	140	60
Barium sulfate	140	60

(continued)

TABLE 6.19 *Continued*

Chemical	Maximum Temperature	
	°F	°C
Barium sulfide	140	60
Benzaldehyde	X	X
Benzene	X	X
Benzene sulfonic acid, 10%	140	60
Benzoic acid	140	60
Benzyl alcohol	X	X
Borax	140	60
Boric acid	140	60
Bromine gas, dry	X	X
Bromine gas, moist	X	X
Bromine liquid	X	X
Butadiene	60	16
Butric acid	X	X
Butyl acetate	X	X
Butyl alcohol	X	X
n-Butylamine	X	X
Calcium bisulfide	140	60
Calcium bisulfite	140	60
Calcium carbonate	140	60
Calcium chlorate	140	60
Calcium chloride	140	60
Calcium hydroxide, 10%	140	60
Calcium hydroxide, sat.	140	60
Calcium hypochlorite	140	60
Calcium nitrate	140	60
Calcium oxide	140	60
Calcium sulfate	140	60
Carbon dioxide, dry	140	60
Carbon dioxide, wet	140	60
Carbon disulfide	X	X
Carbon monoxide	140	60
Carbon tetrachloride	X	X
Carbonic acid	140	60
Cellosolve	X	X
Chloracetic acid	105	40
Chlorine gas, dry	140	60
Chlorine gas, wet	X	X
Chlorine, liquid	X	X
Chlorobenzene	X	X
Chloroform	X	X
Chlorosulfonic acid	60	16
Chromic acid, 10%	140	60
Chromic acid, 50%	X	X
Citric acid, 15%	140	60
Citric acid, conc.	140	60
Copper carbonate	140	60
Copper chloride	140	60

(continued)

TABLE 6.19 *Continued*

Chemical	Maximum Temperature	
	°F	°C
Copper cyanide	140	60
Copper sulfate	140	60
Cresol	X	X
Cyclohexanol	X	X
Dichloroacetic acid	120	49
Dichloroethane (ethylene dichloride)	X	X
Ethylene glycol	140	60
Ferric chloride	140	60
Ferric nitrate, 10–50%	140	60
Ferrous chloride	140	60
Ferrous nitrate	140	60
Fluorine gas, dry	X	X
Fluorine gas, moist	X	X
Hydrobromic acid, dil.	140	60
Hydrobromic acid, 20%	140	60
Hydrobromic acid, 50%	140	60
Hydrochloric acid, 20%	140	60
Hydrochloric acid, 38%	140	60
Hydrocyanic acid, 10%	140	60
Hydrofluoric acid, 30%	120	49
Hydrofluoric acid, 70%	68	20
Hypochlorous acid	140	60
Ketones, general	X	X
Lactic acid, 25%	140	60
Lactic acid, conc.	80	27
Magnesium chloride	140	60
Malic acid	140	60
Methyl chloride	X	X
Methyl ethyl ketone	X	X
Methyl isobutyl ketone	X	X
Muriatic acid	140	60
Nitric acid, 5%	100	38
Nitric acid, 20%	140	60
Nitric acid, 70%	70	23
Nitric acid, anhydrous	X	X
Nitrous acid, conc.	60	16
Oleum	X	X
Perchloric acid, 10%	60	16
Perchloric acid, 70%	60	16
Phenol	X	X
Phosphoric acid, 50–80%	140	60
Picric acid	X	X
Potassium bromide, 30%	140	60
Salicylic acid	X	X
Silver bromide, 10%	105	40
Sodium carbonate	140	60
Sodium chloride	140	60

(continued)

TABLE 6.19 *Continued*

Chemical	Maximum Temperature	
	°F	°C
Sodium hydroxide, 10%	140	60
Sodium hydroxide, 50%	140	60
Sodium hydroxide, conc.	140	60
Sodium hypochlorite, 20%	140	60
Sodium hypochlorite, conc.	140	60
Sodium sulfide, to 50%	140	60
Stannic chloride	140	60
Stannous chloride	140	60
Sulfuric acid, 10%	140	60
Sulfuric acid, 50%	140	60
Sulfuric acid, 70%	140	60
Sulfuric acid, 90%	X	X
Sulfuric acid, 98%	X	X
Sulfuric acid, 100%	X	X
Sulfuric acid, fuming	X	X
Sulfurous acid	140	60
Thionyl chloride	X	X
Toluene	X	X
Trichloroacetic acid	X	X
White liquor	140	60
Zinc chloride	140	60

The chemicals listed are in the pure state or in a saturated solution unless otherwise indicated. Compatibility is shown to the maximum allowable temperature for which data is available. Incompatibility is shown by an X. A blank space indicates that data is unavailable.

Source: From P.A. Schweitzer. 2004. *Corrosion Resistance Tables*, Vols. 1–4, 5th ed., New York: Marcel Dekker.

PVDF is manufactured under various trade names:

Trade Name	Manufacturer
Kynar	Elf Atochem
Solef	Solvay
Hylar	Ausimont USA

Refer to Table 6.21 for the compatibility of PVDF with selected corrodents. Reference [1] provides additional listings.

6.6.4 Polypropylene

The maximum allowable temperature that polypropylene (PP) can be continuously operated under is 180°F (82°C). PP must be attached to

TABLE 6.20

Compatibility of CPVC with Selected Corrodents

Chemical	Maximum Temperature	
	°F	°C
Acetaldehyde	X	X
Acetic acid, 10%	90	32
Acetic acid, 50%	X	X
Acetic acid, 80%	X	X
Acetic acid, glacial	X	X
Acetic anhydride	X	X
Acetone	X	X
Acetyl chloride	X	X
Acrylic acid	X	X
Acrylonitrile	X	X
Adipic acid	200	93
Allyl alcohol, 96%	200	93
Allyl chloride	X	X
Alum	200	93
Aluminum acetate	100	33
Aluminum chloride, aqueous	200	93
Aluminum chloride, dry	180	82
Aluminum fluoride	200	93
Aluminum hydroxide	200	93
Aluminum nitrate	200	93
Aluminum oxychloride	200	93
Aluminum sulfate	200	93
Ammonia gas, dry	200	93
Ammonium bifluoride	140	60
Ammonium carbonate	200	93
Ammonium chloride, 10%	180	82
Ammonium chloride, 50%	180	82
Ammonium chloride, sat.	200	93
Ammonium fluoride, 10%	200	93
Ammonium fluoride, 25%	200	93
Ammonium hydroxide, 25%	X	X
Ammonium hydroxide, sat.	X	X
Ammonium nitrate	200	93
Ammonium persulfate	200	93
Ammonium phosphate	200	93
Ammonium sulfate, 10–40%	200	93
Ammonium sulfide	200	93
Ammonium sulfite	160	71
Amyl acetate	X	X
Amyl alcohol	130	54
Amyl chloride	X	X
Aniline	X	X
Antimony trichloride	200	93
Aqua regia, 3:1	80	27
Barium carbonate	200	93
Barium chloride	180	82
Barium hydroxide	180	82

(continued)

TABLE 6.20 *Continued*

Chemical	Maximum Temperature	
	°F	°C
Barium sulfate	180	82
Barium sulfide	180	82
Benzaldehyde	X	X
Benzene	X	X
Benzene sulfonic acid, 10%	180	82
Benzoic acid	200	93
Benzyl alcohol	X	X
Benzyl chloride	X	X
Borax	200	93
Boric acid	210	99
Bromine gas, dry	X	X
Bromine gas, moist	X	X
Bromine liquid	X	X
Butadiene	150	66
Butyl acetate	X	X
Butyl alcohol	140	60
n-Butylamine	X	X
Butyric acid	140	60
Calcium bisulfide	180	82
Calcium bisulfite	210	99
Calcium carbonate	210	99
Calcium chlorate	180	82
Calcium chloride	180	82
Calcium hydroxide, 10%	170	77
Calcium hydroxide, sat.	210	99
Calcium hypochlorite	200	93
Calcium nitrate	180	82
Calcium oxide	180	82
Calcium sulfate	180	82
Caprylic acid	180	82
Carbon bisulfide	X	X
Carbon dioxide, dry	210	99
Carbon dioxide, wet	160	71
Carbon disulfide	X	X
Carbon monoxide	210	99
Carbon tetrachloride	X	X
Carbonic acid	180	82
Cellosolve	180	82
Chloracetic acid	X	X
Chloracetic acid, 50% water	100	38
Chlorine gas, dry	140	60
Chlorine gas, wet	X	X
Chlorine, liquid	X	X
Chlorobenzene	X	X
Chloroform	X	X
Chlorosulfonic acid	X	X
Chromic acid, 10%	210	99

(continued)

TABLE 6.20 *Continued*

Chemical	Maximum Temperature	
	°F	°C
Chromic acid, 50%	210	99
Chromyl chloride	180	82
Citric acid, 15%	180	82
Citric acid, conc.	180	82
Copper acetate	80	27
Copper carbonate	180	82
Copper chloride	210	99
Copper cyanide	180	82
Copper sulfate	210	99
Cresol	X	X
Cupric chloride, 5%	180	82
Cupric chloride, 50%	180	82
Cyclohexane	X	X
Cyclohexanol	X	X
Dichloroacetic acid, 20%	100	38
Dichloroethane (ethylene dichloride)	X	X
Ethylene glycol	210	99
Ferric chloride	210	99
Ferric chloride, 50% in water	180	82
Ferric nitrate, 10–50%	180	82
Ferrous chloride	210	99
Ferrous nitrate	180	82
Fluorine gas, dry	X	X
Fluorine gas, moist	80	27
Hydrobromic acid, dil.	130	54
Hydrobromic acid, 20%	180	82
Hydrobromic acid, 50%	190	88
Hydrochloric acid, 20%	180	82
Hydrochloric acid, 38%	170	77
Hydrocyanic acid, 10%	80	27
Hydrofluoric acid, 30%	X	X
Hydrofluoric acid, 70%	90	32
Hydrofluoric acid, 100%	X	X
Hypochlorous acid	180	82
Ketones, general	X	X
Lactic acid, 25%	180	82
Lactic acid, conc.	100	38
Magnesium chloride	230	110
Malic acid	180	82
Manganese chloride	180	82
Methyl chloride	X	X
Methyl ethyl ketone	X	X
Methyl isobutyl ketone	X	X
Muriatic acid	170	77
Nitric acid, 5%	180	82
Nitric acid, 20%	160	71
Nitric acid, 70%	180	82

(continued)

TABLE 6.20 *Continued*

Chemical	Maximum Temperature	
	°F	°C
Nitric acid, anhydrous	X	X
Nitrous acid, conc.	80	27
Oleum	X	X
Perchloric acid, 10%	180	82
Perchloric acid, 70%	180	82
Phenol	140	60
Phosphoric acid, 50–80%	180	82
Picric acid	X	X
Potassium bromide, 30%	180	82
Salicylic acid	X	X
Silver bromide, 10%	170	77
Sodium carbonate	210	99
Sodium chloride	210	99
Sodium hydroxide, 10%	190	88
Sodium hydroxide, 50%	180	82
Sodium hydroxide, conc.	190	88
Sodium hypochlorite, 20%	190	88
Sodium hypochlorite, conc.	180	82
Sodium sulfide, to 50%	180	82
Stannic chloride	180	82
Stannous chloride	180	82
Sulfuric acid, 10%	180	82
Sulfuric acid, 50%	180	82
Sulfuric acid, 70%	200	93
Sulfuric acid, 90%	X	X
Sulfuric acid, 98%	X	X
Sulfuric acid, 100%	X	X
Sulfuric acid, fuming	X	X
Sulfurous acid	180	82
Thionyl chloride	X	X
Toluene	X	X
Trichloroacetic acid, 20%	140	60
White liquor	180	82
Zinc chloride	180	82

The chemicals listed are in the pure state or in saturated solution unless otherwise indicated. Compatibilty is shown to the maximum allowable temperature for which data is available. Incompatibility is shown by an X.

Source: From P.A. Schweitzer. 2004. *Corrosion Resistance Tables*, Vols. 1–4, 5th ed., New York: Marcel Dekker.

a backing sheet in order to be secured as a lining. If a backing sheet is not used, the lining will be loose.

PP is resistant to sulfur-bearing compounds, caustics, solvents, acids, and other organic chemicals. It is not resistant to oxidizing-type acids, detergents, low-boiling hydrocarbons, alcohols, aromatics, and some chlorinated organic materials.

TABLE 6.21

Compatibility of PVDF with Selected Corrodents

Chemical	Maximum Temperature	
	°F	°C
Acetaldehyde	150	66
Acetamide	90	32
Acetic acid, 10%	300	149
Acetic acid, 50%	300	149
Acetic acid, 80%	190	88
Acetic acid, glacial	190	88
Acetic anhydride	100	38
Acetone	X	X
Acetyl chloride	120	49
Acrylic acid	150	66
Acrylonitrile	130	54
Adipic acid	280	138
Allyl alcohol	200	93
Allyl chloride	200	93
Alum	180	82
Aluminum acetate	250	121
Aluminum chloride, aqueous	300	149
Aluminum chloride, dry	270	132
Aluminum fluoride	300	149
Aluminum hydroxide	260	127
Aluminum nitrate	300	149
Aluminum oxychloride	290	143
Aluminum sulfate	300	149
Ammonia gas	270	132
Ammonium bifluoride	250	121
Ammonium carbonate	280	138
Ammonium chloride, 10%	280	138
Ammonium chloride, 50%	280	138
Ammonium chloride, sat.	280	138
Ammonium fluoride, 10%	280	138
Ammonium fluoride, 25%	280	138
Ammonium hydroxide, 25%	280	138
Ammonium hydroxide, sat.	280	138
Ammonium nitrate	280	138
Ammonium persulfate	280	138
Ammonium phosphate	280	138
Ammonium sulfate, 10–40%	280	138
Ammonium sulfide	280	138
Ammonium sulfite	280	138
Amyl acetate	190	88
Amyl alcohol	280	138
Amyl chloride	280	138
Aniline	200	93
Antimony trichloride	150	66
Aqua regia, 3:1	130	54
Barium carbonate	280	138
Barium chloride	280	138

(continued)

TABLE 6.21 *Continued*

Chemical	Maximum Temperature	
	°F	°C
Barium hydroxide	280	138
Barium sulfate	280	138
Barium sulfide	280	138
Benzaldehyde	120	49
Benzene	150	66
Benzene sulfonic acid, 10%	100	38
Benzoic acid	250	121
Benzyl alcohol	280	138
Benzyl chloride	280	138
Borax	280	138
Boric acid	280	138
Bromine gas, dry	210	99
Bromine gas, moist	210	99
Bromine liquid	140	60
Butadiene	280	138
Butyl acetate	140	60
Butyl alcohol	280	138
Butyl phthalate	80	27
n-Butylamine	X	X
Butyric acid	230	110
Calcium bisulfide	280	138
Calcium bisulfite	280	138
Calcium carbonate	280	138
Calcium chlorate	280	138
Calcium chloride	280	138
Calcium hydroxide, 10%	270	132
Calcium hydroxide, sat.	280	138
Calcium hypochlorite	280	138
Calcium nitrate	280	138
Calcium oxide	250	121
Calcium sulfate	280	138
Caprylic acid	220	104
Carbon bisulfide	80	27
Carbon dioxide, dry	280	138
Carbon dioxide, wet	280	138
Carbon disulfide	80	27
Carbon monoxide	280	138
Carbon tetrachloride	280	138
Carbonic acid	280	138
Cellosolve	280	138
Chloracetic acid	200	93
Chloracetic acid, 50% water	210	99
Chlorine gas, dry	210	99
Chlorine gas, wet, 10%	210	99
Chlorine, liquid	210	S9
Chlorobenzene	220	104
Chloroform	250	121

(continued)

TABLE 6.21 *Continued*

Chemical	Maximum Temperature	
	°F	°C
Chlorosulfonic acid	110	43
Chromic acid, 10%	220	104
Chromic acid, 50%	250	121
Chromyl chloride	110	43
Citric acid, 15%	250	121
Citric acid, conc.	250	121
Copper acetate	250	121
Copper carbonate	250	121
Copper chloride	280	138
Copper cyanide	280	138
Copper sulfate	280	138
Cresol	210	99
Cupric chloride, 5%	270	132
Cupric chloride, 50%	270	132
Cyclohexane	250	121
Cyclohexanol	210	99
Dichloroacetic acid	120	49
Dichloroethane (ethylene dichloride)	280	138
Ethylene glycol	280	138
Ferric chloride	280	138
Ferric chloride, 50% in water	280	138
Ferrous chloride	280	138
Ferrous nitrate	280	138
Ferrous nitrate, 10–50%	280	138
Fluorine gas, dry	80	27
Fluorine gas, moist	80	27
Hydrobromic acid, dil.	260	127
Hydrobromic acid, 20%	280	138
Hydrobromic acid, 50%	280	138
Hydrochloric acid, 20%	280	138
Hydrochloric acid, 38%	280	138
Hydrocyanic acid, 10%	280	138
Hydrofluoric acid, 30%	260	127
Hydrofluoric acid, 70%	200	93
Hydrofluoric acid, 100%	200	93
Hypochlorous acid	280	138
Iodine solution, 10%	250	121
Ketones, general	110	43
Lactic acid, 25%	130	54
Lactic acid, conc.	110	43
Magnesium chloride	280	138
Malic acid	250	121
Manganese chloride	280	138
Methyl chloride	X	X
Methyl ethyl ketone	X	X
Methyl isobutyl ketone	110	43
Muriatic acid	280	138

(continued)

TABLE 6.21 *Continued*

Chemical	Maximum Temperature	
	°F	°C
Nitric acid, 5%	200	93
Nitric acid, 20%	180	82
Nitric acid, 70%	120	49
Nitric acid, anhydrous	150	66
Nitrous acid, conc.	210	99
Oleum	X	X
Perchloric acid, 10%	210	99
Perchloric acid, 70%	120	49
Phenol	200	93
Phosphoric acid, 50–80%	220	104
Picric acid	80	27
Potassium bromide, 30%	280	138
Salicylic acid	220	104
Silver bromide, 10%	250	121
Sodium carbonate	280	138
Sodium chloride	280	138
Sodium hydroxide, 10%	230	110
Sodium hydroxide, 50%	220	104
Sodium hydroxide, conc.	150	66
Sodium hypochlorite, 20%	280	138
Sodium hypochlorite, conc.	280	138
Sodium sulfide, to 50%	280	138
Stannic chloride	280	138
Stannous chloride	280	138
Sulfuric acid, 10%	250	121
Sulfuric acid, 50%	220	104
Sulfuric acid, 70%	220	104
Sulfuric acid, 90%	210	99
Sulfuric acid, 98%	140	60
Sulfuric acid, 100%	X	X
Sulfuric acid, fuming	X	X
Sulfurous acid	220	104
Thionyl chloride	X	X
Toluene	X	X
Trichloroacetic acid	130	54
White liquor	80	27
Zinc chloride	260	127

The chemicals listed are in the pure state or in saturated solution unless otherwise indicated. Compatibilty is shown to the maximum allowable temperature for which data is available. Incompatibility is shown by an X.

Source: From P.A. Schweitzer. 2004. *Corrosion Resistance Tables*, Vols. 1–4, 5th ed., New York: Marcel Dekker.

Polypropylene has FDA approval for handling of food products. It should not be used with oxidizing type acids, detergents, low boiling hydrocarbons, alcohols, aromatics, and some chlorinated organic materials.

Polypropylene may be subject to environmental stress cracking. The occurrence is difficult to predict. It is dependent upon the process chemistry, operating conditions, and quality of fabrication. In Table 6.22 that gives the compatibility of PP with selected corrodents, certain chemical/liner ratings are identified by an "e" indicating that the liner in contact with that specific chemical may be susceptible to ESC. Reference [1] provides a more comprehensive listing.

6.6.5 Polyethylene

The maximum allowable temperature, at continuous contact that poly-ethylene (PE) can be used is 120°F (49°C). PE is the least expensive of all the plastic materials.

PE has a wide range of corrosion resistance ranging from potable water to corrosive wastes. It exhibits excellent resistance to strong oxidizing chemicals, alkalies, acids, and salt solutions.

Physical and mechanical properties differ by density and molecular weight. The three main classifications of density are low, medium, and high. These specific gravity ranges are 0.91–0.925, 0.925–0.940, and 0.940–0.965. These grades are sometimes referred to as types 1, 11, and 111.

Industry practice breaks the molecular weight of polyethylenes into four distinct classifications:

Medium molecular weight: less than 100,000

High molecular weight: 110,000–250,000

Extra high molecular weight: 250,000–1,500,000

Ultra high molecular weight: 1,500,000 and higher

Usually, the ultra high molecular weight material has a molecular weight of at least 3.1 million.

The two varieties of polyethylene used, for corrosive applications are EHMW and UHMW.

The key properties of these two varieties are:

Good abrasion resistance

Excellent impact resistance

Light weight

Easily heat fused

High tensile strength

Low moisture absorption

Nontoxic

Nonstaining

Corrosion resistant

TABLE 6.22

Compatibility of PP with Selected Corrodents

Chemical	Maximum Temperature	
	°F	°C
Acetaldehyde	120	49
Acetamide	110	43
Acetic acid, 10%	220	104
Acetic acid, 50%	200	93
Acetic acid, 80%	200	93
Acetic acid, glacial	190	88
Acetic anhydride	100	38
Acetone	220	104
Acetyl chloride	X	X
Acrylic acid	X	X
Acrylonitryle	90	32
Adipic acid	100	38
Allyl alcohol	140	60
Allyl chloride	140	60
Alum	220	104
Aluminum acetate	100	38
Aluminum chloride, aqueous	200	93
Aluminum chloride, dry	220	104
Aluminum fluoride	200	93
Aluminum hydroxide	200	93
Aluminum nitrate	200	93
Aluminum oxychloride	220	104
Aluminum sulfate		
Ammonia gas	150	66
Ammonium bifluoride	200	93
Ammonium carbonate	220	104
Ammonium chloride, 10%	180	82
Ammonium chloride, 50%	180	82
Ammonium chloride, sat.	200	93
Ammonium fluoride, 10%	210	99
Ammonium fluoride, 25%	200	93
Ammonium hydroxide, 25%	200	93
Ammonium hydroxide, sat.	200	93
Ammonium nitrate	200	93
Ammonium persulfate	220	104
Ammonium phosphate	200	93
Ammonium sulfate, 10–40%	200	93
Ammonium sulfide	220	104
Ammonium sulfite	220	104
Amyl acetate	X	X
Amyl alcohol	200	93
Amyl chloride	X	X
Aniline	180	82
Antimony trichloride	180	82
Aqua regia, 3:1	X	X
Barium carbonate	200	93
Barium chloride	220	104

(continued)

TABLE 6.22 *Continued*

Chemical	Maximum Temperature	
	°F	°C
Barium hydroxide	200	93
Barium sulfate	200	93
Barium sulfide	200	93
Benzaldehyde	80	27
Benzene	140	60
Benzene sulfonic acid, 10%	180	82
Benzoic acid	190	88
Benzyl alcohol	140	60
Benzyl chloride	80	27
Borax	210	99
Boric acid	220	104
Bromine gas, dry	X	X
Bromine gas, moist	X	X
Bromine liquid	X	X
Butadiene	X	X
Butyl acetate	X	X
Butyl alcohol	200	93
Butyl phthalate	180	82
n-Butylamine	90	32
Butyric acid	180	82
Calcium bisulfide	210	99
Calcium bisulfite	210	99
Calcium carbonate	210	99
Calcium chlorate	220	104
Calcium chloride	220	104
Calcium hydroxide, 10%	200	93
Calcium hydroxide, sat.	220	104
Calcium hypochlorite	210	99
Calcium nitrate	210	99
Calcium oxide	220	104
Calcium sulfate	220	104
Caprylic acid	140	60
Carbon bisulfide	X	X
Carbon dioxide, dry	220	104
Carbon dioxide, wet	140	60
Carbon disulfide	X	X
Carbon monoxide	220	104
Carbon tetrachloride	X	X
Carbonic acid	220	104
Cellosolve	200	93
Chloracetic acid	180	82
Chloracetic acid, 50% water	80	27
Chlorine gas, dry	X	X
Chlorine gas, wet	X	X
Chlorine, liquid	X	X
Chlorobenzene	X	X
Chloroform	X	X

(continued)

TABLE 6.22 *Continued*

Chemical	Maximum Temperature	
	°F	°C
Chlorosulfonic acid	X	X
Chromic acid, 10%	140	60
Chromic acid, 50%	e150	66
Chromyl chloride	140	60
Citric acid, 15%	220	104
Citric acid, conc.	220	104
Copper acetate	80	27
Copper carbonate	200	93
Copper chloride	200	93
Copper cyanide	200	93
Copper sulfate	200	93
Cresol	X	X
Cupric chloride, 5%	140	60
Cupric chloride, 50%	140	60
Cyclohexane	X	X
Cyclohexanol	150	66
Dichloroethane (ethylene dichloride)	80	27
Dictiloroacetic acid	100	38
Ethylene glycol	210	99
Ferric chloride	e210	99
Ferric chloride, 50% in water	210	99
Ferric nitrate, 10–50%	210	99
Ferrous chloride	e210	99
Ferrous nitrate	210	99
Fluorine gas, dry	X	X
Fluorine gas, moist	X	X
Hydrobromic acid, dil.	230	110
Hydrobromic acid, 20%	200	93
Hydrobromic acid, 50%	190	88
Hydrochloric acid, 20%	220	104
Hydrochloric acid, 38%	200	93
Hydrocyanic acid, 10%	150	66
Hydrofluoric acid, 30%	180	82
Hydrofluoric acid, 70%	200	93
Hydrofluoric acid, 100%	200	93
Hypochlorous acid	140	60
Iodine solution, 10%	X	X
Ketones, general	110	43
Lactic acid, 25%	150	66
Lactic acid, conc.	150	66
Magnesium chloride	210	99
Malic acid	130	54
Manganese chloride	120	49
Methyl chloride	X	X
Methyl ethyl ketone	X	X
Methyl isobutyl ketone	80	27
Muriatic acid	200	93

(continued)

TABLE 6.22 *Continued*

Chemical	Maximum Temperature	
	°F	°C
Nitric acid anhydrous	X	X
Nitric acid, 5%	140	60
Nitric acid, 20%	140	60
Nitric acid, 70%	X	X
Nitrous acid, conc.	X	X
Oleum	X	X
Perchloric acid, 10%	140	60
Perchloric acid, 70%	X	X
Phenol	180	82
Phosphoric acid, 50–80%	210	99
Picric acid	140	60
Potassium bromide, 30%	210	99
Salicylic acid	130	54
Silver bromide, 10%	170	77
Sodium carbonate	220	104
Sodium chloride	200	93
Sodium hydroxide, 10%	220	104
Sodium hydroxide, 50%	220	104
Sodium hydroxide, conc.	140	60
Sodium hypochlorite, 20%	120	49
Sodium hypochlorite, conc.	110	43
Sodium sulfide, to 50%	190	88
Stannic chloride	150	66
Stannous chloride	200	93
Sulfuric acid, 10%	200	93
Sulfuric acid, 50%	200	93
Sulfuric acid, 70%	180	82
Sulfuric acid, 90%	180	82
Sulfuric acid, 98%	120	49
Sulfuric acid, 100%	X	X
Sulfuric acid, fuming	X	X
Sulfurous acid	180	82
Thionyl chloride	100	38
Toluene	X	X
Trichloroacetic acid	150	66
White liquor	220	104
Zinc chloride	200	93

The chemicals listed are in the pure state or in saturated solution unless otherwise indicated. Compatibilty is shown to the maximum allowable temperature for which data is available. Incompatibility is shown by an X.

Source: From P.A. Schweitzer. 2004. *Corrosion Resistance Tables*, Vols. 1–4, 5th ed., New York: Marcel Dekker.

Polyethylene exhibits a wide range of corrosion resistance—ranging from potable water to corrosive wastes. It is resistant to most mineral acids, including sulfuric up to 70% concentration, inorganic salts including chlorides, alkalies, and many organic acids.

It is not resistant to bromine, aromatics, or chlorinated hydrocarbons. Refer to Table 6.23 for the compatibility of EHMWPE with selected corrodents and Table 6.24 for HMWPE's compatibility. Reference [1] provides a more comprehensive listing.

6.6.6 Tetrafluoroethylene

Tetrafluoroethylene (PTFE) can be used continuously at 500°F (260°C). It cannot be bonded directly to metal substrates, but it can be bonded with a laminated fiberglass sheet as a backing.

The material tends to creep under stress at elevated temperatures. When the vessel is designed, provisions should be made to retain the PTFE. PTFE liners are subject to permeation by some corrodents. Table 6.25 provides the vapor permeation of PTFE by selected materials.

PTFE is chemically inert in the presence of most corrodents. There are very few chemicals that will attack it within normal use temperatures. These reactants are among the most violent oxidizers and reducing agents known. Elemental sodium in intimate contact with fluorocarbons removes fluorine from the polymer molecule. The other alkali metals (potassium, lithium, etc.) react in a similar manner.

Fluorine and related compounds (e.g., chlorine trifluoride) are absorbed into the PTFE resin with such intimate contact that the mixture becomes sensitive to a source of ignition such as impact.

The handling of 80% sodium hydroxide, aluminum chloride, ammonia, and certain amines at high temperatures may produce the same effect as elemental sodium. Also slow oxidative attack can be produced by 70% nitric acid under pressure at 480°F (250°C).

Refer to Table 6.26 for the compatibility of PTFE with selected corrodents. Reference [1] provides listing of additional corrodents.

6.6.7 Fluorinated Ethylene Propylene

Fluorinated ethylene propylene (FEP) has a lower maximum operating temperature than PTFE. It exhibits changes in physical strength after prolonged exposure above 400°F (204°C). Therefore, the recommended maximum continuous operating temperature is 375°F (190°C).

Permeation of FEP liners can pose a problem. Table 6.27 provides some permeation data relating to the more common chemicals. There is also some absorption of chemicals by FEP. This absorption can also lead to problems. Table 6.28 is a listing of the absorption of selected liquids by FEP.

TABLE 6.23

Compatibility of EHMW PE with Selected Corrodents

Chemical	Maximum Temperature	
	°F	°C
Acetaldehyde, 40%	90	32
Acetamide		
Acetic acid, 10%	140	60
Acetic acid, 50%	140	60
Acetic acid, 80%	80	27
Acetic acid, glacial		
Acetic anhydride	X	X
Acetone	120	49
Acetyl chloride		
Acrylic acid		
Acrylonitrile	150	66
Adipic acid	140	60
Allyl alcohol	140	60
Allyl chloride	80	27
Alum	140	60
Aluminum acetate		
Aluminum chloride, aqueous	140	60
Aluminum chloride, dry	140	60
Aluminum fluoride	140	60
Aluminum hydroxide	140	60
Aluminum nitrate		
Aluminum oxychloride		
Aluminum sulfate	140	60
Ammonia gas	140	60
Ammonium bifluoride		
Ammonium carbonate	140	60
Ammonium chloride, 10%	140	60
Ammonium chloride, 50%	140	60
Ammonium chloride, sat.	140	60
Ammonium fluoride, 10%	140	60
Ammonium fluoride, 25%	140	60
Ammonium hydroxide, 25%	140	60
Ammonium hydroxide, sat.	140	60
Ammonium nitrate	140	60
Ammonium persulfate	140	60
Ammonium phosphate	80	27
Ammonium sulfate, 10–40%	140	60
Ammonium sulfide	140	60
Ammonium sulfite		
Amyl acetate	140	60
Amyl alcohol	140	60
Amyl chloride	X	X
Aniline	130	54
Antimony trichloride	140	60
Aqua regia, 3:1	130	54
Barium carbonate	140	60
Barium chloride	140	60

(continued)

TABLE 6.23 *Continued*

Chemical	Maximum Temperature	
	°F	°C
Barium hydroxide	140	60
Barium sulfate	140	60
Barium sulfide	140	60
Benzaldehyde	X	X
Benzene	X	X
Benzene sulfonic acid, 10%	140	60
Benzoic acid	140	60
Benzyl alcohol	170	77
Benzyl chloride		
Borax	140	60
Boric acid	140	60
Bromine gas, dry	X	X
Bromine gas, moist	X	X
Bromine liquid	X	X
Butadiene	X	X
Butyl acetate	90	32
Butyl alcohol	140	60
Butyl phthalate	80	27
n-Butylamine	X	X
Butyric acid	130	54
Calcium bisulfide	140	60
Calcium bisulfite	80	27
Calcium carbonate	140	60
Calcium chlorate	140	60
Calcium chloride	140	60
Calcium hydroxide, 10%	140	60
Calcium hydroxide, sat.	140	60
Calcium hypochlorite	140	60
Calcium nitrate	140	60
Calcium oxide	140	60
Calcium sulfate	140	60
Caprylic acid		
Carbon bisulfide	X	X
Carbon dioxide, dry	140	60
Carbon dioxide, wet	140	60
Carbon disulfide	X	X
Carbon monoxide	140	60
Carbon tetrachloride	X	X
Carbonic acid	140	60
Cellosolve		
Chloracetic acid	X	X
Chloracetic acid, 50% in water	X	X
Chlorine gas, dry	80	27
Chlorine gas, wet, 10%	120	49
Chlorine, liquid	X	X
Chlorobenzene	X	X
Chloroform	80	27

(continued)

TABLE 6.23 *Continued*

Chemical	Maximum Temperature	
	°F	°C
Chlorosulfonic acid	X	X
Chromic acid, 10%	140	60
Chromic acid, 50%	90	32
Chromyl chloride		
Citric acid, 15%	140	60
Citric acid, conc.	140	60
Copper acetate		
Copper carbonate		
Copper chloride	140	60
Copper cyanide	140	60
Copper sulfate	140	60
Cresol	80	27
Cupric chloride, 5%	80	27
Cupric chloride, 50%		
Cyclohexane	130	54
Cyclohexanol	170	77
Dichloroacetic acid	73	23
Dichloroethane (ethylene dichloride)	X	X
Ethylene glycol	140	60
Ferric chloride	140	60
Ferric chloride, 50% in water	140	60
Ferric nitrate, 10–50%	140	60
Ferrous chloride	140	60
Ferrous nitrate	140	60
Fluorine gas, dry	X	X
Fluorine gas, moist	X	X
Hydrobromic acid, dil.	140	60
Hydrobromic acid, 20%	140	60
Hydrobromic acid, 50%	140	60
Hydrochloric acid, 20%	140	60
Hydrochloric acid, 38%	140	60
Hydrocyanic acid, 10%	140	60
Hydrofluoric acid, 30%	80	27
Hydrofluoric acid, 70%	X	X
Hydrofluoric acid, 100%	X	X
Hypochlorous acid		
Iodine solution, 10%	80	27
Ketones, general	X	X
Lactic acid, 25%	140	60
Lactic acid/conc.	140	60
Magnesium chloride	140	60
Malic acid	100	38
Manganese chloride	80	27
Methyl chloride	X	X
Methyl ethyl ketone	X	X
Methyl isobutyl ketone	80	27
Muriatic acid	140	60

(continued)

TABLE 6.23 *Continued*

Chemical	Maximum Temperature	
	°F	°C
Nitric acid, 5%	140	60
Nitric acid, 20%	140	60
Nitric acid, 70%	X	X
Nitric acid, anhydrous	X	X
Nitrous acid, conc.		
Oleum		
Perchloric acid, 10%	140	60
Perchloric acid, 70%	X	X
Phenol	100	38
Phosphoric acid, 50–80%	100	38
Picric acid	100	38
Potassium bromide, 30%	140	60
Salicylic acid		
Silver bromide, 10%		
Sodium carbonate	140	60
Sodium chloride	140	60
Sodium hydroxide, 10%	170	77
Sodium hydroxide, 50%	170	77
Sodium hydroxide, conc.		
Sodium hypochlorite, 20%	140	60
Sodium hypochlorite, conc.	140	60
Sodium sulfide, to 50%	140	60
Stannic chloride	140	60
Stannous chloride	140	60
Sulfuric acid, 10%	140	60
Sulfuric acid, 50%	140	60
Sulfuric acid, 70%	80	27
Sulfuric acid, 90%	X	X
Sulfuric acid, 98%	X	X
Sulfuric acid, 100%	X	X
Sulfuric acid, fuming	X	X
Sulfurous acid	140	60
Thionyl chloride	X	X
Toluene	X	X
Trichloroacetic acid	140	60
White liquor		
Zinc chloride	140	60

The chemicals listed are in the pure state or in saturated solution unless otherwise indicated. Compatibilty is shown to the maximum allowable temperature for which data is available. Incompatibility is shown by an X.

Source: From P.A. Schweitzer. 2004. *Corrosion Resistance Tables*, Vols. 1–4, 5th ed., New York: Marcel Dekker.

TABLE 6.24

Compatibility of HMWPE with Selected Corrodents

Chemical	Maximum Temperature	
	°F	°C
Acetaldehyde	X	X
Acetamide	140	60
Acetic acid, 10%	140	60
Acetic acid, 50%	140	60
Acetic acid, 80%	80	27
Acetic anhydride	X	X
Acetone	80	27
Acetyl chloride	X	X
Acrylonitrile	150	66
Adipic acid	140	60
Allyl alcohol	140	60
Allyl chloride	110	43
Alum	140	60
Aluminum chloride, aqueous	140	60
Aluminum chloride, dry	140	60
Aluminum fluoride	140	60
Aluminum hydroxide	140	60
Aluminum nitrate	140	60
Aluminum sulfate	140	60
Ammonium bifluoride	140	60
Ammonium carbonate	140	60
Ammonium chloride, 10%	140	60
Ammonium chloride, 50%	140	60
Ammonium chloride, sat.	140	60
Ammonium fluoride, 10%	140	60
Ammonium fluoride, 25%	140	60
Ammonium gas	140	60
Ammonium hydroxide, 25%	140	60
Ammonium hydroxide, sat.	140	60
Ammonium nitrate	140	60
Ammonium persulfate	150	66
Ammonium phosphate	80	27
Ammonium sulfate, to 40%	140	60
Ammonium sulfide	140	60
Ammonium sulfite	140	60
Amyl acetate	140	60
Amyl alcohol	140	60
Amyl chloride	X	X
Aniline	130	44
Antimony trichloride	140	60
Aqua regia, 3:1	130	44
Barium carbonate	140	60
Barium chloride	140	60
Barium hydroxide	140	60
Barium sulfate	140	60
Barium sulfide	140	60
Benzaldehyde	X	X

(continued)

TABLE 6.24 *Continued*

Chemical	Maximum Temperature	
	°F	°C
Benzene	X	X
Benzoic acid	140	60
Benzyl alcohol	X	X
Borax	140	60
Boric acid	140	60
Bromine gas, dry	X	X
Bromine gas, moist	X	X
Bromine, liquid	X	X
Butadiene	X	X
Butyl acetate	90	32
Butyl alcohol	140	60
n-Butylamine	X	X
Butyric acid	X	X
Calcium bisulfide	140	60
Calcium bisulfite	140	60
Calcium carbonate	140	60
Calcium chlorate	140	60
Calcium chloride	140	60
Calcium hydroxide, 10%	140	60
Calcium hydroxide, sat.	140	60
Calcium hypochlorite	140	60
Calcium nitrate	140	60
Calcium oxide	140	60
Calcium sulfate	140	60
Carbon bisulfide	X	X
Carbon dioxide, dry	140	60
Carbon dioxide, wet	140	60
Carbon disulfide	X	X
Carbon monoxide	140	60
Carbon tetrachloride	X	X
Carbonic acid	140	60
Cellosolve	X	X
Chlorine gas, dry	X	X
Chlorine gas, wet	X	X
Chlorine, liquid	X	X
Chloroacetic acid	X	X
Chlorobenzene	X	X
Chloroform	X	X
Chlorosulfonic acid	X	X
Chromic acid, 10%	140	60
Chromic acid, 50%	90	32
Citric acid, 15%	140	60
Citric acid, conc.	140	60
Copper chloride	140	60
Copper cyanide	140	60
Copper sulfate	140	60
Cresol	X	X

(continued)

TABLE 6.24 *Continued*

Chemical	Maximum Temperature	
	°F	°C
Cupric chloride, 5%	140	60
Cupric chloride, 50%	140	60
Cyclohexane	80	27
Cyclohexanol	80	27
Dibutyl phthalate	80	27
Dichloroethane	80	27
Ethylene glycol	140	60
Ferric chloride	140	60
Ferrous chloride	140	60
Ferrous nitrate	140	60
Fluorine gas, dry	X	X
Fluorine gas, moist	X	X
Hydrobromic acid, dil.	140	60
Hydrobromic acid, 20%	140	60
Hydrobromic acid, 50%	140	60
Hydrochloric acid, 20%	140	60
Hydrochloric acid, 38%	140	60
Hydrocyanic acid, 10%	140	60
Hydrofluoric acid, 30%	140	60
Hydrofluoric acid, 70%	X	X
Hydrochlorous acid	150	66
Iodine solution, 10%	80	27
Ketones, general	80	27
Lactic acid, 25%	150	66
Magnesium chloride	140	60
Malic acid	140	60
Manganese chloride	80	27
Methyl chloride	X	X
Methyl ethyl ketone	X	X
Methyl isobutyl ketone	80	27
Nitric acid, 5%	140	60
Nitric acid, 20%	140	60
Nitric acid, 70%	X	X
Nitric acid, anhydrous	X	X
Nitrous acid, conc.	120	49
Perchloric acid, 10%	140	60
Perchloric acid, 70%	X	X
Phenol	100	38
Phosphoric acid, 50–80%	100	38
Picric acid	100	38
Potassium bromide, 30%	140	60
Salicylic acid	140	60
Sodium carbonate	140	60
Sodium chloride	140	60
Sodium hydroxide, 10%	150	66
Sodium hydroxide, 50%	150	66
Sodium hypochlorite, 20%	140	60

(continued)

TABLE 6.24 *Continued*

Chemical	Maximum Temperature	
	°F	°C
Sodium hypochlorite, conc.	140	60
Sodium sulfide, to 50%	140	60
Stannic chloride	140	60
Stannous chloride	140	60
Sulfuric acid, 10%	140	60
Sulfuric acid, 50%	140	60
Sulfuric acid, 70%	80	27
Sulfuric acid, 90%	X	X
Sulfuric acid, 98%	X	X
Sulfuric acid, 100%	X	X
Sulfuric acid, fuming	X	X
Sulfurous acid	140	60
Thionyl chloride	X	X
Toluene	X	X
Trichloroacetic acid	80	27
Zinc chloride	140	60

The chemicals listed are in the pure state or in saturated solution unless otherwise indicated. Compatibilty is shown to the maximum allowable temperature for which data is available. Incompatibility is shown by an X.

Source: From P.A. Schweitzer. 2004. *Corrosion Resistance Tables*, Vols. 1–4, 5th ed., New York: Marcel Dekker.

TABLE 6.25

Vapor Permeation PTFE

Gases	Permeation (g 100 in.^{2}/24 h/mil) at	
	73°F(23°C)	86°F(30°C)
Carbon dioxide		0.66
Helium		0.22
Hydrogen chloride, anhy.		<0.01
Nitrogen		0.11
Acetophenone	0.56	
Benzene	0.36	0.80
Carbon tetrachloride	0.06	
Ethyl alcohol	0.13	
Hydrochloric acid, 20%	<0.01	
Piperdine	0.07	
Sodium hydroxide, 50%	5×10^{-5}	
Sulfuric acid, 98%	1.8×10^{-5}	

TABLE 6.26

Compatibility of PTFE with Selected Corrodents

Chemical	Maximum Temperature	
	°F	°C
Acetaldehyde	450	232
Acetamide	450	232
Acetic acid, 10%	450	232
Acetic acid, 50%	450	232
Acetic acid, 80%	450	232
Acetic acid, glacial	450	232
Acetic anhydride	450	232
Acetone	450	232
Acetyl chloride	450	232
Acrylonitrile	450	232
Adipic acid	450	232
Allyl alcohol	450	232
Allyl chloride	450	232
Alum	450	232
Aluminum chloride, aqueous	450	232
Aluminum fluoride	450	232
Aluminum hydroxide	450	232
Aluminum nitrate	450	232
Aluminum oxychloride	450	232
Aluminum sulfate	450	232
Ammonia gas[a]	450	232
Ammonium bifluoride	450	232
Ammonium carbonate	450	232
Ammonium chloride, 10%	450	232
Ammonium chloride, 50%	450	232
Ammonium chloride, sat.	450	232
Ammonium fluoride, 10%	450	232
Ammonium fluoride, 25%	450	232
Ammonium hydroxide, 25%	450	232
Ammonium hydroxide, sat.	450	232
Ammonium nitrate	450	232
Ammonium persulfate	450	232
Ammonium phosphate	450	232
Ammonium sulfate, 10–40%	450	232
Ammonium sulfide	450	232
Amyl acetate	450	232
Amyl alcohol	450	232
Amyl chloride	450	232
Aniline	450	232
Antimony trichloride	450	232
Aqua regia, 3:1	450	232
Barium carbonate	450	232
Barium chloride	450	232
Barium hydroxide	450	232
Barium sulfate	450	232
Barium sulfide	450	232
Benzaldehyde	450	232

(continued)

TABLE 6.26 *Continued*

Chemical	Maximum Temperature	
	°F	°C
Benzene[a]	450	232
Benzene sulfonic acid, 10%	450	232
Benzoic acid	450	232
Benzyl alcohol	450	232
Benzyl chloride	450	232
Borax	450	232
Boric acid	450	232
Bromine gas, dry[a]	450	232
Bromine liquid[a]	450	232
Butadiene[a]	450	232
Butyl acetate	450	232
Butyl alcohol	450	232
Butyl phthalate	450	232
n-Butylamine	450	232
Butyric acid	450	232
Calcium bisulfide	450	232
Calcium bisulfite	450	232
Calcium carbonate	450	232
Calcium chlorate	450	232
Calcium chloride	450	232
Calcium hydroxide, 10%	450	232
Calcium hydroxide, sat.	450	232
Calcium hypochlorite	450	232
Calcium nitrate	450	232
Calcium oxide	450	232
Calcium sulfate	450	232
Caprylic acid	450	232
Carbon bisulfide[a]	450	232
Carbon dioxide, dry	450	232
Carbon dioxide, wet	450	232
Carbon disulfide	450	232
Carbon monoxide	450	232
Carbon tetrachloride[b]	450	232
Carbonic acid	450	232
Chloracetic acid	450	232
Chloracetic acid, 50% water	450	232
Chlorine gas, dry	X	X
Chlorine gas, wet[a]	450	232
Chlorine, liquid	X	X
Chlorobenzene[a]	450	232
Chloroform[a]	450	232
Chlorosulfonic acid	450	232
Chromic acid, 10%	450	232
Chromic acid, 50%	450	232
Chromyl chloride	450	232
Citric acid, 15%	450	232
Citric acid, conc.	450	232

(continued)

TABLE 6.26 *Continued*

Chemical	Maximum Temperature	
	°F	°C
Copper carbonate	450	232
Copper chloride	450	232
Copper cyanide, 10%	450	232
Copper sulfate	450	232
Cresol	450	232
Cupric chloride, 5%	450	232
Cupric chloride, 50%	450	232
Cyclohexane	450	232
Cyclohexanol	450	232
Dichloroacetic acid	450	232
Dichloroethane (ethylene dichloride)[a]	450	232
Ethylene glycol	450	232
Ferric chloride	450	232
Ferric chloride, 50% in water	450	232
Ferric nitrate, 10–50%	450	232
Ferrous chloride	450	232
Ferrous nitrate	450	232
Fluorine gas, dry	X	X
Fluorine gas, moist	X	X
Hydrobromic acid, dil.[a,b]	450	232
Hydrobromic acid, 20%[b]	450	232
Hydrobromic acid, 50%[b]	450	232
Hydrochloric acid, 20%[b]	450	232
Hydrochloric acid, 38%[b]	450	232
Hydrocyanic acid, 10%	450	232
Hydrofluoric acid, 100%[a]	450	232
Hydrofluoric acid, 30%[a]	450	232
Hydrofluoric acid, 70%[a]	450	232
Hypochlorous acid	450	232
Iodine solution, 10%[a]	450	232
Ketones, general	450	232
Lactic acid, 25%	450	232
Lactic acid, conc.	450	232
Magnesium chloride	450	232
Malic acid	450	232
Methyl chloride[a]	450	232
Methyl ethyl ketone[a]	450	232
Methyl isobutyl ketone[b]	450	232
Muriatic acid[a]	450	232
Nitric acid, 5%[a]	450	232
Nitric acid, 20%[a]	450	232
Nitric acid, 70%[a]	450	232
Nitric acid, anhydrous[a]	450	232
Nitrous acid, 10%	450	232
Oleum	450	232
Perchloric acid, 10%	450	232
Perchloric acid, 70%	450	232

(continued)

TABLE 6.26 *Continued*

Chemical	Maximum Temperature	
	°F	°C
Pheno[a]	450	232
Phosphoric acid, 50–80%	450	232
Picric acid	450	232
Potassium bromide, 30%	450	232
Salicylic acid	450	232
Sodium carbonate	450	232
Sodium chloride	450	232
Sodium hydroxide, 10%	450	232
Sodium hydroxide, 50%	450	232
Sodium hydroxide, conc.	450	232
Sodium hypochlorite, 20%	450	232
Sodium hypochlorite, conc.	450	232
Sodium sulfide, to 50%	450	232
Stannic chloride	450	232
Stannous chloride	450	232
Sulfuric acid, 10%	450	232
Sulfuric acid, 50%	450	232
Sulfuric acid, 70%	450	232
Sulfuric acid, 90%	450	232
Sulfuric acid, 98%	450	232
Sulfuric acid, 100%	450	232
Sulfuric acid, fuming[a]	450	232
Sulfurous acid	450	232
Thionyl chloride	450	232
Toluene[a]	450	232
Trichloroacetic acid	450	232
White liquor	450	232
Zinc chloride[c]	450	232

The chemicals listed are in the pure state or in saturated solution unless otherwise indicated. Compatibilty is shown to the maximum allowable temperature for which data is available. Incompatibility is shown by an X. A blank space indicates that the data is unavailable.

[a] Material will permeate.
[b] Material will cause stress cracking.
[c] Material will be absorbed.

Source: From P.A. Schweitzer. 2004. *Corrosion Resistance Tables*, Vols. 1–4, 5th ed., New York: Marcel Dekker.

FEP basically has the same corrosion resistance properties as PTFE but at lower maximum temperatures. It is resistant to practically all chemicals, the exception being extremely potent oxidizers such as chlorine trifluoride and related compounds. Some chemicals will attack FEP when present in high concentrations, at or near the service temperature limit.

Refer to Table 6.29 for the compatibility of FEP with selected corrodents. Reference [1] provides additional listings.

TABLE 6.27

Vapor Permeation of FEP

	Permeation (g 100 in.^{2}/24 h/mil) at:		
	73°F(23°C)	95°F(35°C)	122°F(50°C)
Gases			
Nitrogen	0.18		
Oxygen	0.39		
Vapors			
Acetic acid		0.42	
Acetone	0.13	0.95	3.29
Acetophenone	0.47		
Benzene	0.15	0.64	
n-Butyl ether	0.08		
Carbon tetrachloride	0.11	0.31	
Decane	0.72		1.03
Ethyl acetate	0.06	0.77	2.9
Ethyl alcohol	0.11	0.69	
Hexane		0.57	
Hydrochloric acid, 20%	0.01		
Methanol			5.61
Sodium hydroxide, 50%	4×10^{-5}		
Sulfuric acid, 98%	8×10^{-6}		
Toluene	0.37		2.93
Water	0.09	0.45	0.89

TABLE 6.28

Absorption of Selected Liquid[a] by FEP

Chemical	Temperature (°F/°C)	Range of Weight Gains (%)
Aniline	365/185	0.3–0.4
Acetophenone	394/201	0.6–0.8
Benzaldehyde	354/179	0.4–0.5
Benzyl alcohol	400/204	0.3–0.4
n-Butylamine	172/78	0.3–0.4
Carbon tetrachloride	172/78	2.3–2.4
Dimethylsulfoxide	372/190	0.1–0.2
Nitrobenzene	410/210	0.7–0.9
Perchlorethylene	250/121	2.0–2.3
Sulfuryl chloride	154/68	1.7–2.7
Toluene	230/110	0.7–0.8
Tributyl phosphate[b]	392/200	1.8–2.0

[a] 16 h exposure at their boiling points.
[b] Not boiling.

TABLE 6.29

Compatibility of FEP with Selected Corrodents

Chemical	Maximum Temperature	
	°F	°C
Acetaldehyde	200	93
Acetamide	400	204
Acetic acid, 10%	400	204
Acetic acid, 10%	400	204
Acetic acid, 50%	400	204
Acetic acid, glacial	400	204
Acetic anhydride	400	204
Acetone	400	204
Acetyl chloride	400	204
Acrylic acid	200	93
Acrylonitrile	400	204
Adipic acid	400	204
Allyl alcohol	400	204
Allyl chloride	400	204
Alum	400	204
Aluminum acetate	400	204
Aluminum chloride, aqueous	400	204
Aluminum chloride, dry	300	149
Aluminum fluoride[a]	400	204
Aluminum hydroxide	400	204
Aluminum nitrate	400	204
Aluminum oxychloride	400	204
Aluminum sulfate	400	204
Ammonia gas[a]	400	204
Ammonium bifluoride[a]	400	204
Ammonium carbonate	400	204
Ammonium chloride, 10%	400	204
Ammonium chloride, 50%	400	204
Ammonium chloride, sat.	400	204
Ammonium fluoride, 10%[a]	400	204
Ammonium fluoride, 25%[a]	400	204
Ammonium hydroxide, 25%	400	204
Ammonium hydroxide, sat.	400	204
Ammonium nitrate	400	204
Ammonium persulfate	400	204
Ammonium phosphate	400	204
Ammonium sulfate, 10–40%	400	204
Ammonium sulfide	400	204
Ammonium sulfite	400	204
Amyl acetate	400	204
Amyl alcohol	400	204
Amyl chloride	400	204
Aniline[b]	400	204
Antimony trichloride	250	121
Aqua regia, 3:1	400	204
Barium carbonate	400	204
Barium chloride	400	204

(continued)

TABLE 6.29 *Continued*

Chemical	Maximum Temperature	
	°F	°C
Barium hydroxide	400	204
Barium sulfate	400	204
Barium sulfide	400	204
Benzaldehyde[b]	400	204
Benzene sulfonic acid, 10%	400	204
Benzene[b]	400	204
Benzoic acid	400	204
Benzyl alcohol	400	204
Benzyl chloride	400	204
Borax	400	204
Boric acid	400	204
Bromine gas, dry[a]	200	93
Bromine gas, moist[a]	200	93
Bromine liquid[a,b]	400	204
Butadiene[a]	400	204
Butyl acetate	400	204
Butyl alcohol	400	204
Butyl phthalate	400	204
n-Butylamine[b]	400	204
Butyric acid	400	204
Calcium bisulfide	400	204
Calcium bisulfite	400	204
Calcium carbonate	400	204
Calcium chlorate	400	204
Calcium chloride	400	204
Calcium hydroxide, 10%	400	204
Calcium hydroxide, sat.	400	204
Calcium hypochlorite	400	204
Calcium nitrate	400	204
Calcium oxide	400	204
Calcium sulfate	400	204
Caprylic acid	400	204
Carbon bisulfide[a]	400	204
Carbon dioxide, dry	400	204
Carbon dioxide, wet	400	204
Carbon disulfide	400	204
Carbon monoxide	400	204
Carbon tetrachloride[a,b,c]	400	204
Carbonic acid	400	204
Cellosolve	400	204
Chloracetic acid	400	204
Chloracetic acid, 50% water	400	204
Chlorine gas, dry	X	X
Chlorine gas, wet[a]	400	204
Chlorine, liquid[b]	400	204
Chlorobenzene[a]	400	204
Chloroform[a]	400	204

(continued)

TABLE 6.29 *Continued*

Chemical	Maximum Temperature	
	°F	°C
Chlorosulfonic acid[b]	400	204
Chromic acid, 50%[b]	400	204
Chromic acid, 10%	400	204
Chromyl chloride	400	204
Citric acid, 15%	400	204
Citric acid, conc.	400	204
Copper acetate	400	204
Copper carbonate	400	204
Copper chloride	400	204
Copper cyanide	400	204
Copper sulfate	400	204
Cresol	400	204
Cupric chloride, 5%	400	204
Cupric chloride, 50%	400	204
Cyclohexane	400	204
Cyclohexanol	400	204
Dichloroacetic acid	400	204
Dichloroethane (ethylene dichloride)[a]	400	204
Ethylene glycol	400	204
Ferric chloride	400	204
Ferric chloride, 50% in water[b]	260	127
Ferric nitrate, 10–50%	260	127
Ferrous chloride	400	204
Ferrous nitrate	400	204
Fluorine gas, dry	200	93
Fluorine gas, moist	X	X
Hydrobromic acid, dil.	400	204
Hydrobromic acid, 20%[a,c]	400	204
Hydrobromic acid, 50%[a,c]	400	204
Hydrochloric acid, 20%[a,c]	400	204
Hydrochloric acid, 38%[a,c]	400	204
Hydrocyanic acid, 10%	400	204
Hydrofluoric acid, 30%[a]	400	204
Hydrofluoric acid, 70%[a]	400	204
Hydrofluoric acid, 100%[a]	400	204
Hypochlorous acid	400	204
Iodine solution, 10%	400	204
Ketones, general	400	204
Lactic acid, 25%	400	204
Lactic acid, conc.	400	204
Magnesium chloride	400	204
Malic acid	400	204
Manganese chloride	300	149
Methyl chloride[a]	400	204
Methyl ethyl ketone[a]	400	204
Methyl isobutyl ketone[a]	400	204
Muriatic acid[a]	400	204

(continued)

TABLE 6.29 *Continued*

Chemical	Maximum Temperature	
	°F	°C
Nitric acid, 5%[a]	400	204
Nitric acid, 20%[a]	400	204
Nitric acid, 70%[a]	400	204
Nitric acid, anhydrous[a]	400	204
Nitrous acid, conc.	400	204
Oleum	400	204
Perchloric acid, 10%	400	204
Perchloric acid, 70%	400	204
Phenol[a]	400	204
Phosphoric acid, 50–80%	400	204
Picric acid	400	204
Potassium bromide, 30%	400	204
Salicylic acid	400	204
Silver bromide, 10%	400	204
Sodium carbonate	400	204
Sodium chloride	400	204
Sodium hydroxide, 10%[b]	400	204
Sodium hydroxide, 50%	400	204
Sodium hydroxide, conc.	400	204
Sodium hypochlorite, 20%	400	204
Sodium hypochlorite, conc.	400	204
Sodium sulfide, to 50%	400	204
Stannic chloride	400	204
Stannous chloride	400	204
Sulfuric acid, 50%	400	204
Sulfuric acid, 10%	400	204
Sulfuric acid, 70%	400	204
Sulfuric acid, 90%	400	204
Sulfuric acid, 98%	400	204
Sulfuric acid, 100%	400	204
Sulfuric acid, fuming[c]	400	204
Sulfurous acid	400	204
Thionyl chloride[a]	400	204
Toluene	400	204
Trichloroacetic acid	400	204
White liquor	400	204
Zinc chloride[c]	400	204

The chemicals listed are in the pure state or in saturated solution unless otherwise indicated. Compatibilty is shown to the maximum allowable temperature for which data is available. Incompatibility is shown by an X. A blank space indicates that the data is unavailable.

[a] Material will cause stress cracking.
[b] Material will permeate.
[c] Material will be absorbed.

Source: From P.A. Schweitzer. 2004. *Corrosion Resistance Tables*, Vols. 1–4, 5th ed., New York: Marcel Dekker.

TABLE 6.30

Permeation of Various Gases in PFA at 77°F (25°C)

Gas	Permeation (cm^3 mil thickness/100 in.^{2} 24 h atm)
Carbon dioxide	2260
Nitrogen	291
Oxygen	881

6.6.8 Perfluoralkoxy

Perfluoralkoxy (PFA) can be continuously used at 500°F (260°C). It is subject to permeation by certain gases and will absorb liquids. Table 6.30 illustrates the permeability of PFA, and Table 6.31 lists the absorption of representative liquids by PFA.

Perfluoralkoxy is inert to strong mineral acids, inorganic bases, inorganic oxidizers, aromatics, some aliphatic hydrocarbons, alcohols, aldehydes, ketones, esters, ethers, chlorocarbons, fluorocarbons, and mixtures of these compounds.

TABLE 6.31

Absorption of Representative Liquids in PFA

Liquids[a]	Temperature, °F/°C	Range of Weight Gains, %
Aniline	365/185	0.3–0.4
Acetophenone	394/201	0.6–0.8
Benzaldehyde	354/179	0.4–0.5
Benzyl alcohol	400/204	0.3–0.4
n-Butylamine	172/78	0.3–0.4
Carbon tetrachloride	172/78	2.3–2.4
Dimethyl sulfoxide	372/190	0.1–0.2
Freon 113	117/47	1.2
Isooctane	210/99	0.7–0.8
Nitrobenzene	410/210	0.7–0.9
Perchlorethylene	250/121	2.0–2.3
Sulfuryl chloride	154/68	1.7–2.7
Toluene	230/110	0.7–0.8
Tributyl phosphate	392/200[b]	1.8–2.0
Bromine, anhydrous	−5/−22	0.5
Chlorine, anhydrous	248/120	0.5–0.6
Chlorosulfonic acid	302/150	0.7–0.8
Chromic acid, 50%	248/120	0.00–0.01
Ferric chloride	212/100	0.00–0.01
Hydrochloric acid, 37%	248/120	0.00–0.03
Phosphoric acid, conc.	212/100	0.00–0.01
Zinc chloride	212/100	0.00–0.03

[a] Liquids were exposed for 168 hours at the boiling point of the solvents. The acidic reagents were exposed for 168 hours.
[b] Not boiling.

PFA will be attacked by certain halogenated complexes containing fluorine. These include chlorine trifluoride, bromine trifluoride, iodine pentafluoride, and fluorine. PFA can also be attacked by such metals as sodium or potassium, particularly in the molten state.

PFA has the advantage over PTFE in that it can be thermoformed, heat sealed, welded, and heat laminated. This makes the material much easier to produce lining from than that of PTFE. PFA can be bonded to a metal substrate. Refer to Table 6.32 for the compatibility of PFA with selected corrodents. Reference [1] provides additional listings.

6.6.9   Ethylene Tetrafluoroethylene

Ethylene Tetrafluoroethylene (ETFE) is sold under the trade name of Tefzel by DuPont. It is a partially fluorinated copolymer of ethylene and tetrafluoroethylene with a maximum service temperature of 300°F/149°C. Tefzel can be melt bonded to untreated aluminum, steel, and copper. It can also be melt bonded to itself. To adhesive bond Tefzel with polyester or epoxy compounds, the surface must be chemically etched or subjected to corona or flame treatments.

ETFE is a rugged thermoplastic. It is less dense, tougher, stiffer, and exhibits a higher tensile strength and creep resistance than PTFE and FEP fluorocarbon resins.

Vapor permeation of ETFE is as follows:

Gas	Permeation (cm^3 mil thickness/100 in.^{2} 24 h atm)
Carbon dioxide	250
Nitrogen	30
Oxygen	100
Helium	900

Tefzel is inert to strong mineral acids, inorganic bases, halogens, and metal salt solutions. Even carboxylic acids, anhydrides, aromatic and aliphatic hydrocarbons, alcohols, ketones, ethers, esters, chlorocarbons, and classic polymer solvents have little effect on ETFE.

Very strong oxidizing acids near their boiling points, such as nitric acid, at high concentration will affect ETFE in varying degrees as well as organic bases such as amines and sulfonic acids. Refer to Table 6.33 or the compatibility of ETFE with selected corrodents. Reference [1] provides a more comprehensive listing.

6.6.10   Chlorotrifluoroethylene

Chlorotrifluoroethylene (ECTFE) is sold under the tradename of Halar by Ausimont USA, Inc. Halar has many of the desirable properties of PTFE

TABLE 6.32

Compatibility of PFA with Selected Corrodents

Chemical	Maximum Temperature	
	°F	°C
Acetaldehyde	450	232
Acetamide	450	232
Acetic acid, 10%	450	232
Acetic acid, 50%	450	232
Acetic acid, 80%	450	232
Acetic acid, glacial	450	232
Acetic anhydride	450	232
Acetone	450	232
Acetyl chloride	450	232
Acrylonitrile	450	232
Adipic acid	450	232
Allyl alcohol	450	232
Allyl chloride	450	232
Alum	450	232
Aluminum chloride, aqueous	450	232
Aluminum fluoride	450	232
Aluminum hydroxide	450	232
Aluminum nitrate	450	232
Aluminum oxychloride	450	232
Aluminum sulfate	450	232
Ammonia gas[b]	450	232
Ammonium bifluoride[b]	450	232
Ammonium carbonate	450	232
Ammonium chloride, 10%	450	232
Ammonium chloride, 50%	450	232
Ammonium chloride, sat.	450	232
Ammonium fluoride, 10%[b]	450	232
Ammonium fluoride, 25%[b]	450	232
Ammonium hydroxide, 250%	450	232
Ammonium hydroxide, sat.	450	232
Ammonium nitrate	450	232
Ammonium persulfate	450	232
Ammonium phosphate	450	232
Ammonium sulfate, 10–40%	450	232
Ammonium sulfide	450	232
Amyl acetate	450	232
Amyl alcohol	450	232
Amyl chloride	450	232
Aniline[a]	450	232
Antimony trichloride	450	232
Aqua regia, 3:1	450	232
Barium carbonate	450	232
Barium chloride	450	232
Barium hydroxide	450	232
Barium sulfate	450	232
Barium sulfide	450	232
Benzaldehyde[a]	450	232

(continued)

TABLE 6.32 *Continued*

Chemical	Maximum Temperature	
	°F	°C
Benzene[b]	450	232
Benzene sulfonic acid, 10%	450	232
Benzoic acid	450	232
Benzyl alcohol	450	232
Benzyl chloride[b]	450	232
Borax	450	232
Boric acid	450	232
Bromine gas, dry	450	232
Bromine liquid[a,b]	450	232
Butadiene[b]	450	232
Butyl acetate	450	232
Butyl alcohol	450	232
Butyl phthalate	450	232
n-Butylamine[a]	450	232
Butyric acid	450	232
Calcium bisulfide	450	232
Calcium bisulfite	450	232
Calcium carbonate	450	232
Calcium chlorate	450	232
Calcium chloride	450	232
Calcium hydroxide, 10%	450	232
Calcium hydroxide, sat.	450	232
Calcium hypochlorite	450	232
Calcium nitrate	450	232
Calcium oxide	450	232
Calcium sulfate	450	232
Caprylic acid	450	232
Carbon bisulfide[b]	450	232
Carbon dioxide, dry	450	232
Carbon dioxide, wet	450	232
Carbon disulfide[b]	450	232
Carbon monoxide	450	232
Carbon tetrachloride[a,b,c]	450	232
Carbonic acid	450	232
Chloracetic acid	450	232
Chloracetic acid, 50% water	450	232
Chlorine gas, dry	X	X
Chlorine gas, wet[b]	450	232
Chlorine, liquid[a]	X	X
Chlorobenzene[b]	450	232
Chloroform[b]	450	232
Chlorosulfonic acid[a]	450	232
Chromic acid, 10%	450	232
Chromic acid, 50%[a]	450	233
Chromyl chloride	450	232
Citric acid, 15%	450	232
Citric acid, conc.	450	232

(continued)

TABLE 6.32 *Continued*

Chemical	Maximum Temperature	
	°F	°C
Copper carbonate	450	232
Copper chloride	450	232
Copper cyanide	450	232
Copper sulfate	450	232
Cresol	450	232
Cupric chloride, 5%	450	232
Cupric chloride, 50%	450	232
Cyclohexane	450	232
Cyclohexanol	450	232
Dichloroacetic acid	450	232
Dichloroethane (ethylene dichloride)[b]	450	232
Ethylene glycol	450	232
Ferric chloride	450	232
Ferric chloride, 50% in water	450	232
Ferric nitrate, 10–50%	450	233
Ferrous chloride	450	232
Ferrous nitrate	450	232
Fluorine gas, dry	X	X
Fluorine gas, moist	X	X
Hydrochloric acid, 20%[b,c]	450	232
Hydrobromic acid, 20%[b,c]	450	232
Hydrobromic acid, 50%[b,c]	450	232
Hydrobromic acid, 20%[b,c]	450	232
Hydrochloric acid, 38%[b,c]	450	232
Hydrocyanic acid, 10%	450	232
Hydrofluoric acide, 30%[b]	450	232
Hydrofluoric acid, 70%[b]	450	232
Hydrofluoric acid, 100%[b]	450	232
Hypochlorous acid	450	232
Iodine solution, 10%[b]	450	232
Ketones, general	450	232
Lactic acid, 25%	450	232
Lactic acid, conc.	450	232
Magnesium chloride	450	232
Malic acid	450	232
Methyl chloride[b]	450	232
Methyl ethyl ketone[b]	450	232
Methyl isobutyl ketone[b]	450	232
Muriatic acid[b]	450	232
Nitric acid, 5%[b]	450	232
Nitric acid, 20%[b]	450	232
Nitric acid, 70%[b]	450	232
Nitric acid, anhydrous[b]	450	232
Nitrous acid, 10%	450	232
Oleum	450	232
Perchloric acid, 10%	450	232
Perchloric acid, 70%	450	232

(continued)

TABLE 6.32 *Continued*

Chemical	Maximum Temperature	
	°F	°C
Phenol[b]	450	232
Phosphoric acid, 50–80%[a]	450	232
Picric acid	450	232
Potassium bromide, 30%	450	232
Salicylic acid	450	232
Sodium carbonate	450	232
Sodium chloride	450	232
Sodium hydroxide, 10%	450	232
Sodium hydroxide, 50%	450	232
Sodium hydroxide, conc.	450	232
Sodium hypochlorite, 20%	450	232
Sodium hypochlorite, conc.	450	232
Sodium sulfide, to 50%	450	232
Stannic chloride	450	232
Stannous chloride	450	232
Sulfuric acid, 10%	450	232
Sulfuric acid, 50%	450	232
Sulfuric acid, 70%	450	232
Sulfuric acid, 90%	450	232
Sulfuric acid, 98%	450	232
Sulfuric acid, 100%	450	232
Sulfuric acid, fuming	450	232
Sulfurous acid	450	232
Thionyl chloride[b]	450	232
Toluene[b]	450	232
Trichloroacetic acid	450	232
White liquor	450	232
Zinc chloride[a]	450	232

The chemicals listed are in the pure state or in a saturated solution unless otherwise indicated. Compatibility is shown to the maximum allowable temperature for which data is available. Incompatibility is shown by an X. A blank space indicates that the data is unavailable.

[a] Material will be absorbed.
[b] Material will permeate.
[c] Material will cause stress cracking.

Source: From P.A. Schweitzer. 2004. *Corrosion Resistance Tables*, Vols. 1–4, 5th ed., New York: Marcel Dekker.

without some of the disadvantages. ECTFE can be welded with ordinary thermoplastic welding equipment. It is used to line vessels and pipe. Liner thicknesses are usually 0.160 in. thick, that resists permeation. The water absorption rate of ECTFE is low, being less than 0.1%.

Halar has a broad use temperature range from cryogenic to 340°F/171°C with continuous service to 300°F/149°C. It exhibits excellent impact strength

TABLE 6.33

Compatibility of ETFE with Selected Corrodents

Chemical	Maximum Temperature	
	°F	°C
Acetaldehyde	200	93
Acetamide	250	121
Acetic acid, 10%	250	121
Acetic acid, 50%	250	121
Acetic acid, 80%	230	110
Acetic acid, glacial	230	110
Acetic anhydride	300	149
Acetone	150	66
Acetyl chloride	150	66
Acrylonitrile	150	66
Adipic acid	280	138
Allyl alcohol	210	99
Allyl chloride	190	88
Alum	300	149
Aluminum chloride, aqueous	300	149
Aluminum chloride, dry	300	149
Aluminum fluoride	300	149
Aluminum hydroxide	300	149
Aluminum nitrate	300	149
Aluminum oxychloride	300	149
Aluminum sulfate	300	149
Ammonium bifluoride	300	149
Ammonium carbonate	300	149
Ammonium chloride, 10%	300	149
Ammonium chloride, 50%	290	143
Ammonium chloride, sat.	300	149
Ammonium fluoride, 10%	300	149
Ammonium fluoride, 25%	300	149
Ammonium hydroxide, 25%	300	149
Ammonium hydroxide, sat.	300	149
Ammonium nitrate	230	110
Ammonium persulfate	300	149
Ammonium phosphate	300	149
Ammonium sulfate, 10–40%	300	149
Ammonium sulfide	300	149
Amyl acetate	250	121
Amyl alcohol	300	149
Amyl chloride	300	149
Aniline	230	110
Antimony trichloride	210	99
Aqua regia, 3:1	210	99
Barium carbonate	300	149
Barium chloride	300	149
Barium hydroxide	300	149
Barium sulfate	300	149
Barium sulfide	300	149
Benzaldehyde	210	99
Benzene	210	99

(continued)

TABLE 6.33 *Continued*

Chemical	Maximum Temperature	
	°F	°C
Benzene sulfonic acid, 10%	210	99
Benzoic acid	270	132
Benzyl alcohol	300	149
Benzyl chloride	300	149
Borax	300	149
Boric acid	300	149
Bromine gas, dry	150	66
Bromine water, 10%	230	110
Butadiene	250	121
Butyl acetate	230	110
Butyl alcohol	300	149
Butyl phthalate	150	66
n-Butylamine	120	49
Butyric acid	250	121
Calcium bisulfide	300	149
Calcium carbonate	300	149
Calcium chlorate	300	149
Calcium chloride	300	149
Calcium hydroxide, 10%	300	149
Calcium hydroxide, sat.	300	149
Calcium hypochlorite	300	149
Calcium nitrate	300	149
Calcium oxide	260	127
Calcium sulfate	300	149
Caprylic acid	210	99
Carbon bisulfide	150	66
Carbon dioxide, dry	300	149
Carbon dioxide, wet	300	149
Carbon disulfide	150	66
Carbon monoxide	300	149
Carbon tetrachloride	270	132
Carbonic acid	300	149
Cellosolve	300	149
Chloracetic acid, 50%	230	110
Chloracetic acid, 50% water	230	110
Chlorine gas, dry	210	99
Chlorine gas, wet	250	121
Chlorine, water	100	38
Chlorobenzene	210	99
Chloroform	230	110
Chlorosulfonic acid	80	27
Chromic acid, 10%	150	66
Chromic acid, 50%	150	66
Chromic chloride	210	99
Citric acid, 15%	120	49
Copper chloride	300	149
Copper cyanide	300	149
Copper sulfate	300	149
Cresol	270	132

(continued)

TABLE 6.33 *Continued*

Chemical	Maximum Temperature	
	°F	°C
Cupric chloride, 5%	300	149
Cyclohexane	300	149
Cyclohexanol	250	121
Dichloroacetic acid	150	66
Ethylene glycol	300	149
Ferric chloride, 50% in water	300	149
Ferric nitrate, 10–50%	300	149
Ferrous chloride	300	149
Ferrous nitrate	300	149
Fluorine gas, dry	100	38
Fluorine gas, moist	100	38
Hydrobromic acid, dil.	300	149
Hydrobromic acid, 20%	300	149
Hydrobromic acid, 50%	300	149
Hydrochloric acid, 20%	300	149
Hydrochloric acid, 38%	300	149
Hydrocyanic acid, 10%	300	149
Hydrofluoric acid, 30%	270	133
Hydrofluoric acid, 70%	250	121
Hydrofluoric acid, 100%	230	110
Hypochlorous acid	300	149
Lactic acid, 25%	250	121
Lactic acid, conc.	250	121
Magnesium chloride	300	149
Malic acid	270	132
Manganese chloride	120	49
Methyl chloride	300	149
Methyl ethyl ketone	230	110
Methyl isobutyl ketone	300	149
Muriatic acid	300	149
Nitric acid, 5%	150	66
Nitric acid, 20%	150	66
Nitric acid, 70%	80	27
Nitric acid, anhydrous	X	X
Nitrous acid, conc.	210	99
Oleum	150	66
Perchloric acid, 10%	230	110
Perchloric acid, 70%	150	66
Phenol	210	99
Phosphoric acid 50–80%	270	132
Picric acid	130	54
Potassium bromide, 30%	300	149
Salicylic acid	250	121
Sodium carbonate	300	149
Sodium chloride	300	149
Sodium hydroxide, 10%	230	110
Sodium hydroxide, 50%	230	110
Sodium hypochlorite, 20%	300	149
Sodium hypochlorite, conc.	300	149

(continued)

TABLE 6.33 *Continued*

Chemical	Maximum Temperature	
	°F	°C
Sodium sulfide, to 50%	300	149
Stannic chloride	300	149
Stannous chloride	300	149
Sulfuric acid, 10%	300	149
Sulfuric acid, 50%	300	149
Sulfuric acid, 70%	300	149
Sulfuric acid, 90%	300	149
Sulfuric acid, 98%	300	149
Sulfuric acid, 100%	300	149
Sulfuric acid, fuming	120	49
Sulfurous acid	210	99
Thionyl chloride	210	99
Toluene	250	121
Trichloroacetic acid	210	99
Zinc chloride	300	149

The chemicals listed are in the pure state or in a saturated solution unless otherwise indicated. Compatibility is shown to the maximum allowable temperature for which data is available. Incompatibility is shown by an X. A blank space indicates that the data is unavailable.

Source: Material extracted from P.A. Schweitzer. 2004. *Corrosion Resistance Tables*, Vols. 1–4, 5th ed., New York: Marcel Dekker.

and abrasion resistance over its entire operating range, and good tensile, flexural, and wear related properties.

Halar is very similar in its corrosion resistance to PTFE, but it does not have the permeation problem associated with PTFE, PFA, and FEP. It is resistant to strong mineral and oxidizing acids, alkalies, metal etchants, liquid oxygen, and practically all organic solvents except hot amines, aniline, dimethylamine, etc. ECTFE is not subject to chemically induced stress cracking from strong acids, bases, or solvents. Some halogenated solvents can cause ECTFE to become slightly plasticized when it comes into contact with them. Under normal circumstances, this does not affect the usefulness of the polymer because upon removal of the solvent from contact and upon drying, its mechanical properties return to their original values, indicating that no chemical attack has taken place.

Like other fluoropolymers, ECTFE will be attacked by metallic sodium and potassium. Table 6.34 provides the compatibility of ECTFE with selected corrodents. Reference [1] provides a more comprehensive listing of corrosion resistance.

TABLE 6.34

Compatibility of ECTFE with Selected Corrodents

Chemical	Maximum Temperature	
	°F	°C
Acetic acid, 10%	250	121
Acetic acid, 50%	250	121
Acetic acid, 80%	150	66
Acetic acid, glacial	200	93
Acetic anhydride	100	38
Acetone	150	66
Acetyl chloride	150	66
Acrylonitrile	150	66
Adipic acid	150	66
Allyl chloride	300	149
Alum	300	149
Aluminum chloride, aqueous	300	149
Aluminum chloride, dry		
Aluminum fluoride	300	149
Aluminum hydroxide	300	149
Aluminum nitrate	300	149
Aluminum oxychloride	150	66
Aluminum sulfate	300	149
Ammonia gas	300	149
Ammonium bifluoride	300	149
Ammonium carbonate	300	149
Ammonium chloride, 10%	290	143
Ammonium chloride, 50%	300	149
Ammonium chloride, sat.	300	149
Ammonium fluoride, 10%	300	149
Ammonium fluoride, 25%	300	149
Ammonium hydroxide, 25%	300	149
Ammonium hydroxide, sat.	300	149
Ammonium nitrate	300	149
Ammonium persulfate	150	66
Ammonium phosphate	300	149
Ammonium sulfate, 10–40%	300	149
Ammonium sulfide	300	149
Amyl acetate	160	71
Amyl alcohol	300	149
Amyl chloride	300	149
Aniline	90	32
Antimony trichloride	100	38
Aqua regia, 3:1	250	121
Barium carbonate	300	149
Barium chloride	300	149
Barium hydroxide	300	149
Barium sulfate	300	149
Barium sulfide	300	149
Benzaldehyde	150	66
Benzene	150	66
Benzene sulfonic acid, 10%	150	66

(continued)

TABLE 6.34 *Continued*

Chemical	Maximum Temperature	
	°F	°C
Benzoic acid	250	121
Benzyl alcohol	300	149
Benzyl chloride	300	149
Borax	300	149
Boric acid	300	149
Bromine gas, dry	X	X
Bromine liquid	150	66
Butadiene	250	121
Butyl acetate	150	66
Butyl alcohol	300	149
Butyric acid	250	121
Calcium bisulfide	300	149
Calcium bisulfite	300	149
Calcium carbonate	300	149
Calcium chlorate	300	149
Calcium chloride	300	149
Calcium hydroxide, 10%	300	149
Calcium hydroxide, sat.	300	149
Calcium hypochlorite	300	149
Calcium nitrate	300	149
Calcium oxide	300	149
Calcium sulfate	300	149
Caprylic acid	220	104
Carbon bisulfide	80	27
Carbon dioxide, dry	300	149
Carbon dioxide, wet	300	149
Carbon disulfide	80	27
Carbon monoxide	150	66
Carbon tetrachloride	300	149
Carbonic acid	300	149
Cellosolve	300	149
Chloracetic acid	250	121
Chloracetic acid, 50% water	250	121
Chlorine gas, dry	150	66
Chlorine gas, wet	250	121
Chlorine, liquid	250	121
Chlorobenzene	150	66
Chloroform	250	121
Chlorosulfonic acid	80	27
Chromic acid, 10%	250	121
Chromic acid, 50%	250	121
Citric acid, 15%	300	149
Citric acid, conc.	300	149
Copper carbonate	150	66
Copper chloride	300	149
Copper cyanide	300	149
Copper sulfate	300	149

(continued)

TABLE 6.34 *Continued*

Chemical	Maximum Temperature	
	°F	°C
Cresol	300	149
Cupric chloride, 5%	300	149
Cupric chloride, 50%	300	149
Cyclohexane	300	149
Cyclohexanol	300	149
Ethylene glycol	300	149
Ferric chloride	300	149
Ferric chloride, 50% in water	300	149
Ferric nitrate, 10–50%	300	149
Ferrous chloride	300	149
Ferrous nitrate	300	149
Fluorine gas, dry	X	X
Fluorine gas, moist	80	27
Hydrobromic acid, dil.	300	149
Hydrobromic acid, 20%	300	149
Hydrobromic acid, 50%	300	149
Hydrochloric acid, 20%	300	149
Hydrochloric acid, 38%	300	149
Hydrocyanic acid, 10%	300	149
Hydrofluoric acid, 30%	250	121
Hydrofluoric acid, 70%	240	116
Hydrofluoric acid, 100%	240	116
Hypochlorous acid	300	149
Iodine solution, 10%	250	121
Lactic acid, 25%	150	66
Lactic acid, conc.	150	66
Magnesium chloride	300	149
Malic acid	250	121
Methyl chloride	300	149
Methyl ethyl ketone	150	66
Methyl isobutyl ketone	150	66
Muriatic acid	300	149
Nitric acid, 5%	300	149
Nitric acid, 20%	250	121
Nitric acid, 70%	150	66
Nitric acid, anhydrous	150	66
Nitrous acid, conc.	250	121
Oleum	X	X
Perchloric acid, 10%	150	66
Perchloric acid, 70%	150	66
Phenol	150	66
Phosphoric acid, 50–80%	250	121
Picric acid	80	27
Potassium bromide, 30%	300	149
Salicylic acid	250	121
Sodium carbonate	300	149
Sodium chloride	300	149

(continued)

TABLE 6.34 *Continued*

Chemical	Maximum Temperature	
	°F	°C
Sodium hydroxide, 10%	300	149
Sodium hydroxide, 50%	250	121
Sodium hydroxide, conc.	150	66
Sodium hypochlorite, 20%	300	149
Sodium hypochlorite, conc.	300	149
Sodium sulfide, to 50%	300	149
Stannic chloride	300	149
Stannous chloride	300	149
Sulfuric acid, 10%	250	121
Sulfuric acid, 50%	250	121
Sulfuric acid, 70%	250	121
Sulfuric acid, 90%	150	66
Sulfuric acid, 98%	150	66
Sulfuric acid, 100%	80	27
Sulfuric acid, fuming	300	149
Sulfurous acid	250	121
Thionyl chloride	150	66
Toluene	150	66
Trichloroacetic acid	150	66
White liquor	250	121
Zinc chloride	300	149

The chemicals listed are in the pure state or in a saturated solution unless otherwise indicated. Compatibility is shown to the maximum allowable temperature for which data is available. Incompatibility is shown by an X. A blank space indicates that the data is unavailable.

Source: From P.A. Schweitzer. 2004. *Corrosion Resistance Tables*, Vols. 1–4, 5th ed., New York: Marcel Dekker.

6.6.11 Polyvinylidene Chloride (Saran)

Saran is the trademark of Dow Chemical for their proprietary polyvinylidene chloride (PVDC) resin. It has a maximum allowable operating temperature of 175°F/81°C. It is used to line piping as well as vessels.

PVDC has found wide application in the plating industry and for handling deionized water, Pharmaceuticals, food processing, and other applications where stream purity protection is critical. The material complies with FDA regulations for food processing and potable water and also with regulations prescribed by the Meat Inspection Division of the Department of Agriculture for transporting fluids used in meat production. In applications such as plating solutions, chlorines, and certain other chemicals, Saran is superior to polypropylene and finds many applications in the handling of municipal water supplies and waste waters. Refer to Table 6.35 for the compatibility of PVDC with selected corrodents. Reference [1] provides a more comprehensive listing.

TABLE 6.35

Compatibility of Polyvinylidene Chloride (Saran) with Selected Corrodents

	Maximum Temperature	
Chemical	°F	°C
Acetaldehyde	150	66
Acetic acid, 10%	150	66
Acetic acid, 50%	130	54
Acetic acid, 80%	130	54
Acetic acid, glacial	140	60
Acetic anhydride	90	32
Acetone	90	32
Acetyl chloride	130	54
Acrylonitrile	90	32
Adipic acid	150	66
Allyl alcohol	80	27
Alum	180	82
Aluminum chloride, aqueous	150	66
Aluminum fluoride	150	66
Aluminum hydroxide	170	77
Aluminum nitrate	180	82
Aluminum oxychloride	140	60
Aluminum sulfate	180	82
Ammonia gas	X	X
Ammonium bifluoride	140	60
Ammonium carbonate	180	82
Ammonium chloride, sat.	160	71
Ammonium fluoride, 10%	90	32
Ammonium fluoride, 25%	90	32
Ammonium hydroxide, 25%	X	X
Ammonium hydroxide, sat.	X	X
Ammonium nitrate	120	49
Ammonium persulfate	90	32
Ammonium phosphate	150	66
Ammonium sulfate, 10–40%	120	49
Ammonium sulfide	80	27
Amyl acetate	120	49
Amyl alcohol	150	66
Amyl chloride	80	27
Aniline	X	X
Antimony trichloride	150	66
Aqua regia, 3:1	120	49
Barium carbonate	180	82
Barium chloride	180	82
Barium hydroxide	180	82
Barium sulfate	180	82
Barium sulfide	150	66
Benzaldehyde	X	X
Benzene	X	X
Benzene sulfonic acid, 10%	120	49
Benzoic acid	120	49
Benzyl chloride	80	27

(continued)

TABLE 6.35 *Continued*

Chemical	Maximum Temperature	
	°F	°C
Boric acid	170	77
Bromine liquid	X	X
Butadiene	X	X
Butyl acetate	120	49
Butyl alcohol	150	66
Butyl phthalate	180	82
Butyric acid	80	27
Calcium bisulfite	80	27
Calcium carbonate	180	82
Calcium chlorate	160	71
Calcium chloride	180	82
Calcium hydroxide, 10%	160	71
Calcium hydroxide, sat.	180	82
Calcium hypochlorite	120	49
Calcium nitrate	150	66
Calcium oxide	180	82
Calcium sulfate	180	82
Caprylic acid	90	32
Carbon bisulfide	90	32
Carbon dioxide, dry	180	82
Carbon dioxide, wet	80	27
Carbon disulfide	80	27
Carbon monoxide	180	82
Carbon tetrachloride	140	60
Carbonic acid	180	82
Cellosolve	80	27
Chloracetic acid	120	49
Chloracetic acid, 50% water	120	49
Chlorine gas, dry	80	27
Chlorine gas, wet	80	27
Chlorine, liquid	X	X
Chlorobenzene	80	27
Chloroform	X	X
Chlorosulfonic acid	X	X
Chromic acid, 10%	180	82
Chromic acid, 50%	180	82
Citric acid, conc.	180	82
Citric, acid, 15%	180	82
Copper carbonate	180	82
Copper chloride	180	82
Copper cyanide	130	54
Copper sulfate	180	82
Cresol	150	66
Cupric chloride, 5%	160	71
Cupric chloride, 50%	170	77
Cyclohexane	120	49
Cyclohexanol	90	32

(continued)

TABLE 6.35 *Continued*

Chemical	Maximum Temperature	
	°F	°C
Dichloroacetic acid	120	49
Dichloroethane (ethylene dichloride)	80	27
Ethylene glycol	180	82
Ferric chloride	140	60
Ferric chloride, 50% in water	140	60
Ferric nitrate, 10–50%	130	54
Ferrous chloride	130	54
Ferrous nitrate	80	27
Fluorine gas, dry	X	X
Fluorine gas, moist	X	X
Hydrobromic acid, dil.	120	49
Hydrobromic acid, 20%	120	49
Hydrobromic acid, 50%	130	54
Hydrochloric acid, 20%	180	82
Hydrochloric acid, 38%	180	82
Hydrocyanic acid, 10%	120	49
Hydrofluoric acid, 30%	160	71
Hydrofluoric acid, 100%	X	X
Hypochlorous acid	120	49
Ketones, general	90	32
Lactic acid, conc.	80	27
Magnesium chloride	180	82
Malic acid	80	27
Methyl chloride	80	27
Methyl ethyl ketone	X	X
Methyl isobutyl ketone	80	27
Muriatic acid	180	82
Nitric acid, 5%	90	32
Nitric acid, 20%	150	66
Nitric acid, 70%	X	X
Nitric acid, anhydrous	X	X
Oleum	X	X
Perchloric acid, 10%	130	54
Perchloric acid, 70%	120	49
Phenol	X	X
Phosphoric acid, 50–80%	130	54
Picric acid	120	49
Potassium bromide, 30%	110	43
Salicylic acid	130	54
Sodium carbonate	180	82
Sodium chloride	180	82
Sodium hydroxide, 0%	90	32
Sodium hydroxide, 50%	150	66
Sodium hydroxide, conc.	X	X
Sodium hypochlorite, 10%	130	54
Sodium hypochlorite, conc.	120	49
Sodium sulfide, to 50%	140	60

(continued)

TABLE 6.35 *Continued*

Chemical	Maximum Temperature	
	°F	°C
Stannic chloride	180	82
Stannous chloride	180	82
Sulfuric acid, 10%	120	49
Sulfuric acid, 50%	X	X
Sulfuric acid, 70%	X	X
Sulfuric acid, 90%	X	X
Sulfuric acid, 98%	X	X
Sulfuric acid, 100%	X	X
Sulfuric acid, fuming	X	X
Sulfurous acid	80	27
Thionyl chloride	X	X
Toluene	80	27
Trichloroacetic acid	80	27
Zinc chloride	170	77

The chemicals listed are in the pure state or in a saturated solution unless otherwise indicated. Compatibility is shown to the maximum allowable temperature for which data is available. Incompatibility is shown by an X. A blank space indicates that the data is unavailable.

Source: From P.A. Schweitzer. 2004. *Corrosion Resistance Tables*, Vols. 1–4, 5th ed., New York: Marcel Dekker.

Reference

1.   P.A. Schweitzer. 2004. *Corrosion Resistance Tables*, Vols. 1–4, 5th ed., New York: Marcel Dekker.

7

Coatings

7.1 Introduction to Coatings

Construction metals are selected because of their mechanical properties and machineability at a low price, although, they should also be corrosion resistant. Very seldom can these properties be met in one and the same material. This is where coatings come into play. By applying an appropriate coating, the base metal with the good mechanical properties can be utilized while the appropriate coating provides corrosion protection.

The majority of coatings are applied on external surfaces to protect the metal from natural atmospheric corrosion, and atmospheric pollution. On occasion, it may also be necessary to provide protection from accidental spills and/or splashes. In some instances, coatings are applied internally in vessels for corrosion resistance.

There are basically four different classes of coatings:

Organic	Inorganic	Conversion	Metallic[a]
Coal tars	Silicates	Anodizing	Galvanizing
Phenolics	Ceramics	Phosphating	Vacuum vapor deposition
Vinyls	Glass	Chromate	Electroplating
Acrylics		Molybdate	Diffusion
Epoxy			
Alkyds			
Urethanes			

[a] These are processes rather than individual coatings because many metals may be applied by each process. The process and item to be coated will determine which metal will be used.

7.1.1 Principles of Corrosion Protection

Most metals used for construction purposes are unsuitable in the atmosphere. These unstable metals are produced by reducing ores artificially; therefore, they will return to their original ores or to similar metallic compounds when exposed to the atmosphere. For example, metallic iron is oxidized to ferric oxyhydride in a thermodynamically stable state (iron in the higher level of free energy is changed to lepidocrocite, γ-FeOOH, in the lower level):

$$4Fe + 3O_2 + 2H_2O \rightarrow 4FeOOH$$

This reaction of a metal in a natural environment is called corrosion. By means of a coating a longer period of time is required for rust to form on the substrate, as shown in Figure 7.1. Therefore, it is important that the proper coating material be selected for application in the specific environment.

For a coating to be effective, it must isolate the base metal from the environment. The service life of a coating is dependent upon the thickness and the chemical properties of the coating layer. The latter determines the durability of a coating material in a specific environment that is the corrosion resistance of a metal coating or the stability of its organic and inorganic compounds. In order to be effective, the coating's durability must be greater than that of the base metal's, or it must be maintained by some means. In addition, a coating is often required to protect the base metal with its original pore and crack or with a defect that may have resulted from mechanical damage and/or pitting corrosion.

Coatings are classified according to the electrochemical principle that they operate on to provide protection. These categories are:

1. EMF control
2. Cathodic control protection
3. Anodic control protection
4. Mixed control protection
5. Resistance control protection

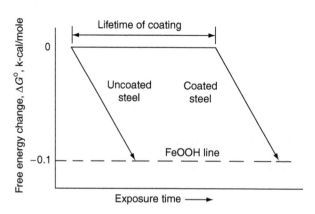

FIGURE 7.1
Role of corrosion resistant coating.

The mechanism of the corrosion cell can explain the theories that these five categories operate on.

7.1.2 Corrosion Cell

A corrosion cell is formed on a metal surface when oxygen and water are present. Refer to Figure 7.2. The electrochemical reactions taking place in the corrosion cell are:

Anodic reaction (M = metal):

$$M \rightarrow M^{n+} + ne \qquad (7.1)$$

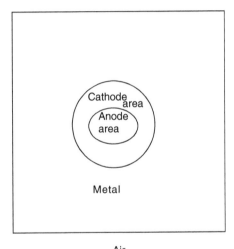

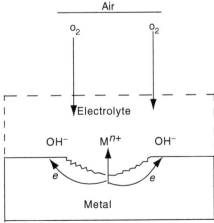

FIGURE 7.2
Structure of corrosion cell.

Cathodic reaction in acidic solution:

$$2H^+ + 2e \rightarrow H_2 \tag{7.2}$$

Cathodic reaction in neutral and alkaline solutions:

$$O_2 + 2H_2O + 4e \rightarrow 4OH^- \tag{7.3}$$

The Evans diagram shown in Figure 7.3 represents the mechanism of the corrosion cell. The cathodic current is expressed in the same direction as the anodic current.

In Figure 7.3, E_a shows the single potential for H_2/H^+ or for O_2/OH^- at the cathode and E_a shows the single potential for metal/metal in equilibria at the anode. The single potential is given by the Nernst equation:

$$E = E^\circ + \frac{RT}{nF} \ln a \tag{7.4}$$

where

$E =$ single potential
$E^\circ =$ standard single potential
$R =$ absolute temperature
$n =$ charge on an ion
$F =$ Faraday constant
$a =$ activity of the ion

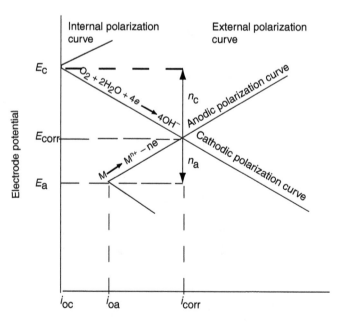

FIGURE 7.3
Mechanism of corrosion cell.

When $a=1$, E is equal to $E°$. The standard single potential $E°$ shows the degree of activity of the metal and gas.

The electrochemical series consists of the arrangement of metals in order of electrode potential. The more negative the single potential is, the more active the metal. Table 7.1 provides the standard single potentials of various metals and nonmetal reactants.

When the electromotive force (E_c-E_a) is supplied, the corrosion cell is formed with the current flowing between the anode and the cathode. The cathodic electrode potential is shifted to the less noble direction. The shifting of potentials is called cathodic and anodic polarization. The reaction rate curves $(E-i)$ are known as cathodic or anodic polarization curves. The corrosion potential E^c_{corr} and the corrosion current i_{corr} are given by the intersection of the cathodic and anodic polarization curves—indication that both electrodes react at the same rate in the corrosion process.

The polarization curves in the current density range greater than i_{corr} are called external polarization curves, and those in the current density range less than i_{corr} are called internal polarization curves. By sweeping the electrode potential from the corrosion potential to the cathodic or anodic

TABLE 7.1

Standard Single Potentials, $E°$ (V, SHE, 25°C)

Active		Inert	
Electrode	$E°$	**Electrode**	$E°$
Li/Li^+	-3.01	Mo/Mo^{3+}	-0.2
Rb/Rb^+	-0.298	Sn/Sn^{2+}	-0.140
Cs/Cs^+	-0.292	Pb/Pb^{2+}	-0.126
K/K^+	-2.92	H_2/H^+	±0
Ba/Ba^{2+}	-2.92	Bi/BiO	$+0.32$
Sr/Sr^{2+}	-2.89	Cu/Cu^{2+}	$+0.34$
Ca/Ca^+	-2.84	Rh/Rh^{2+}	$+0.6$
Na/Na^+	-2.71	Hg/Hg^+	$+0.798$
Mg/Mg^{2+}	-2.38	Ag/Ag^+	$+0.799$
Th/Th^{4+}	-2.10	Pd/Pd^{2+}	$+0.83$
Ti/Ti^{2+}	-1.75	Ir/Ir^{3+}	$+1.0$
Be/Be^{2+}	-1.70	Pt/Pt^{2+}	$+1.2$
Al/Al^{3+}	-1.66	Au/Au^{3+}	$+1.42$
V/V^{2+}	-1.5	Au/Au^+	$+1.7$
Mn/Mn^{2+}	-1.05		
Zn/Zn^{2+}	-0.763	O_2/OH	$+0.401$
Cr/Cr^{3+}	-0.71	I_2/I^-	$+0.536$
Fe/Fe^{2+}	-0.44	Br_2/Br^-	$+1.066$
Cd/Cd	-0.402	Cl_2/Cl^-	$+1.256$
In/In^{3+}	-0.34	F_2/F^-	$+2.85$
Ti/Ti^+	-0.335	S/S^{2-}	-0.51
Co/Co^{2+}	-0.27	Se/Se^{2+}	-0.78
Ni/Ni^{2+}	-0.23	Te/Te^{2+}	-0.92

direction, the external polarization curve can be determined. The internal polarization curve cannot be directly measured by the electrochemical technique because it is impossible to pick up the current separately from the anode and cathode that exist in the electrode. By analyzing the metallic ions dissolved and the oxidizer reaction, the internal polarization curves can be determined.

Anodic or cathodic overpotential is represented by the difference in potential between E_{corr} and E_a or E_{corr} and E_c and is expressed as n_a or n_c, where

$$n_a = E_{corr} - E_a \qquad n_a > 0 \tag{7.5}$$
$$n_c = E_{corr} - E_c \qquad n_c < 0. \tag{7.6}$$

The anodic and cathodic resistance is represented by n_a/i_{corr}

As soon as the cell circuit is formed, the corrosion reaction starts.

$$E_c - E_a = [n_c] - i_{corr}R, \tag{7.7}$$

where R is the resistance of the electrolyte between the anode and cathode. As the current passes through three processes (the anodic process, the cathodic process, and the transit process in the electrolytes), the electromotive force of a corrosion cell is dissipated.

When the electrode is polarized, the overpotential n is composed of the activation overpotential n^a and the concentration overpotential n^c.

$$n = n^a + n^c. \tag{7.8}$$

The activation overpotential n^a results from the potential energy barrier to be overcome for a charge to cross the double layer at the interface $(M = M^{n+} + ne)$ and is given as follows:

In the anodic reaction:

$$n_a^a = \beta_a \log \frac{i_a}{i_{oa}} \quad \text{(Tafel equation)} \tag{7.9}$$

$$\beta_a = 2.3 \frac{RT}{\alpha nF}. \tag{7.10}$$

In the cathodic reaction:

$$n_c^a = \beta_c \log \frac{i_c}{i_{oc}} \quad \text{(Tafel equation)} \tag{7.11}$$

$$\beta_c = \frac{2.3RT}{(1 - \alpha)nF}, \tag{7.12}$$

where

n_a^a = activation overpotential in the anodic reaction
n_c^a = activation overpotential in the cathodic reaction
β_a = anodic Tafel coefficient
β_c = cathodic Tafel coefficient
α = transfer coefficient
i_a = anodic current density
i_c = cathodic current density
i_{oa} = exchange current density of anode
i_{oc} = exchange current density of cathode

The degree of contribution of electrical energy for the activation energy in the electrode reaction $(0<\alpha<1)$ is indicated by the energy transfer factor α, that, in most cases, is in the range of 0.3–0.7. The exchange current density i_a or i_c is the flux of charge that passes through the electrical double layer at the single-equilibrium potential E_a or E_c. A linear relationship exists between n^a and $\log i_a$ or $\log i_c$. The Tafel coefficient β_a or β_c is the slope dn^a/d ($\log i_a$ or $\log i_c$) of the polarization curve. Therefore, β is one of the important factors controlling the corrosion rate.

The electrode reaction at the low reaction rate is controlled by the activation overpotential. One of the processes controlled by the activation overpotential is the cathodic reaction (in the acid solution):

$$2H^+ + 2e = H_2. \tag{7.13}$$

Table 7.2 shows the hydrogen overpotentials of various metals. The activation overpotential varies with the kind of metal and the electrolyte

TABLE 7.2

Hydrogen Overpotentials of Various Metals

Metal	Temperature (°C)	Solutions	Hydrogen Overpotential $|n°|$ (V/mA/ cm^{-2})	Tatel Coefficient $|\beta_c|$ (V)	Exchange Current Density $|i_{oc}|$ (A/cm^2)
Pt smooth	20	1 N HCl	0.00	0.03	10^{-3}
Mo	20	1 N HCl	0.12	0.04	10^{-6}
Au	20	1 N HCl	0.15	0.05	10^{-6}
Ag	20	0.1 N HCl	0.30	0.09	5×10^{-7}
Ni	20	0.1 N HCl	0.31	0.10	8×10^{-7}
Bi	20	1 N HCl	0.40	0.10	10^{-7}
Fe	16	1 N HCl	0.45	0.15	10^{-6}
Cu	20	0.1 N HCl	0.44	0.12	2×10^{-7}
Al	20	2 N H$_2$SO$_4$	0.70	0.10	10^{-10}
Sn	20	1 N HCl	0.75	0.15	10^{-8}
Cd	16	1 N HCl	0.80	0.20	10^{-7}
Zn	20	1 N H$_2$SO$_4$	0.94	0.12	1.6×10^{-11}
Pb	20	0.01–8 N HCl	1.16	0.12	2×10^{13}

condition. Metal dissolution and metal ion deposition are usually controlled by the activation overpotential.

The anodic overpotential is given by

$$n_a = \beta_a \log \frac{i_a}{i_{oa}}.$$ (7.14)

At high reaction rates, the concentration overpotential n^c becomes the controlling factor in the electrode reaction. In this case, the electrode reaction is controlled by mass transfer process that is the diffusion rate of reactive species. The diffusion current i is given as follows:

$$i = \frac{nFD(C - C_0)}{\delta},$$ (7.15)

where

$i =$ current density
$D =$ diffusion coefficient
$C =$ concentration of reactive species in the bulk solution
$C_0 =$ concentration of reactive species at the interface
$\delta =$ thickness of the diffusion layer.

When the concentration of the reactive species at the interface is zero, $C_0 = 0$, and the current density reaches a critical value, i_L called the limiting current density:

$$i_L = \frac{nFDC}{\delta}.$$ (7.16)

From Equation 7.15 and Equation 7.16,

$$\frac{C_0}{C} = 1 = \frac{i}{i_L}.$$ (7.17)

The concentration overpotential is given as follows:

$$n^c = \left[2.3 \frac{RT}{nF}\right] \log \left[\frac{C_0}{C}\right].$$ (7.18)

From Equation 7.17 and Equation 7.18,

$$n^c = \left[\frac{2.3RT}{nF}\right] \log \left[1 - \frac{i}{i_L}\right].$$ (7.19)

As seen in Equation 7.19, the concentration overpotential increases rapidly as i approaches i_L.

The cathodic reaction is controlled by the activation overpotential n_c^a and the concentration overpotential n_c^a.

The cathodic overpotential is

$$n_c = n_c^a + n_c^c. \tag{7.20}$$

The cathodic overpotential n_c can be written in the form:

$$n_c = \beta_c \log\left[\frac{i_c}{i_{oc}} + \frac{2.3RT}{nF}\right]\log\left[1 - \frac{i_c}{i_{cL}}\right]. \tag{7.21}$$

See Equation 7.12, Equation 7.19, and Equation 7.20. In most cases, the corrosion rate can be determined from the anodic and cathodic overpotentials since the rate determining process is determined by the slopes of the two polarization curves.

As previously mentioned, the role of the coating is to isolate the substrate from the environment. The coating accomplishes this based on two characteristics of the coating material: (1) the corrosion resistance of the coating material when the coating is formed by the defect free continuous layer and (2) the electrochemical action of the coating material when the coating layer has some defect such as a pore or crack. The mechanism of the corrosion cell can explain the action required of the coating layer. For better understanding, Equation 7.7 is rewritten as follows:

$$i_{corr} = \frac{(E_c - E_a) - |n_c|n_a}{R}. \tag{7.22}$$

A corrosion resistant coating is achieved by one of the five different methods to decrease i_{corr} based on Equation 7.22:

1. *EMF control protection*: Decrease in electromotive force $(E_c - E_a)$
2. *Cathodic control protection*: Increase in cathodic overpotential $|n_c|$
3. *Anodic control protection*: Increase in anodic overpotential $|n_a|$
4. *Mixed control protection*: Increase in both anodic overpotential $|n_a|$ and cathodic overpotential $|n_c|$
5. *Resistance control protection*: Increase in resistance of corrosion cell R

7.1.3 EMF Control Protection

The difference in potential between the anode and the cathode $(E_c - E_a)$ is the EMF of the corrosion cell. It is also the degree of thermodynamic instability of the surface metal for the environment. In other words, the less the EMF, the lower the corrosion rate. By covering the surface of the active metal with a continuous layer of a more stable metal, the active metal surface becomes more thermodynamically stable. Dissolved oxygen and hydrogen ions, that are the reactants in the cathodic reaction, are normal oxidizers found in natural environments. In the natural atmosphere, the single-potential of dissolved oxygen is nearly constant. Because of this, metals with more noble electrode potentials are used as coating materials. These include copper, silver, platinum, gold, and their alloys. A copper coating system provides excellent corrosion resistance under the condition that the defect-free continuous layer covers the surface of the iron substrate. In doing so, the EMF of the iron surface is decreased by the copper coating. The corrosion potential is changed from E_{corr} of uncoated iron to E_{corr} of copper by coating with copper. Under this condition, the iron corrodes at the low rate of i_{corr} of copper. However, if the iron substrate is exposed to the environment, as the result of mechanical damage, the substrate is corroded predominantly at the rate of i_{corr} of exposed copper by coupling iron and copper (galvanic corrosion).

Organic coatings and paints are also able to provide EMF control protection. Surface conditions are converted to more stable states by coating with organic compounds. These coatings delay the generation of electromotive force, causing the corrosion of the substrate.

How long an organic coating will be serviceable is dependent upon the durability of the coating and its adhesive ability on the base metal. The former is the stability of the coating layer as exposed to various environmental factors, and the latter is determined by the condition of the interface between the organic film and the substrate.

EMF can also be decreased by the use of glass lining, porcelain enameling, and temporary coating with greases and oils.

7.1.4 Cathodic Control Protection

Cathodic control protection protects the substrate by coating with a less noble metal, for which the slopes of the cathodic polarization curves are steep. The cathodic overpotential of the surface is increased by the coating; therefore, the corrosion potential becomes more negative than that of the substrate. Coating materials used for this purpose are zinc, aluminum, manganese, cadmium, and their alloys. The electrode potential of these metals are more negative than those of iron and steel. When exposed to the environment, these coatings act as sacrificial anodes for the iron and steel substrates.

The protective abilities of coatings are as follows:

1. Original barrier action of coating layer
2. Secondary barrier action of corrosion product layer
3. Galvanic action of coating layer as sacrificial anode

Barrier coatings 1 and 2 predominate as the protective ability even though a sacrificial metal coating is characterized by galvanic action.

Initially, the substrate is protected against corrosion by the barrier action of the coating, followed by the barrier action of the corrosion product layer.

Upon exposure to air, aluminum forms a chemically inert Al_2O_3 oxide film that is a rapidly forming self-healing film. Therefore, the passive film on aluminum, as well as the corrosion product layer, is a main barrier and leads to a resistant material in natural environments.

On the other hand, the surface oxide film that forms on zinc is not as an effective barrier as the oxide film on aluminum.

Upon exposure in the natural atmosphere, many corrosion cells are formed on the surface of a sacrificial metal coating, thereby, accelerating the corrosion rate. During this period, corrosion products are gradually formed and converted to a stable layer. This period may last for several months, after this period, the corrosion rate becomes constant. These corrosion products form the second barrier and are amorphous Al_2O_3 on aluminum, and $Zn(OH)_2$ and basic zinc salts on zinc. ZnO, being electrically conductive, loosens the corrosion product layer and, therefore, does not contribute to formation of the barrier. When materials such as CO_2, NaCl, and SO_x are present, basic zinc salts are formed, for example, $2ZnCO_3 \cdot 3Zn(OH)_2$ in mild atmospheres, $ZnCl_2 \cdot 6Zn(OH)_2$ in chloride atmospheres, and $ZnSO_4 \cdot 4Zn(OH)_2$ in SO_x atmospheres. How stable each basic zinc salt will be is dependent on the pH and anion concentration of the electrolyte on the zinc. Zinc carbonate forms an effective barrier on steel in mild atmospheres, whereas, basic zinc sulfate and chloride dissolve with decreasing pH of the electrolyte. The basic zinc sulfate is restricted under atmospheric conditions in its effort to act as a barrier since the pH value of rain in an SO_x atmosphere is usually low, in the area of less than 5. In a chloride environment, the value of pH in the electrolyte is not as low as in the SO_x atmosphere; therefore, a secondary barrier will form. However, in a severe chloride environment, the zinc coating layer will corrode in spite of the existence of basic zinc chloride on the surface.

7.1.4.1 Galvanic Action of Coating Layer

Sacrificial metal coatings protect the substrate metal by means of galvanic action. When the base metal is exposed to the atmosphere as a result of mechanical damage or the like, the exposed portion of the base metal is polarized cathodically to the corrosion potential of the coating layer. As a result, little corrosion takes place on the exposed metal. A galvanic couple is formed between the exposed part of the base metal and the surrounding coating metal. Sacrificial metals are more negative in electrochemical potential than other metals such as iron or steel. Therefore, the sacrificial metal coating acts as a cathode. This type of reaction of sacrificial metal coatings is known as galvanic or cathodic protection. In addition, the defects are protected by a second barrier of corrosion products of the coating layer. Figure 7.4 schematically illustrates the galvanic action of a sacrificial metal coating.

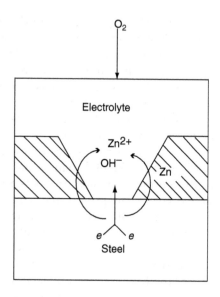

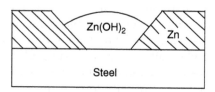

FIGURE 7.4
Schematic illustration of galvanic action of sacrificial metal coating.

7.1.5   Anodic Control Protection

Noble metal coatings provide anodic control protection. They are usually used where corrosion protection and decorative appearance are required. Nickel, chromium, tin, lead, and their alloys are the coating metals that provide anodic protection.

7.1.5.1   *Single Layer Coatings*

Single layer metal coatings provide corrosion protection as a result of the original barrier action of the noble metal. With the exception of lead, a second barrier of corrosion products is not formed. Noble metals do not provide cathodic protection to steel substances in natural atmospheres because the corrosion potential of the noble metal is more noble than that of the steel. Refer to Table 7.3.

TABLE 7.3

Corrosion Potentials of Noble Metals

pH	Corrosion Potentials (V, SCE)	
	2.9	6.5
Chromium	−0.119	−0.186
Nickel	−0.412	−0.430
Tin	−0.486	−0.554
Lead	−0.435	−0.637
Steel	−0.636	−0.505

The service life of a single layer coating is affected by any discontinuity in the coating such as that caused by pores and cracks. For metals to form a protective barrier, the coating thickness must be greater than 30 µm to ensure absence of defects. The surface of a bright nickel coating will remain bright in a clean atmosphere but will change to a dull color when exposed in an SO_x atmosphere.

Chromium coatings are applied as a thin layer to maintain a bright, tarnish-free surface. Cracking of chromium coatings begins at a thickness of 0.5 µm after which a network of fine cracks is formed.

For protection of steel in an SO_x atmosphere, lead and its alloys (5–10% tin) coatings are employed. Pitting will occur in the lead coating at the time of initial exposure, but the pits are self-healing, and the lead surface is protected by the formation of insoluble lead sulfate.

7.1.5.2 Multilayer Coatings

There are three types of nickel coatings: bright nickel, semibright, and dull. The difference is the quantity of sulfur contained in them as shown below:

Bright nickel deposits	>0.04% Sulfur
Semibright nickel deposits	<0.005% Sulfur
Dull nickel deposits	<0.001% Sulfur

The corrosion potentials of the nickel deposits are dependent on the sulfur content. Figure 7.5 shows the effect of sulfur content on the corrosion potential of a nickel deposit.

As the sulfur content increases, the corrosion potential of a nickel deposit becomes more negative. A bright nickel coating is less protective than a semibright or dull nickel coating. The difference in potential of bright nickel and semibright deposits is more than 50 mV.

Use is made in the differences in the potential in the application of multilayer coatings. The more negative bright nickel deposits are used as sacrificial intermediate layers. When bright nickel is used as an intermediate

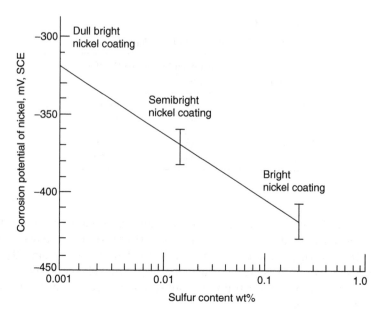

FIGURE 7.5
Effect of sulfur content on the corrosion potential of nickel deposit.

layer, the corrosion behavior is characterized by a sideways diversion. Pitting corrosion is diverted laterally when it reaches the more noble semibright deposit. Thus, the behavior of bright nickel prolongs the time for pitting penetration to reach the base metal.

The most negative of all nickel deposits is trinickel. In the triplex layer coating system, a coating of trinickel approximately 1 μm thick, containing 0.1–0.25% sulfur is applied between bright nickel and semibright nickel deposits. The high sulfur nickel layer dissolves preferentially, even when pitting corrosion reaches the surface of the semibright deposit. Since the high sulfur layer reacts with the bright nickel layer, pitting corrosion does not penetrate the high sulfur nickel layer in the tunneling form. The application of a high sulfur nickel strike definitely improves the protective ability of a multistage nickel coating.

7.1.6 Resistance Control Protection

Resistance control protection is achieved by using organic compounds, such as some paints, as coating materials. The coating layer delays the transit of ions to the substrate thereby, inhibiting the formation of corrosion cells. Figure 7.6 illustrates the principles of resistance control protection by an organic coating. The corrosion rate of iron is inhibited by the coating from i_{corr} of uncoated iron to i_{corr} of coated iron.

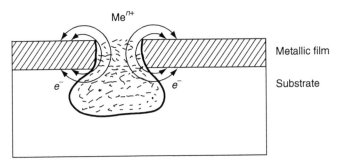

FIGURE 7.6
Dissolution of substrate in coating metal defect.

7.2 Metallic Coatings

Metallic coatings are applied to metal substrates for several purposes. Typical purposes include improved corrosion resistance, wear resistance, and appearance. Of primary concern is corrosion resistance.

By providing a barrier between the substrate and the environment or by cathodically protecting the substrate, metallic coatings protect the substrate from corrosion. Coatings of chromium, copper, and nickel provide increased wear resistance and good corrosion resistance. However, these noble metals make the combination of the substrate (mostly steel or an aluminum alloy) with the protective layer sensitive to galvanically induced local corrosion. Nonnoble metallic layers such as zinc or cadmium provide good cathodic protection but show poor wear resistance.

A coating of a corrosion resistant metal on a corrosion prone substrate can be formed by various methods. The choice of coating material and selection of an application method are determined by the end use.

7.2.1 Methods of Producing Coatings

7.2.1.1 Electroplating

This is one of the most versatile methods. The metal to be coated is made the cathode in an electrolytic cell. A potential is applied between the cathode that the plating occurs on and the anode that may be the same metal or an inert material such as graphite. This method can be used for all metals that can be electrolytically reduced from the ionic state to the metallic state when present in an electrolyte. Aluminum, titanium, sodium, magnesium, and calcium cannot be electrodeposited from aqueous solution because the competing cathodic reaction, $2H^+ + 2e^- = H_2$, is strongly thermodynamically favored and takes place in preference to the reduction of the metal ion. These metals can be electrodeposited from conducting organic solutions or molten salt solutions in which the H^+ ion concentration is negligible.

Many alloys may be electrodeposited, including copper–zinc, copper–tin, lead–tin, cobalt–tin, nickel–cobalt, nickel–iron, and nickel–tin. The copper–zinc alloys are used to coat steel wire used in tire-cord. Lead–tin alloys are known as terneplate and have many corrosion resistant applications.

The thickness of coating can be accurately controlled since the amount deposited is a function of the number of coulombs passed.

7.2.1.2 Electroless Plating

This method is also known as chemical plating or immersion plating. It is based on the formation of metal coatings, resulting from chemical reduction of metal ions from solution. The surface that is to be coated must be catalytically active and remain catalytically active as the deposition proceeds in a solution that must contain a reducing agent. If the catalyst is a reduction product (metal), autocatalysis is ensured. In this case, it is possible to deposit a coating, in principle, of unlimited thickness. The advantages of electroless plating are:

1. Deposits have fewer pinholes.
2. Electric power supply is not required.
3. Nonconductive materials are metallized.
4. A functional layer is deposited.
5. A uniform layer is deposited, even on complex parts.
6. The equipment for electroless plating is simple.

Electroless plating is limited by the fact that:

1. It is more expensive than electroplating because the reducing agents cost more than an equivalent amount of electricity.
2. It is less intensive because the metal deposition rate is limited by metal ion reduction in the bulk of the solution.

Copper, silver, cobalt, and palladium are the most commonly plated metals using this process. The silvering of mirrors falls into this category.

Hypophosphite, amine boranes, formaldehyde, borohydride, and hydrozine are typical reducing agents. Deposits of nickel formed with hypophosphite as a reducing agent contain phosphorus. This alloying constituent determines many of the properties. Table 7.4 shows coatings obtained by electroless plating.

7.2.1.3 Electrophoretic Deposition

Finely divided materials suspended in an electrolyte develop a charge as a result of asymmetry in the charge distribution caused by the selective

TABLE 7.4

Coatings Obtained by Electroless Plating

	Reducing Agent						
Metal	$H_2PO_4^-$	N_2H_2	CH_2O	BH_4^-	RBH_3	Me ions	Others
Ni	Ni–P	Ni		Ni–B	Ni–B		
Co	Co–P	Co	Co	Co–B	Co–B		
Fe				Fe–B			
Cu	Cu	Cu	Cu	Cu	Cu	Cu	
Ag		Ag	Ag	Ag	Ag	Ag	Ag
Au		Au	Au	Au	Au		Au
Pd	Pd–P	Pd	Pd	Pd–B	Pd–B		
Rh		Rh					Rh
Ru			Ru				
Pt		Pt	Pt				Pt
Sn						Sn	
Pb			Pb				

adsorption of one of the constituent ions. When the substrate metal is immersed in the electrolyte, and a potential is applied, a coating will be formed. If the particles have a negative charge, they will be deposited on the anode, and if they have a positive charge, they will be deposited on the cathode. Commercial applications of this method in the case of metals is limited.

7.2.1.4 Cathodic Sputtering

This method is carried out under a partial vacuum. The substrate to be coated is attached to the anode. Argon, or a similar inert gas, is admitted at low pressure. A discharge is initiated, and the positively charged gas ions are attracted to the cathode. Atoms are dislodged from the cathode as the gas ions collide with the cathode. These atoms are attracted to the anode and coat the substrate. This method can be used for nonconducting as well as conducting materials. The major disadvantages are the heating of the substrate and low deposition rates.

Some of the most commonly used metals deposited by sputtering are aluminum, copper, chromium, gold, molybdenum, nickel, platinum, silver, tantalum, titanium, tungsten, vanadium, and zirconium.

Sputtered coatings are used for a wide variety of applications. Some examples are:

1. Metals and alloys are used as conductors, contacts, and resistors and in other components such as capacitors.
2. Some high performance magnetic data storage media are deposited via sputtering.

3. Thin metal and dielectric coatings are used to construct mirrors, antireflection coatings, light valves, laser optics, and lens coatings.

4. Hard coatings such as titanium carbide, nitride, and carbon produce wear-resistant coatings for cutting tools.

5. Thin film coatings can be used to provide high temperature environmental corrosion resistance for aerospace and engine parts, gas barrier layers, and lightweight battery components.

6. Titanium nitride is deposited on watch bands and jewelry as a hard gold colored coating.

7.2.1.5 Diffusion Coating

This method requires a preliminary step followed by thermal treatment and diffusion of the coating metal into the substrate. A commercial material known as galvannealed steel is made by coating steel with zinc followed by heat treatment and the formation of an iron–zinc intermetallic coating by diffusion.

7.2.1.6 Flame Spraying

The coating metal is melted and kept in the molten condition until it strikes the substrate to be coated. Aluminum and zinc are applied in this manner. Flame sprayed aluminum has a lower density than pure aluminum because of voids in the coating.

7.2.1.7 Plasma Spraying

This method is similar to flame spraying except that forms of heating other than a flame are used.

7.2.1.8 Hot Dipping

Zinc coatings are applied to steel sheets by immersion of the steel in a molten bath of zinc to form galvanized steel. A small amount of aluminum is added to the bath to establish a good bond at the zinc–steel interface. The thickness of the zinc coating is controlled by rigid control of the galvanizing bath temperature, the speed of passage through the bath, the temperature of the steel sheet before it enters the bath, and the use of air jets that exert a wiping action as the steel sheet exits the bath.

7.2.1.9 Vacuum and Vapor Deposition

This method is used primarily for the formation of metallic coatings on nonconductive substrates. Common deposited coatings using this method include aluminum coatings on plastics and rhodium coatings on mirrors.

7.2.1.10 Gas Plating

Some metal compounds can be decomposed by heat to form the metal. Typical examples are metal carbonyls, metal halides, and metal methyl compounds.

7.2.1.11 Fusion Bonding

Coatings of low melting metals such as tin, lead, zinc, and aluminum may be applied by cementing the metal as a powder on the substrate then heating the substrate to a temperature above the melting point of the coating metal.

7.2.1.12 Explosion Bonding

This method produces a bond between two metals by the exertion of a strong force that compresses the two metals sufficiently to develop a strong interfacial interaction.

7.2.1.13 Metal Cladding

The most common method is roll bonding that produces full-sized sheets of clad (coated) material. The bond formed is partly mechanical and partly metallurgical; consequently, metallurgically incompatible materials cannot be produced.

7.3 Noble Coatings

Because of the high corrosion resistance of the noble metals, these materials are used where a high degree of corrosion resistance and decorative appearance are requirements. They find application in domestic appliances, window frames, bicycles. motorbikes, parts for car bodies, furniture, tools, flanges, hydraulic cylinders, shock absorbers in cars, and parts of equipment for the chemical and food processing industries.

Noble metal coatings protect substrates from corrosion by means of anodic control or EMF control. Coating metals that provide protection by means of anodic control include nickel, chromium, tin, lead, and their alloys.

They can protect the substrate metal as a result of their resistance to corrosion insofar as they form a well-adhering and nonporous barrier layer. However, when the coating is damaged, galvanically induced corrosion will lead to severe attack. This corrosion process is extremely fast for coated systems because of the high current density in the defect as a result of the large ratio between the surface areas of the cathodic outer surface and the anodic defect, as shown in Figure 7.7. To compensate for these defects in the coating, multilayer coating systems have been developed. The corrosion resistance of a single layer noble metal coating results from the original barrier action of the noble metal, the surface of the noble metal being

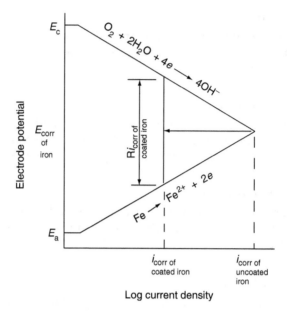

FIGURE 7.7
Resistance control protection. (From I. Suzuki. 1989. *Corrosion Resistant Coatings Technology,* New York: Marcel Dekker.)

passivated. With the exception of lead, a secondary barrier of corrosion products is not formed. Noble metals do not provide cathodic protection for the steel substrate because their corrosion potential is more noble than those of iron and steel in a natural environment. Refer to Table 7.5. In multilayer coating systems, a small difference in potential between coating layers results in galvanic action in coating layers.

Noble coating metals that provide corrosion protection by means of EMF control include copper, silver, platinum, gold, and their alloys. The standard single potentials of these metals are more noble than those of hydrogen (refer to Table 7.6). Therefore, the oxidizer in corrosion cells formed on these metals in a natural environment, containing no other particular oxidizers,

TABLE 7.5

Corrosion Potential of Noble Metals

	Corrosion Potentials (V, SCE)	
pH	2.9	6.5
Chromium	-0.119	-0.186
Nickel	-0.412	-0.430
Tin	-0.486	-0.554
Lead	-0.435	-0.637
Steel	-0.636	-0.505

TABLE 7.6

Standard Single Potentials,
$E°$ (V, SHE, 25°C)

Inert	
Electrode	$E°$
H_2/H^+	±0
Cu/Cu^{2+}	+0.34
Cu/Cu^+	+0.52
Ag/Ag^+	+0.799
Pt/Pt^{2+}	+1.2
Au/Au^{3+}	+1.42
Au/Au^+	+1.7

is dissolved oxygen. Consequently, the electromotive force that causes corrosion is so small that coating with noble metals is an effective means of providing corrosion protection. With the exception of copper, the other members of this group are precious metal and are primarily used for electrical conduction and decorative appearance.

7.3.1 Nickel Coatings

There are three types of nickel coatings: bright, semibright, and dull bright. The difference between the coatings is in the quantity of sulfur contained in them as shown below:

Bright nickel deposits	>0.04% Sulfur
Semibright nickel deposits	<0.005% Sulfur
Dull bright nickel deposits	<0.001% Sulfur

The corrosion potentials of the nickel deposits are dependent on the sulfur content. Figure 7.8 shows the effect of sulfur content on the corrosion potential of a nickel deposit. A single layer nickel coating must be greater than 30 μm to ensure absence of defects.

As the sulfur content increases, the corrosion potential of a nickel deposit becomes more negative. A bright nickel coating is less protective than a semi-bright or dull nickel coating. The difference in the potential of bright nickel and semibright nickel deposits is more than 50 mV.

Use is made of the differences in the potential in the application of multilayer coatings. The more negative bright nickel deposits are used as sacrificial intermediate layers. When bright nickel is used as an intermediate layer, the corrosion behavior is characterized by a sideways diversion. Pitting corrosion is laterally diverted when it reaches the more noble semi-bright nickel deposit. Thus, the corrosion behavior of bright nickel prolongs the time for pitting penetration to reach the base metal.

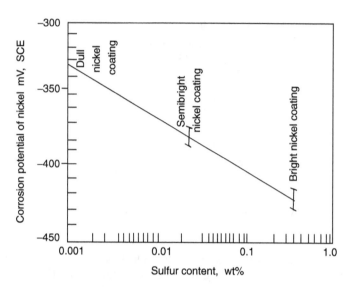

FIGURE 7.8
Effect of sulfur content on corrosion protection of nickel.

The most negative of all nickel deposits is trinickel. In the triplex layer coating system, a coating of trinickel approximately 1 μm thick, containing 0.1–0.25% sulfur, is applied between bright nickel and semibright nickel deposits. The high sulfur nickel layer dissolves preferentially, even when pitting corrosion reaches the surface of the semibright nickel deposit. Because the high sulfur layer reacts with the bright nickel layer, pitting corrosion does not penetrate the high sulfur nickel layer in the tunneling form. The application of a high sulfur nickel strike definitely improves the protective ability of a multilayer nickel coating.

In the duplex nickel coating system, the thickness ratio of semibright nickel deposit to bright nickel deposit is nominally 3:1, and a thickness of 20–25 μm is required to provide high corrosion resistance. The properties required for a semibright nickel deposit are as follows:

1. The deposit contains little sulfur.
2. Internal stress must be slight.
3. Surface appearance is semibright and extremely level.

For a trinickel (high sulfur) strike, the following properties are required:

1. The deposit contains a stable 0.1–0.25% sulfur.
2. The deposit provides good adhesion for semibright nickel deposits.

Nickel coatings can be applied by electrodeposition or electrolessly from an aqueous solution without the use of an externally applied current.

Depending on the production facilities and the electrolyte composition, electrodeposited nickel can be relatively hard (120–400 HV). Despite competition from hard chromium and electroless nickel, electrodeposited nickel is still being used as an engineering coating because of its relatively low price. Some of its properties are:

1. Good general corrosion resistance
2. Good protection from fretting corrosion
3. Good machineability
4. The ability of layers of 50–75 μm to prevent scaling at high temperatures
5. Mechanical properties, including the internal stress and hardness, that are variable and that can be fixed by selecting the manufacturing parameters
6. Excellent combination with chromium layers
7. A certain porosity
8. A tendency for layer thicknesses below 10–20 μm on steel to give corrosion spots because of porosity

The electrodeposition can be either directly on steel or over an intermediate coating of copper. Copper is used as an underlayment to facilitate buffing, because it is softer than steel, and to increase the required coating thickness with a material less expensive than nickel.

The most popular electroless nickel plating process is the one where hypophosphite is used as the reducer. Autocatalytic nickel ion reduction by hypophosphite takes place in both acid and alkaline solutions. In a stable solution with a high coating quality, the deposition rate may be as high as 20–25 μm/h. However, a relatively high temperature of 194°F/90°C is required. Because hydrogen ions are formed in the reduction reaction,

$$Ni^{2+} + 2H_2PO_2^- + 2H_2O \rightarrow Ni + 2H_2 + 2H^-,$$

a high buffering capacity of the solution is necessary to ensure a steady state process. For this reason, acetate, citrate, propionate, glycolate, lactate, or aminoacetate is added to the solutions. These substances along with buffering may form complexes with nickel ions. Binding Ni^{2+} ions into a complex is required in alkaline solutions (here ammonia and pyrophosphate may be added in addition to citrate and amino-acetate). In addition, such binding is desirable in acid solutions because free nickel ions form a compound with the reaction product (phosphate) that precipitates and prevents further use of the solution.

When hypophosphite is used as the reducing agent, phosphorus will be present in the coating. Its amount (in the range of 2–15 mass percent) depends on pH, buffering capacity, ligands, and other parameters of electroless solutions.

Borohydride and its derivatives may also be used as reducing agents. When borohydride is used in the reduction, temperatures of 140°F/60°C to 194°F/90°C are required. The use of dimethylaminoborane (DMAB) enables the deposition of Ni–B coatings with a small amount of boron (0.5–1.0 mass percent) at temperatures in the range of 86°F/30°C to 104°F/60°C. Neutral and alkaline solutions may be used.

Depending upon exposure conditions, certain minimum coating thicknesses to control porosity are recommended for the coating to maintain its appearance and have a satisfactory life:

Indoor exposures	0.3–0.5 mil (0.008–0.013 mm)
Outdoor exposures	0.5–1.5 mil (0.013–0.04 mm)
Chemical industry	1–10 mil (0.025–0.25 mm)

For applications near the seacoast, thicknesses in the range of 1.5 mil (0.04 mm) should be considered. This also applies to automobile bumpers and applications in general industrial atmospheres.

Nickel is sensitive to attack by industrial atmospheres and forms a film of basic nickel sulfate that causes the surface to "fog" or lose its brightness. To overcome this fogging, a thin coating of chromium (0.01–0.03 mil/0.003–0.0007 mm) is electrodeposited over the nickel. This finish is applied to all materials for which continued brightness is desired.

Single-layer coatings of nickel exhibit less corrosion resistance than multilayer coatings because of their discontinuities. The electroless plating process produces a coating with fewer discontinuous deposits. Therefore, the single layer deposited by electroless plating provides more corrosion resistance than does an electroplated single layer.

Most electroless plated nickel deposits contain phosphorus that enhances corrosion resistance. In the same manner, an electroplated nickel deposit containing phosphorus will also be more protective.

7.3.2 Satin-Finish Nickel Coating

A satin-finish nickel coating consists of nonconductive materials such as aluminum oxide, kaolin, and quartz that are codeposited with chromium on the nickel deposit. Some particles are exposed on the surface of the chromium deposit, so the deposit has a rough surface. Because the reflectance of the deposit is decreased to less than half of that of a level surface, the surface appearance looks like satin.

A satin finish nickel coating provides good corrosion resistance because of the discontinuity of the top coat of chromium.

7.3.3 Nickel–Iron Alloy Coating

To reduce production costs of bright nickel, the nickel–iron alloy coating was developed. The nickel–iron alloy deposits full brightness, high leveling, and excellent ductility and good reception for chromium.

This coating has the disadvantage of forming red rust when immersed in water; consequently, nickel–iron alloy coating is suitable for use in mild atmospheres only. Typical applications include kitchenware and tubular furniture.

7.3.4 Chromium Coatings

In the northern parts of the United States, immediately after World War II, it was not unusual for the chromium plated bumpers on the most expensive cars to show severe signs of rust within a few months of winter exposure. This was partially the result of trying to extend the short supply of strategic metals by economizing on the amount used. However, the more basic reason was the lack of sufficient knowledge of the corrosion process in order to control the attack by the atmosphere. Consequently, an aggressive industrial program was undertaken to obtain a better understanding of the corrosion process and ways to control it.

Chromium-plated parts on automobiles consist of steel substrates with an intermediate layer of nickel, or in some cases, layered deposits of copper and nickel. The thin chromium deposit provides bright appearance and stain free surface while the nickel layer provides the corrosion protection to the steel substrate. With this system, it is essential that the nickel completely cover the steel substrate because the iron will be the anode and nickel the cathode. Any breaks or pores in the coating will result in the condition shown in Figure 7.9. This figure illustrates the reason for the corrosion of chrome trim on automobiles after World War II.

The corrosion problem was made worse by the fact that addition agents used in the plating bath resulted in a bright deposit. Bright deposits contain sulfur that makes the nickel more active from a corrosion standpoint. From a corrosion standpoint, this is discouraging. However, it occurred to the investigators that this apparent disadvantage of bright nickel could be put to good use.

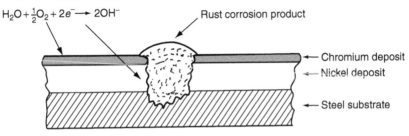

Cathodic reaction takes place on chromium or nickel

$$H_2O + \tfrac{1}{2}O_2 + 2e^- \longrightarrow 2OH^-$$ Rust corrosion product

← Chromium deposit
← Nickel deposit

← Steel substrate

Anodic reaction takes place on iron exposed through coating
$$Fe - 2e^- \longrightarrow Fe^{++}$$

FIGURE 7.9
Corrosion of steel at breaks in a nickel/chromium coating when exposed to the atmosphere.

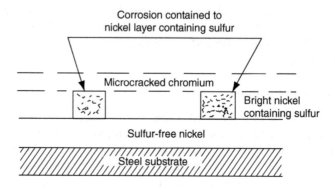

FIGURE 7.10
Duplex nickel electrode deposit to prevent corrosion of steel substrate.

To solve this problem, a duplex nickel coating was developed as shown in Figure 7.10. An initial layer of sulfur-free nickel is applied to the steel substrate, followed by an inner layer of a bright nickel containing sulfur, with an outer layer of microcracked chromium.

Any corrosion that takes place is limited to the bright nickel layer containing sulfur. The corrosion laterally spreads between the chromium and sulfur free nickel deposits because the outer members of this sandwich (chromium and sulfur free nickel) are cathodic to the sulfur containing nickel.

A potential problem that could result from this system of corrosion control would be the undermining of the chromium and the possibility that the brittle chromium deposits could flake off the surface. This potential problem was prevented by the development of a microcracked or microporous chromium coating. These coatings contain microcracks or micropores that do not detract from the bright appearance of the chromium. They are uniformly formed over the exterior of the plated material and serve to distribute the corrosion process over the entire surface. The result has been to extend the life of chromium-plated steel exposed to outdoor atmospheric conditions.

Microcracked chromium coatings are produced by first depositing a high stress nickel strike on a sulfur free nickel layer and then a decorative chromium deposit. The uniform crack network results from the interaction of the thin chromium layer and the high stress nickel deposit. The result is a mirror surface as well as a decorative chromium coating.

Microporous chromium coatings are produced by first electroplating a bright nickel layer containing suspended nonconductive fine particles. Over this, a chromium layer is deposited which results in a mirror finish. As the chromium thickness increases the number of pores decrease. For a chromium deposit of $0.25\ \mu m$ thickness, a porosity of more than $10,000\ \text{pores/cm}^2$ are required. A porosity of $40,000\ \text{pores/cm}^2$ provides the best corrosion resistance.

Hard (engineering) chromium layers are also deposited directly on a variety of metals. The purpose in applying these layers is to obtain wear resistant surfaces with a high hardness or to restore original dimensions to a workpiece. In addition, the excellent corrosion resistance resulting from these layers make them suitable for outdoor applications.

Thick chromium deposits have high residual internal stress and may be brittle because of the electrodeposition process that hydrogen can be incorporated in the deposited layer. Cracks result during plating when the stress exceeds the tensile stress of the chromium. As plating continues, some cracks are filled. This led to the development of controlled cracking patterns that produce wet-table porous surfaces that can spread oil that is important for engine cylinders, liners, etc.

Some of the properties of engineering chromium layers are:

1. Excellent corrosion resistance
2. Wear resistance
3. Hardness up to 950 HV
4. Controlled porosity is possible

7.3.5 Tin Coatings (Tinplate)

Tinplate is mainly produced by the electroplating process. Alkaline and acid baths are used in the production line. The acid baths are classified as either ferrostan or halogen baths.

A thermal treatment above the melting point of tin follows the electrolytic deposition. The intermetallic compound $FeSn_2$ forms at the interface between the iron and tin during this thermal processing. The corrosion behavior of the tinplate is determined by the quality of the $FeSn_2$ formed, particularly when the amount of the free tin is small. The best performing tinplate is one where the $FeSn_2$ uniformly covers the steel so that the area of iron exposed is very small in case the tin should dissolve. Good coverage requires good and uniform nucleation of $FeSn_2$. Many nuclei form when electrodeposition of tin is carried out from the alkaline stannate bath.

Compared to either iron or tin, $FeSn_2$ is chemically inert in all but the strongest oxidizing environments.

Most of the tinplate (tin coating on steel) produced is used for the manufacture of food containers (tin cans). The nontoxic nature of tin salts makes tin an ideal material for the handling of foods and beverages.

An inspection of the galvanic series will indicate that tin is more noble than steel and, consequently, the steel would corrode at the base of the pores. On the outside of a tinned container this is what happens—the tin is cathodic to the steel. However, on the inside of the container there is a reversal of polarity because of the complexing of the stannous ions by many food products. This greatly reduces the activity of the stannous ions, resulting in a change in the potential of tin in the active direction.

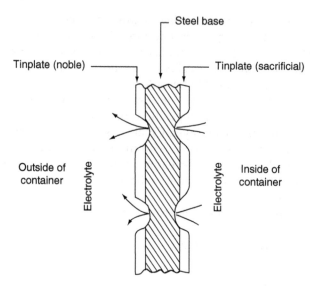

FIGURE 7.11
Tin acting as both a noble and sacrificial coating.

This change in polarity is absolutely necessary because most tin coatings are thin and, therefore, porous. To avoid perforation of the can, the tin must act as a sacrificial coating. Figure 7.11 illustrates this reversal of activity between the outside and inside of the can.

The environment inside a hermetically sealed can varies depending upon the contents that include general foods, beverages, oils, aerosol products, liquid gases, etc. For example, pH values vary for different contents as shown below:

Acidic beverage	2.4–4.5
Beer and wine	3.5–4.5
Meat, fish, marine products, and vegetables	4.1–7.4
Fruit juices, fruit products	3.1–4.3
Nonfood products	1.2–1.5

The cans' interiors are subject to general corrosion, localized corrosion, and discoloring. The coating system for tinplate consists of tin oxide, metallic tin, and alloy. The dissolution of the tin layer in acid fruit products is caused by acids such as citric acid. In acid fruit products, the potential reversal occurs between the tin layer and the steel substrate as shown in Figure 7.12. The potential reversal of a tin layer for steel substrate occurs at a pH range <3.8 in a citric acid solution. This phenomenon results from the potential shift of the tin layer to a more negative direction. Namely, the activity of the stannous ion, Sn^{2+}, is reduced by the formation of soluble tin complexes, and, thereby, the corrosion potential of the tin layer becomes

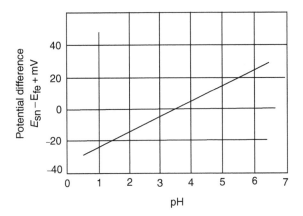

FIGURE 7.12
Potential reversal in tinplate.

more negative than that of steel. Thus, the tin layer acts as a sacrificial anode for steel so that the thickness and density of the pores in the tin layer are important factors affecting the service life of the coating. A thicker tin layer prolongs the service life of a tin can. The function of the alloy layer (FeSn) is to reduce the active area of steel by covering it since it is inert in acid fruit products. When some parts of the steel substrate are exposed, the corrosion of the tin layer is accelerated by galvanic coupling with the steel. The corrosion potential of the alloy layer is between that of the tin layer and that of the steel. A less defective layer exhibits potential closer to that of the tin layer. Therefore, the covering with alloy layer is important to decrease the dissolution of the tin layer.

In carbonated beverages, the potential reversal does not take place; therefore, the steel preferentially dissolves at the defects in the tin layer. Under such conditions, pitting corrosion sometimes results in perforation. Consequently, except for fruit cans, almost all tinplate cans are lacquered.

When tinplate is to be used for structural purposes such as roofs, an alloy of 12–25 parts of tin to 88–75 parts of lead is frequently used. This is called terneplate. It is less expensive and more resistant to the weather than a pure tin coating. Terneplate is used for fuel tanks of automobiles and is also used in the manufacture of fuel lines, brake lines, and radiators in automobiles.

7.3.6 Lead Coatings

Coatings of lead and its alloy (5–10% Sn) protect steel substrate especially in industrial areas having an SO_x atmosphere. At the time of initial exposure, pitting occurs on the lead surface; however, the pits are self-healed and then the lead surface is protected by the formation of insoluble lead sulfate. Little protection is provided by these coatings when in contact with the soil.

Lead coatings are usually applied by either hot dipping or by electro-deposition. When the coating is to be applied by hot dipping, a small percentage of tin is added to improve the adhesion to the steel plate. If 25% or more of tin is added, the resulting coating is termed terneplate.

Caution: Do not use lead coatings where they will come into contact with drinking water or food products. Lead salts can be formed that are poisonous.

7.3.7 Terneplate

Terneplate is a tin-lead alloy coated sheet steel and is produced either by hot dipping or electrodeposition. The hot dipping process with a chloride flux is used to produce most terneplates. The coating layer, whose electrode potential is more noble than that of the steel substrate, contains 8–16% Sn. Since the electrode potential of the coating layer is more noble than the steel substrate, it is necessary to build a uniform and dense alloy layer ($FeSn_2$) in order to form a pinhole free deposit.

Terneplate exhibits excellent corrosion resistance, especially under wet conditions, excellent weldability and formability, with only small amounts of corrosion products forming on the surface. A thin nickel deposit can be applied as an undercoat for the terne layer. Nickel reacts rapidly with the tin-lead alloy to form a nickel–tin alloy layer. This alloy layer provides good corrosion resistance and inhibits localized corrosion.

The main application for terneplate is in the production of fuel tanks for automobiles.

7.3.8 Gold Coatings

Gold electrodeposits are primarily used to coat copper in electronic applications to protect the copper connectors and other copper components from corrosion. It is desirable to obtain the corrosion protection with the minimum thickness of gold because of the cost of the gold. As the thickness of the electrodeposit is decreased, there is a tendency for the deposit to provide inadequate coverage of the copper. For this reason, it is necessary that there be a means whereby the coverage of the copper can be determined. Such a test, using corrosion principles as a guide, has been developed. In a 0.1 M NH_4Cl solution, gold serves as the cathode and copper serves as the anode. At a high cathode/anode surface area fraction, the corrosion potential is linearly related to the area fraction of copper exposed, as shown in Figure 7.13. By measuring the corrosion potential of the gold plated copper in a 0.1 M NH_4Cl solution, the area fraction of copper exposed is determined.

Gold coatings can also be deposited by means of electroless plating. Borohydride or DMAB are used as reducers with a stable gold cyanide complex. Thin gold coatings may be deposited on plastics by an aerosol spray method using gold complexes with amines and hydrazine as a reducer. A relatively thick coat may be obtained.

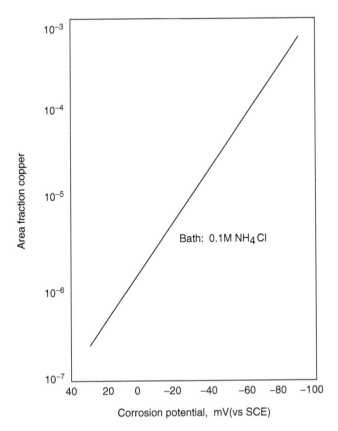

FIGURE 7.13
Data showing that the fractional exposed area of copper in copper/gold system is linearly related to the corrosion potential at low-exposed copper areas.

7.3.9 Copper Coatings

Even though copper is soft, it has many engineering applications in addition to its decorative function. One such application is the corrosion protection of steel. It can be used as an alternative to nickel to prevent fretting and scaling corrosion. Copper can be electrochemically deposited from various aqueous solutions. The properties of the deposit will depend on the chosen bath and the applied procedures. The hardness of the layers varies from 40 to 160 HV.

Since copper is very noble, it causes extreme galvanically induced local corrosion of steel and aluminum substrates. Because of this, extreme care must be taken to produce well-adhering nonporous layers.

The corrosion protection provided by a copper coating is twofold, consisting of an original barrier action of the coating layer and a secondary barrier action of corrosion products. The low EMF of copper is responsible for the formation of the original barrier action. The electrochemical reactions in the corrosion cells on copper are as follows:

$$\text{Anodic reaction:} \quad Cu \rightarrow Cu^+ + e$$

$$Cu \rightarrow Cu^{2+} + 2e$$

$$\text{Cathodic reaction:} \quad O_2 + H_2O \rightarrow 4e + 4OH^-$$

Chloride ions in a natural environment stabilize cuprous ions. Cupric ions are more stable. Since the EMF of corrosion on copper is less than that on iron, the reactivity of a steel surface is decreased by coating it with copper.

Over a period of time, corrosion products gradually build up a secondary layer against corrosion. Initially, a cuprous oxide layer is formed, followed by the copper surface covered with basic copper salts. Pollutants in the atmosphere determine the formation of basic copper salts as follows:

Mild atmosphere	Malachite $CuCO_3{:}Cu(OH)_2$
SO_x atmosphere	Brochanite $CuSO_4{:}3Cu(OH)_2$
Chloride atmosphere	Atacamite $CuCl_2{:}3Cu(OH)_2$

In most coastal areas, the amount of sulfates in the atmosphere exceeds the amount of chlorides. As a layer of copper salt grows on the surface of the corrosion product layer, the protective ability of the corrosion layer is increased. As the exposure time increases, the average corrosion rate of copper gradually decreases. After twenty years, the corrosion rate of copper is reduced to half the value of the first year as a result of the secondary barrier of corrosion products.

The initial corrosion rate of a copper coating is dependent on atmospheric conditions such as time of wetness and type and amount of pollutants. Time of wetness is the most important factor affecting the corrosion rate of copper. The corrosion rate of copper usually obeys parabolic law:

$$M^2 = kt,$$

where

$M =$ mass increase
$k =$ a constant
$t =$ exposure time

Accordingly, the average corrosion rate decreases with increased exposure time, that means the surface of the copper is covered with basic salts by degrees, and thereafter, the corrosion rate approaches a constant value.

Copson[6] conducted twenty year exposure tests and found the average corrosion rate of copper to be as follows:

0.0034 mil/year in dry rural atmospheres
0.143 mil/year in rural atmospheres

0.0476–0.515 mil/year in industrial atmospheres

0.0198–0.0562 mil/year in marine atmospheres

Until the base metal is exposed, the corrosion process of a copper coated layer is similar to that of copper plate. Galvanic corrosion of copper coated steel is induced when the steel substrate is exposed. However, in the case of copper coated stainless steel, the occurrence of galvanic action is dependent on the composition of the stainless steel.

In chloride atmospheres galvanic pitting takes place at the pores in copper layers and galvanic tunneling at cut edges on types 409 and 430 stainless steels, whereas, in SO_x atmospheres uniform corrosion takes place on the copper coating.

Copper coatings are used both for decorative and for corrosion protection from the atmosphere. Copper coated steels are used as roofs, flashings, leaders, gutters, and architectural trim. Copper undercoats also improve the corrosion resistance of multilayered coatings, specifically in the plating of nickel and chromium.

7.4 Nonnoble Coatings

Nonnoble metals protect the substrate by means of cathodic control. Cathodic overpotential of the surface is increased by coating which makes the corrosion potential more negative than that of the substrate. The coating metals used for cathodic control protection are zinc, aluminum, manganese, and cadmium-and their alloys, of which the electrode potentials are more negative than those of iron and steel. Consequently, the coating layers of these metals act as sacrificial anodes for iron and steel substrates, when the substrates are exposed to the atmospheres. The coating layer provides cathodic protection for the substrate by galvanic action. These metals are called sacrificial metals.

The electrical conductivity of the electrolyte, the temperature, and the surface condition determines the galvanic action of the sacrificial metal coating. An increase in the cathodic overpotential is responsible for the corrosion resistance of the coating layer. Figure 7.14 shows the principle of cathodic control protection by a sacrificial metal coating.

The corrosion rate of zinc coated iron $i_{corr\ of\ zinc\ coating}$ becomes lower than that of uncoated iron $i_{corr\ of\ zinc\ uncoated\ iron}$ because the cathodic overpotential of the surface is increased by zinc coating and the exchange current density of dissolved oxygen on zinc $i_{oc\ on\ zinc}$ is lower than that on iron $i_{oc\ on\ iron}$. If a small part of iron is exposed to the atmosphere, the electrode potential of the exposed iron is equal to the corrosion potential of zinc $E_{corr\ of\ zinc\ coating}$ because the exposed iron is polarized cathodically by the surrounding zinc so that little corrosion occurs on the exposed iron

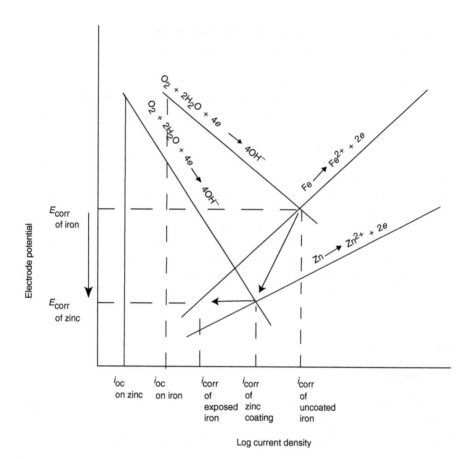

FIGURE 7.14
Cathodic control protection.

$i_{corr \ of \ exposed \ iron}$. Zinc ions predominately dissolved from the zinc coating form the surrounding barrier of corrosion products at the defect and, thereby, protect the exposed iron.

Sacrificial metal coatings protect iron and steel by two or three protective abilities such as:

1. Original barrier action of coating layer

2. Secondary barrier action of corrosion product layer

3. Galvanic action of coating layer

The surface oxide film and the electrochemical properties based on the metallography of the coating metal provide the original barrier action.

An air formed film of Al_2O_3 approximately 25 Å thick forms on aluminum. This film is chemically inert, and its rapid formation of oxide film by a self-healing ability leads to satisfactory performance in natural environments.

Zinc, however, does not produce a surface oxide film that is as effective a barrier as the oxide film on aluminum. The original barriers of zinc and zinc alloy coatings result from electrochemical properties based on the structure of the coating layer.

Nonuniformity of the surface condition generally induces the formation of a corrosion cell. Such nonuniformity results from defects in the surface oxide film, localized distribution of elements, and the difference in crystal face or phase. These nonuniformities of surface cause the potential difference between portions of the surface, thereby, promoting the formation of a corrosion cell.

Many corrosion cells are formed on the surface, accelerating the corrosion rate, as a sacrificial metal and its alloy-coated materials are exposed in the natural atmosphere. During this time, corrosion products are gradually formed and converted to a stable layer after a few months of exposure. Typical corrosion products formed are shown in Table 7.7. Once the stable layer has been formed, the corrosion rate becomes constant. This secondary barrier of corrosion protection continuously regenerates over a long period of time. In most cases, the service life of a sacrificial metal coating depends on the secondary barrier action of the corrosion product layer.

Sacrificial metal coatings are characterized by their galvanic action. Exposure of the base metal, as a result of mechanical damage, polarizes the base metal cathodically to the corrosion potential of the coating layer, as shown in Figure 7.14 so that little corrosion takes place on the exposed base metal. A galvanic couple is formed between the exposed part of the base metal and the surrounding coating metal. Because sacrificial metals are more negative in electrochemical potential than iron or steel, a sacrificial metal acts as an anode and the exposed base metal behaves as a cathode. Table 7.8 shows the corrosion potentials of sacrificial metals and steel in a 3% NaCl solution. Consequently, the dissolution of the coating layer around the defect is accelerated, and the exposed part of base metal is protected against corrosion. Figure 7.15 shows a schematic illustration of the galvanic action of a sacrificial metal coating.

TABLE 7.7

Corrosion Products Formed on Various Sacrificial Metal Coatings

Metal	Corrosion Product
Al	Al_2O_3, $\beta Al_2O_3 \cdot 3H_2O$, $\alpha AlOOH$, $Al(OH)_3$, amorphous Al_2O_3
Zn	ZnO, $Zn(OH)_2$, $2ZnCO_3 \cdot 3Zn(OH)_2$, $ZnSO_4 \cdot 4Zn(OH)_2$, $ZhCl_2 \cdot 4Zn(OH)_2$, $ZnCl2 \cdot 6Zn(OH)_2$
Mn	$\gamma > Mn_2O_3$, $MnCO_3$, γ-MnnOOH
Cd	CdO, $CdOH_2$, $2CdCO_3 \cdot 3Cd(OH)_2$

TABLE 7.8

Corrosion Potentials of Sacrificial Metals in a
3% NaCl Solution

Metal	Corrosion Potential (V, SCE)
Mn	−1.50
Zn	−1.03
Al	−0.79
Cd	−0.70
Steel	−0.61

The loss of metal coating resulting from corrosion determines the service life of the coating. The degree of loss is dependent on the time of wetness on the metal surface and the type of concentration of pollutants in the atmosphere. Table 7.9 shows the average corrosion losses of zinc, aluminum, and 55% Al–Zn coatings in various locations and atmospheres. The losses were calculated from the mean values of time of wetness and the average corrosion rate during wet duration. The time of wetness of walls is 40% of that of roofs. Coating metals and coating thicknesses can be decided from Table 7.9 since the corrosion losses of zinc, aluminum, and Al–Zn alloy are proportional to exposure time.

As can be seen from the table, a G90 sheet that has a 1 mil zinc coating, cannot be used for a roof having a durability of ten years in any atmosphere except in a rural area. Were this sheet to be used in an urban, marine, or industrial atmosphere, it would have to be painted for protection.

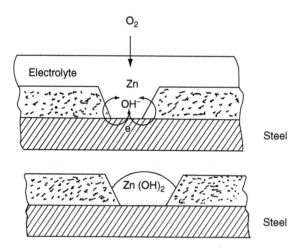

FIGURE 7.15
Schematic illustration of galvanic action of sacrificial metal coating.

TABLE 7.9

Average Corrosion Losses of Sacrificial Metal Coatings for 10 Years

| Location | Atmosphere | Average Corrosion Loss (mil[a]/10 years) | | | | | |
| | | Zinc | | 55% AlZn | | Aluminum | |
		Root	Wall	Root	Wall	Root	Wall
Inland	Rural	0.42	0.17	0.15	0.06	0.06	0.02
	Urban	1.48	0.59				
	Industrial	1.40	0.56	0.25	0.06	0.06	0.02
	Severe industrial	1.59	0.64	0.20	0.08	0.07	0.03
Inland shore of lake or marsh	Rural	0.59	0.24				
	Urban	1.97	0.79				
	Industrial	1.40	0.56				
	Severe industrial	2.12	0.85	0.20	0.08	0.08	0.03
Coast	Rural	0.74	0.23				
	Urban	2.47	0.99				
	Industrial	1.75	0.70	0.25	0.10	0.08	0.04
	Severe industrial	2.65	1.06	0.46	0.18	0.10	0.04
Seashore	Severe industrial	2.06	0.82			0.19	0.07

[a] 1 mil = 25.4 μm.

Source: I. Suzuki. 1989. *Corrosion Resistant Coatings Technology,* New York: Marcel Dekker.

Aluminum and 55% Al–Zn alloy provide galvanic protection for the steel substrate. In rural and industrial atmospheres, an aluminum coating does not act as a sacrificial anode. However, in a chloride atmosphere such as a marine area, it does act as a sacrificial anode.

The choice of what sacrificial metal coating to use will be based on the environment that it will be exposed to and the service life required. The service life required will also determine the coating thickness to be applied that, in turn, will influence the coating process to be used. Sacrificial metal coatings have been used successfully for roofs, walls, ducts, shutters, doors, and window frames in the housing industry; and on structural materials such as transmission towers, structural members of a bridge, antennae, chimney structures, grand-stands, steel frames, high-strength steel bolts, guard rails, corrugated steel pipe, stadium seats, bridge I-beams, footway bridges, road bridges, and fencing.

7.4.1   Zinc Coatings

Approximately half of the world's production of zinc is used to protect steel from rust. Zinc coatings are probably the most important type of metallic coating for corrosion protection of steel. The reasons for the wide application are:

1. Prices are relatively low.
2. Because of large reserves, an ample supply of zinc is available.
3. There is great flexibility in application procedures resulting in many different qualities with well-controlled layer thicknesses.
4. Steel provides good cathodic protection.
5. Many special alloy systems have been developed with improved corrosion protection properties.

The ability to select a particular alloy or to specify a particular thickness of coating depends on the type of coating process used. Zinc coatings can be applied in many ways. The six most commonly used procedures follow.

7.4.1.1   Hot Dipping

This is a process where cleaned steel is immersed in molten zinc or zinc alloy, and a reaction takes place to form a metallurgically bonded coating.

The coating is integral with the steel because the formation process produces zinc–iron alloy layers overcoated with zinc. Continuity and uniformity is good since any discontinuities are readily visible as "black spots."

Coating thicknesses can be varied from approximately 50–125 μm on tube and products. Thicker coatings up to 250 μm can be obtained by grit-blasting before galvanizing. Sheet and wire normally receive thicknesses of 10–30 μm.

Conventional coatings that are applied to finished articles are not formable. The alloy layer is abrasion resistant but brittle on bending. Special coatings with little or no alloy layer are readily formed (e.g., on sheet) and resistance welded.

A chromate conversion coating over the zinc coating prevents wet storage stain, whereas, phosphate conversion coatings provide a good base (on a new sheet) for paints. Weathered coatings are often painted after 10–30 years for longer service.

Hot-dip galvanizing is the most important zinc coating process. All mild steels and cast iron can be coated by this process. The thickness and structure of the coating will depend on the alloying elements. Approximately half of the steel that is coated is in the form of sheet, approximately one quarter is fabricated work, and the remainder is tube or wire. Metallurgically, the processes used for tubes and fabricated work is similar, whereas, the process used for sheet has small additions to the zinc that reduces the quantity of iron–zinc alloy in the coating which provides flexibility.

7.4.1.2 *Zinc Electroplating*

This process is sometimes mistakenly referred to as electrogalvanizing. In this process, zinc salt solutions are used in the electrolytic deposition of a layer of zinc on a cleaned steel surface.

This process provides good adhesion, comparable with other electro-plated coatings. The coating is uniform within the limitations of "throwing power" of the bath. Pores are not a problem as exposed steel is protected by the adjacent zinc.

Coating thickness can be varied at will but is usually 2.5–15 μm. Thicker layers are possible but are not generally economical.

Electroplated steel has excellent formability and can be spot welded. Small components are usually finished before being plated.

Chromate conversion coatings are used to prevent wet storage stain, whereas, phosphate conversion coatings are used as a base for paint.

The process is normally used for simple, fairly small components. It is suitable for barrel plating or for continuous sheet and wire. No heating is used in this process except for hydrogen embrittlement relief on high strength steels.

Electroplated zinc is very ductile, and consequently, this process is widely used for the continuous plating of strip and wire where severe deformation may be required.

The coating on steel from this process gives a bright and smooth finish. It is used for decorative effect to protect delicate objects where rough or uneven finishes cannot be tolerated (e.g., instrument parts). It is also used for articles that cannot withstand the pretreatment or temperatures required in other coating processes.

It was previously mentioned that the term *electrogalvanizing* is sometimes used to describe this process. This is misleading since the chief characteristic of galvanizing is the formation of a metallurgical bond at the zinc–iron interface. This does not occur in electroplating.

7.4.1.3 Mechanical Coating

This process involves the agitating of suitably prepared parts to be coated with a mixture of nonmetallic impactors (e.g., glass beads), zinc powder, a chemical promoter, and water. All types of steel can be coated. However, this process is less suitable for parts heavier than 1/2 lb (250 g) because the tumbling process reduces coating thickness at the edges.

The adhesion is good compared with electroplated coatings. Thickness can be varied as desired from 5 μm to more than 70 μm. However, the coating is not alloyed with the steel nor does it have the hard, abrasion resistance iron–zinc alloy layers of galvanized or sherardized coatings. Conversion coatings can be applied.

7.4.1.4 Sherardizing

The articles to be coated are tumbled in a barrel containing zinc dust at a temperature just below the melting point of zinc, usually around 716°F/380°C. In the case of spring steels, the temperature used is somewhat lower. By means of a diffusion process, the zinc bonds to the steel forming a hard, even coating of zinc–iron compounds. The coating is dull gray in color and can readily be painted if necessary.

The finish is continuous and very uniform, even on threaded and irregular parts. This is a very useful finish for nuts and bolts that, with proper allowance for thickness of coats, can be sherardized after manufacture and used without retapping the threads.

The thickness of coating can be controlled. Usually, a thickness of 30 μm is used for outdoor applications while 15 μm is used indoors.

7.4.1.5 Thermally Sprayed Coatings

In the process, droplets of semimolten zinc are sprayed from a special gun that is fed with either wire or powder onto a grit-blasted surface. The seimmolten droplets coalesce with some zinc oxide present at each interface between droplets. Electrical continuity is maintained both throughout the coating and with the iron substrate so that full cathodic protection can be obtained since the zinc oxide forms only a small percentage of the coating.

The sprayed coating contains voids (typically 10–20% by volume) between coalesced particles. These voids have little effect on the corrosion protection because they soon fill up with zinc corrosion products and are thereafter impermeable. However, the use of a sealer to fill the voids improves appearance in service and adds to life expectancy, but more importantly, it provides a better surface for subsequent application of paints.

There are no size or shape limitations regarding the use of this process.

7.4.1.6 *Zinc Dust Painting*

Zinc dust paints may be used alone for protection or as a primer followed by conventional top coats. More details will be found in Chapter 8.

7.4.1.7 *Corrosion of Zinc Coatings*

In general, zinc coatings corrode in a similar manner as solid zinc. However, there are some differences. For example, the iron–zinc alloy present in most galvanized coatings has a higher corrosion resistance than solid zinc in neutral and acid solutions. At points where the zinc coating is defective, the bare steel is cathodically protected under most conditions.

The corrosion of zinc coatings in air is an approximate straight line relationship between weight loss and time. Because the protective film on zinc increases with time in rural and marine atmospheres of some types, under these conditions, the life of the zinc may increase more than in proportion to thickness. However, this does not always happen.

Zinc coatings are primarily used to protect ferrous parts against atmospheric corrosion. These coatings have good resistance to abrasion by solid pollutants in the atmosphere. General points to consider are:

1. Corrosion increases with time of wetness.
2. The corrosion rate increases with an increase in the amount of sulfur compounds in the atmosphere. Chlorides and nitrogen oxides usually have a lesser effect but are often very significant in combination with sulfates.

Zinc coatings resist atmospheric corrosion by forming protective films consisting of basic salts, notably carbonate. The most widely accepted formula is $3Zn(OH)_2 \cdot 2ZnCO_3$. Environmental conditions that prevent the formation of such films, or conditions that lead to the formation of soluble films, may cause rapid attack on the zinc.

The duration and frequency of moisture contact is one such factor. Another factor is the rate of drying because a thin film of moisture with high oxygen concentration promotes reaction. For normal exposure conditions, the films dry quite rapidly. It is only in sheltered areas that drying times are slow so that the attack on zinc is accelerated significantly.

The effect of atmospheric humidity on the corrosion of a zinc coating is related to the conditions that may cause condensation of moisture on the metal surface and to the frequency and duration of the moisture contact. If the air temperature drops below the dew point, moisture will be deposited. The thickness of the piece, its surface roughness, and its cleanliness also influence the amount of dew deposited. Lowering the temperature of a metal surface below the air temperature in a humid atmosphere will cause moisture to condense on the metal. If the water evaporates quickly, corrosion is usually not severe, and a protective film is formed on the surface. If water from rain or snow remains in contact with zinc when access to air is restricted and the humidity is high, the resulting corrosion can appear to be severe (wet-storage stain known as "white rust") because the formation of a protective basic zinc carbonate is prevented.

In areas having atmospheric pollutants, particularly sulfur oxides and other acid forming pollutants, time of wetness becomes of secondary importance. These pollutants can also make rain more acid. However, in less corrosive areas, time of wetness assumes a greater proportional significance.

In the atmospheric corrosion of zinc, the most important atmospheric contaminent to be considered is sulfur dioxide. At relative humidities of about 70% or above, it usually controls the corrosion rate.

Sulfur oxides and other corrosive species react with the zinc coating in two ways: dry deposition and wet deposition. Sulfur dioxide can deposit on a dry surface of galvanized steel panels until a monolayer of SO_2 is formed. In either case, the sulfur dioxide that deposits on the surface of the zinc forms a sulfurous or other strong acid that reacts with the film of zinc oxide, hydroxide, or basic carbonate to form zinc sulfate. The conversion of sulfur dioxide to sulfur-based acids may be catalyzed by nitrogen compounds in the air (NO_x compounds). This factor may affect corrosion rates in practice. The acids partially destroy the film of corrosion products that will then reform from the underlying metal, thereby, causing continuous corrosion by an amount equivalent to the film dissolved, hence the amount of SO_2 absorbed.

Chloride compounds have less effect than sulfur compounds in determining the corrosion rate of zinc. Chloride is most harmful when combined with acidity because of sulfur gases. This is prevalent on the coast in highly industrial areas.

Atmospheric chlorides will lead to the corrosion of zinc, but to a lesser degree than the corrosion of steel, except in brackish water and flowing seawater. Any salt deposit should be removed by washing. The salt content of the atmosphere will usually decrease rapidly inland farther away from the coast. Corrosion also decreases with distance from the coast, but the change is more gradual and erratic because chloride is not the primary pollutant affecting zinc corrosion. Chloride is most harmful when combined with acidity resulting from sulfur gases.

Other pollutants also have an effect on the corrosion of galvanized surfaces. Deposits of soot or dust can be detrimental because they have the potential to increase the risk of condensation onto the surface and hold more water in position. This is prevalent on upward-facing surfaces. Soot (carbon) absorbs large quantities of sulfur that are released by rainwater.

In rural areas, overmanuring of agricultural land tends to increase the ammonia content of the air. The presence of normal atmospheric quantities of ammonia does not accelerate zinc corrosion, and petrochemical plants where ammonium salts are present show no accelerated attack on galvanized steel. However, ammonia will react with atmospheric sulfur oxides, producing ammonium sulfate that accelerates paint film corrosion as well as zinc corrosion. When ammonium reacts with NO_x^- compounds in the atmosphere, ammonium nitrite and nitrate are produced. Both compounds increase the rate of zinc corrosion, but less than SO_2 or SO_3.

Because of the Mears effect (wire corrodes faster per unit of area than more massive materials) galvanized wire corrodes some 10–80% faster than galvanized sheet. However, the life of rope made from galvanized steel wires is greater than the life of the individual wire. This is explained by the fact that the parts of the wire that lie on the outside are corroded more rapidly and when the zinc film is penetrated in those regions, the uncorroded zinc inside the rope provides cathodic protection for the outer regions.

Table 7.10 lists the compatibility of zinc coatings with various corrodents.

TABLE 7.10

Compatibility of Galvanized Steel with Selected Corrodents

Acetic Acid	U
Acetone	G
Acetonitrile	G
Acrylonitrile	G
Acrylic latex	U
Aluminum chloride, 26%	U
Aluminum hydroxide	U
Aluminum nitrate	U
Ammonia, dry vapor	U
Ammonium acetate solution	U
Ammonium bisulfate	U
Ammonium bromide	U
Ammonium carbonate	U
Ammonium chloride, 10%	U
Ammonium dichloride	U
Ammonium hydroxide	
Vapor	U
Reagent	U
Ammonium molybdate	G
Ammonium nitrate	U
Argon	G
Barium hydroxide	
Barium nitrate solution	S
Barium sulfate solution	S
Beeswax	U
Borax	S
Bromine, moist	U
2-Butanol	G
Butyl acetate	G
Butyl chloride	G
Butyl ether	G
Butylphenol	G
Cadmium chloride solution	U
Cadmium nitrate solution	U
Cadmium sulfate solution	U
Calcium hydroxide	
Sat. solution	U
20% Solution	S
Calcium sulfate, sat. solution	U

(continued)

TABLE 7.10 *Continued*

Cellosolve acetate	G
Chloric acid, 20%	U
Chlorine, dry	G
Chlorine water	U
Chromium chloride	U
Chromium sulfate solution	U
Copper chloride solution	U
Decyl acrylate	G
Diamylamine	G
Dibutylamine	G
Dibutyl cellosolve	G
Dibutyl phthalate	G
Dichloroethyl ether	G
Diethylene glycol	G
Dipropylene glycol	G
Ethanol	G
Ethyl acetate	G
Ethyl acrylate	G
Ethyl amine, 69%	G
N-Ethyl butylamine	G
2-Ethyl butyric acid	G
Ethyl ether	G
Ethyl hexanol	G
Fluorine, dry, pure	G
Formaldehyde	G
Fruit juices	S
Hexanol	G
Hexylamine	G
Hexylene glycol	G
Hydrochloric acid	U
Hydrogen peroxide	S
Iodine, gas	U
Isohexanol	G
Isooctanol	G
Isopropyl ether	G
Lead sulfate	U
Lead sulfite	S
Magnesium carbonate	S
Magnesium chloride, 42.5%	U
Magnesium fluoride	G
Magnesium hydroxide, sat.	S
Magnesium sulfate	
2% Solution	S
10% Solution	U
Methyl amyl alcohol	G
Methyl ethyl ketone	G
Methyl propyl ketone	G
Methyl isobutyl ketone	G
Nickel ammonium sulfate	U
Nickel chloride	U
Nickel sulfate	S
Nitric acid	U
Nitrogen, dry, pure	G
Nonylphenol	G
Oxygen	
Dry, pure	G
Moist	U

(*continued*)

TABLE 7.10 *Continued*

Paraldehyde	G
Perchloric acid solution	S
Permanganate solution	S
Peroxide	
Pure, dry	S
Moist	U
Phosphoric acid, 0.3–3%	G
Polyvinyl acetate latex	U
Potassium carbonate	
10% Solution	U
50% Solution	U
Potassium chloride solution	U
Potassium bichromate	
14.7%	G
20%	S
Potassium disulfate	S
Potassium fluoride, 5–20%	G
Potassium hydroxide	U
Potassium nitrate	
5–10% Solution	S
Potassium peroxide	U
Potassium persulfate, 10%	U
Propyl acetate	G
Propylene glycol	G
Propionaldehyde	G
Propionic acid	U
Silver bromide	U
Silver chloride	
Pure, dry	S
Moist, wet	U
Silver nitrate solution	U
Sodium acetate	S
Sodium aluminum sulfate	U
Sodium bicarbonate solution	U
Sodium bisulfate	U
Sodium carbonate solution	U
Sodium chloride solution	U
Sodium hydroxide solution	U
Sodium nitrate solution	U
Sodium sulfate solution	U
Sodium sulfide	U
Sodium sulfite	U
Styrene, monomeric	G
Styrene oxide	G
Tetraethylene glycol	G
1,1,2 Trichloroethane	G
1,2,3 Trichloropropane	G
Vinyl acetate	G
Vinyl ethyl ether	G
Vinyl butyl ether	G
Water	
Potable, hard	G

G, suitable application; S, borderline application; U, not suitable.

7.5   Zinc–5% Aluminum Hot-Dip Coatings

This zinc alloy coating is known as Galfan. Galfan coatings have a corrosion resistance up to three times that of galvanized steel. The main difference between these two coatings lies in the degree of cathodic protection they afford. This increase in corrosion protection is evident in both a relatively mild urban industrial atmosphere as can be seen in Table 7.11. The latter is particularly significant, because, unlike galvanizing, the corrosion rate appears to be slowing after about four years, and conventional galvanized steel would show rust in Five years. See Figure 7.16. The slower rate of corrosion also means that the zinc–5% aluminum coatings provide full cathodic protection to cut edges over a longer period. Refer to Table 7.12.

Because Galfan can be formed with much smaller cracks than can be obtained in conventional galvanized coatings, it provides excellent protection at panel bulges. This reduced cracking means that less zinc is exposed to the environment that increases the relative performance factor compared with galvanized steel.

7.6   Zinc–55% Aluminum Hot-Dip Coatings

These coatings are known as Galvalume and consist of zinc–55% aluminum–1.5% silicon. This alloy is sold under such tradenames as Zaluite, Aluzene, Alugalva, Algafort, Aluzink, and Zincalume. Galvalume exhibits superior corrosion resistance over galvanized coatings in rural, industrial, marine, and severe marine environments. However, this alloy has limited cathodic protection and less resistance to some alkaline conditions and is subject to weathering discoloration and wet storage staining. The latter two disadvantages can be overcome by chromate passivation that also improves its atmospheric corrosion resistance.

TABLE 7.11

Five-Year Outdoor Exposure Results of Galfan Coating

Atmosphere	Thickness Loss (μm)		Ratio of Improvement
	Galvanized	Galfan	
Industrial	15.0	5.2	2.9
Severe marine	>20.0	9.5	>2.1
Marine	12.5	7.5	1.7
Rural	10.5	3.0	3.5

Source: From F.C. Porter. 1994. *Corrosion Resistance of Zinc and Zinc Alloys*, New York: Marcel Dekker.

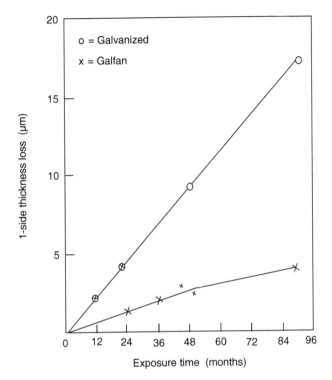

FIGURE 7.16
Seven year exposure of Galfan and galvanized steel in a severe marine atmosphere.

Initially, a relatively high corrosion loss is observed for Galvalume sheet as the zinc-rich portion of the coating corrodes and provides sacrificial protection at cut edges. This takes place in all environments, whereas, aluminum provides adequate galvanic protection only in marine chloride

TABLE 7.12

Comparison of Cathodic Protection for Galfan and Galvanized Coatings

| | Amount (mm) of Bare Edges Exposed after 3 Years (Coating Recession from Edge) | |
Environment	Galvanized	Galfan
Severe marine	1.6	0.1
Marine	0.5	0.06
Industrial	0.5	0.05
Rural	0.1	0

Source: From F.C. Porter. 1994. *Corrosion Resistance of Zinc and Zinc Alloys*, New York: Marcel Dekker.

environments. After approximately three years, the corrosion-time curves take on a more gradual slope reflecting a change from active, zinclike behavior to passive, aluminumlike behavior as the interdentric regions fill with corrosion products. It has been predicted that Galvalume sheets should outlast galvanized sheets of equivalent thickness by at least two to four times over a wide range of environments. Figures comparing the corrosion performance of galvanized sheet and Galvalume sheet are shown in Figure 7.17.

Galvalume sheets provide excellent cut-edge protection in very aggressive conditions where the surface does not remain too passive. However, it does not offer as good a protection on the thicker sheets in mild rural conditions

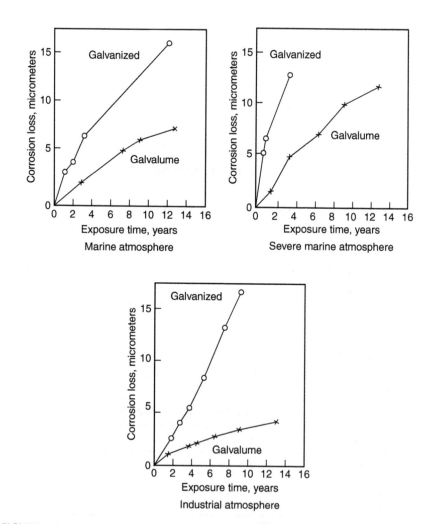

FIGURE 7.17

Thirteen year exposure of Galvalume in marine and industrial atmospheres.

where zinc–5% aluminum coatings provide good general corrosion resistance, and when sheared edges are exposed or localized damage to the coating occurs during fabrication or service, the galvanic protection is retained for a longer period.

7.7 Zinc–15% Aluminum Thermal Spray

Zinc–15% aluminum coatings are available as thermally sprayed coatings. These coatings have a two-phase structure consisting of a zinc-rich and an aluminum-rich phase. The oxidation products formed are encapsulated in the porous layer formed by the latter and do not build up a continuous surface layer as with pure zinc coatings. As a result, no thickness or weight loss is observed even after several years of exposure in atmospheric field testing.

It is normally recommended that thermally sprayed coatings be sealed to avoid initial rust stains, to improve appearance, and to facilitate maintenance painting. Sealing is designed to fill pores and give only a thin overall coating, too thin to be directly measurable. Epoxy or acrylic system resin, having a low viscosity, is used as a sealer.

7.8 Zinc–Iron Alloy Coatings

As compared with pure zinc, the zinc–iron alloy coatings provide increased corrosion resistance in acid atmospheres, but slightly reduced corrosion resistance in alkaline atmospheres.

Electroplated zinc–iron alloy layers containing more than 20% iron provide a corrosion resistance 30% higher than zinc in industrial atmospheres. In other atmospheres, the zinc–iron galvanized coatings provide as good a coating as coatings with an outer zinc layer. Sherardized coatings are superior to electroplated coatings and equal to galvanized coatings of the same thickness. However, the structure of the alloy layer, and its composition, affects the corrosion resistance.

If the zinc layer of a galvanized coating has weathered, or the zinc–iron alloy layer forms the top layer after galvanizing, brown areas may form. Brown staining can occur on sherardized or hot dip galvanized coatings in atmospheric corrosion through the oxidation of iron from the zinc–iron alloy layers or from the substrate. Such staining is usually a dull brown rather than the bright red–brown of uncontrolled rust. Usually, there is a substantial intact galvanized layer underneath, leaving the life of the coating unchanged. Unless the aesthetic appearance is undesirable, no action need be taken.

7.9 Aluminum Coatings

Aluminum coatings protect steel substrates by means of cathodic control, with an original barrier action of an air formed film which is chemically inert, and the rapid formation of oxide film by a self-healing ability. Aluminum coatings are excellent in general corrosion resistance. However, they do not act as a sacrificial anode in rural and industrial atmospheres but do in a chloride area such as a marine environment. In a nonchloride environment, the formation of red rust occurs at sheared edges and in other defects of an aluminum coating layer. However, the growth of red rust is slow.

Aluminum coatings, sealed with organic or composite layers, such as etch primer, zinc chromate, etc., will provide long service in seawater environments. Recommended coating thickness plus sealing for the splash zone and submerged zone is 150 μm.

7.10 Cadmium Coatings

Cadmium coatings are produced almost exclusively by electrodeposition. A cadmium coating on steel does not provide as much cathodic protection to the steel as does a zinc coating because the potential between cadmium and iron is not as great as between zinc and iron. Therefore, it becomes important to minimize defects in the cadmium coating.

Unlike zinc, a cadmium coating will retain a bright metallic appearance. It is more resistant to attack by salt spray and atmospheric condensate than zinc. In aqueous solutions, cadmium will resist attack by strong alkalies, but will be corroded by dilute acids and aqueous ammonia.

Cadmium coatings should not be allowed to come into contact with food products because cadmium salts are toxic. This coating is commonly used on nuts and bolts, but because of its toxicity, usage is declining.

7.11 Manganese Coatings

Manganese is very active, having an electrode potential more negative than zinc (Mn: -1.5 m, Zn -1.03 V, SCE). In a natural atmosphere, a dense corrosion layer builds on the surface of manganese during a very short period. However, defects in the coating accelerate the anodic dissolution of manganese, thereby, shortening the life of the coating. Therefore, manganese is combined with zinc to form a duplex Mn/Zn alloy coating. The types of corrosion products found on these coatings were shown in Table 7.7. The compound γ-Mn_2O_3 is effective for the formation of a barrier. The more γ-Mn_2O_3 in the corrosion products, the denser the layer on a Mn/Zn coating.

Manganese is so negative in electrochemical potential, and active, that its alloy and duplex coatings provide galvanic protection. Mn/Zn alloy coating exhibits high corrosion resistance, and the corrosion potential of manganese is more negative than that of zinc; therefore, this alloy coating provides cathodic protection for a steel substrate. The structure of Mn/Zn alloy coating is composed of the single phase of ε in the manganese content range <20%, and ε and γ phases in the range above 20%. As the manganese content in the deposit is increased, the percentage of γ-Mn phase is increased.

7.12   Conversion Coatings

The term *conversion coating* is used to describe coatings where the substrate metal provides ions that become part of the protective coating. The coating layers are composed of inorganic compounds that are chemically inert. These inert compounds on the surface reduce both anodic and cathodic areas and delay the transit of reactive species to the base metal. This results in increases in the slopes of anodic and cathodic polarization curves, thereby, decreasing the rate of corrosion of the substrate.

Conversion layers are used for various reasons:

1. To improve the adherence of the organic layers
2. To obtain electrically insulating barrier layers
3. To provide a uniform grease-free surface
4. To provide active corrosion inhibition by reducing the rate of the oxygen reduction reaction or by passivating the metallic substrate

Conversion coatings belonging in this group are phosphate, chromate, oxide, and anodized coatings. These coatings are composed of corrosion products that have been artificially formed by chemical or electrochemical reactions in selected solutions. The corrosion products formed build a barrier protection for the substrate metal. This barrier reduces the active surface area on the base metal, thereby, delaying the transport of oxidizers and aggressive species. By doing so the coating inhibits the formation of corrosion cells. The degree of secondary barrier action depends on the compactness, continuity, and stability of the corrosion product layer.

Each conversion coating protects the base metal against corrosion with two or three of the following protective abilities:

1. Secondary barrier action of corrosion products
2. Inhibiting action of soluble compounds contained in the corrosion products
3. Improvement in paint adhesion by the formation of a uniform corrosion product layer

7.12.1 Phosphate Coating

When a metal surface is treated with a weak phosphoric acid solution of iron, zinc, or manganese phosphate, phosphate layers are formed. These phosphate coatings are applied to iron and steel, zinc, aluminum, and their alloys.

Phosphate films are formed by the dissolution of base metal and the precipitation of phosphate films. The metal surface must be free of greases, oils, and other carbonaceous material before immersion in the phosphating solution or before spray application. Baths operating at 120°F/50°C have pH values of approximately 2, whereas, those operating below 120°F/50°C have pH values of approximately 3.

The zinc phosphate coating is basically the result of a corrosion process. Reactions of iron and steel in a zinc phosphate solution are as follows:

1. The dissolution of base metal at the anodic sites:

$$Fe + 2H_3PO_4 \rightarrow Fe(H_2PO_4)_2 + H_2 \qquad (7.23)$$

Promotion by the activator:

$$2Fe + 2H_2PO_4^- + 2H^+ + 3NO_2 \rightarrow 2FePO_4 + 3H_2O + 3NO \quad (7.24)$$

$$4Fe + 3Zn^{2+} + 6H_2PO_4^- + 6NO_2 \rightarrow 4FePO_4 + Zn_3(PO_4)_2$$

$$+ 6H_2O + 6NO \qquad (7.25)$$

2. Precipitation of phosphate films at the cathodic sites:

$$2Zn(H_2PO_4)_2 + Fe(H_2PO_4)_2 + 4H_2O$$

$$\rightarrow \underbrace{Zn_2Fe(PO_4)_2 \cdot 4H_2O}_{\text{Phosphophyllite}} + 4H_3PO_4 \qquad (7.26)$$

$$3Zn(H_2PO_4)_2 + 4H_2O \rightarrow \underbrace{Zn_3(PO_4)_2 \cdot 4H_2O}_{\text{Hopeite}} + 4H_3PO_4 \qquad (7.27)$$

With these reactions, the phosphate film consists of phosphophyllite and hopeite. Phosphate solubilities are lowest in the pH range of 6–8. They are stable in neutral environments and are nonelectric conductive compounds. Phosphate film deposits on cathodic areas and anodic sites remain in the form of pinholes. Consequently, the continuity of phosphate films is not as good as those of anodic oxide and chromate films.

Since the barrier action of a conversion film is dependent on its solubility and continuity, it is evident that the phosphate films provide only limited protection. However, they do provide an excellent base for paint, plastic, and rubber coatings.

The chemical effects of phosphating on the surface is to convert the surface to a nonalkaline condition, protecting the surface from reactions with oils in paint and to protect against the spread of corrosion from defects. Alkaline residues on the surface of the base metal lead to underfilm corrosion.

Phosphating increases the uniformity in the surface texture and surface area that improves paint adhesion, that in turn, increases the service life of a paint film.

7.12.2 Chromate Coating

Chromate conversion coatings are formed on aluminum and its alloys—magnesium, zinc, and cadmium. These coatings provide good corrosion protection and improve the adhesion of organic layers. A chromate coating is composed of a continuous layer consisting of insoluble trivalent chromium compounds and soluble hexavalent chromium compounds. The coating structure provides a secondary barrier, inhibiting action, and also good adhesion for lacquer films.

Chromate coatings provide their corrosion resistance based on the following three properties:

1. Cr(III) oxide, that is formed by the reduction of Cr(IV) oxide, has poor solubility in aqueous media and, thereby, provides a barrier layer.
2. Cr(VI) will be included in the conversion coating and will be reduced to Cr(III) to passivate the surface when it is damaged, preventing hydrogen gas from developing.
3. The rate of cathodic oxygen reactions is strongly reduced.

Most chromate conversion coatings are amorphous gel-like precipitates, so they are excellent in continuity. The service life is dependent on thickness, characteristics of the base metal, coating conditions—particularly dry heat—and the environmental conditions that the chromated products are used under.

When a chromated product is exposed to the atmosphere, hexavalent chromium slowly leaches from the film, with the result that the surface appearance changes from irridescent yellow to either a green color or to clear. The structure of the film consists of more of the insoluble trivalent chromium compounds. Passivation is provided for any damaged areas by the leached hexavalent chromium.

The longer the time of wetness, the shorter the service life of the coating because chromate coatings absorb moisture, and moisture results in the leaching of hexavalent chromium. The leaching behavior of a chromate film is also affected by its aging process, drying process, and long-term storage. Aging of a chromate coating reduces its protective ability.

Chrome baths always contain a source of hexavalent chromium ion (e.g., chromate, dichromate. or chromic acid) and an acid to produce a low pH that usually is in the range of 0–3. A source of fluoride ions is also usually present. These fluoride ions will attack the original (natural) aluminum oxide film, exposing the base metal substrate to the bath solution. Fluoride also prevents the aluminum ions (that are released by the dissolution of the oxide layer) from precipitating by forming complex ions. The fluoride concentration is critical. If the concentration is too low, a conversion layer will not form because of the failure of the fluoride to attack the natural oxide layer, whereas, too high a concentration results in poor adherence of the coating because of reaction of the fluoride with the aluminum metal substrate.

During the reaction, hexavalent chromium is partially reduced to trivalent chromium, forming a complex mixture consisting largely of hydrated hydroxides of both chromium and aluminum:

$$6H^+ + H_2Cr_2O_7 + 6e^- \rightarrow 2Cr(OH)_3 + H_2O \qquad (7.28)$$

There are two types of processes that can produce conversion coatings: chromic acid processes and chromic acid–phosphoric acid processes. In the formation of the chromic acid-based conversion coating, the following overall equation governs:

$$6H_2Cr_2O_7 + 3OHF + 12Al + 18HNO_3 \rightarrow 3Cr_2O_3 + Al_2O_3$$
$$+ 10AlF_3 + 6Cr(NO_3)_3 + 30H_2O \qquad (7.29)$$

The oxide Cr_2O_3 is better described as an amorphous chromium hydroxide, $Cr(OH)_3$. The conversion coating is yellow to brown in color.

In the chromic acid–phosphoric acid process the following reaction governs:

$$2H_2Cr_2O_7 + 10H_3PO_4 + 12HF + 4Al \rightarrow CrPO_4 + 4AlF$$
$$+ 3Cr(H_2PO_4)_3 + 14H_2O$$

This conversion coating is greenish in color and primarily consists of hydrated chromium phosphate with hydrated chromium oxide concentrated toward the metal.

The barrier action of a chromate coating increases with its thickness. Chromate conversion coatings can be used as a base for paint or alone for corrosion protection. Previously, it was described how the leached hexavalent chromium acts as an anodic inhibitor, by forming passive films over defects in the coating. Since the films formed on aluminum by the chromic acid–phosphoric acid process contain no hexavalent chromium, they do not provide self-healing from defects.

The service life of a chromate coating depends on the coating thickness. Chromate coatings absorb moisture, and moisture results in the leaching of hexavalent chromium. Therefore, the longer the time of wetness, the shorter the life of the coating. However, as long as the leaching of the hexavalent chromium continues, the base metal is protected.

Environmental conditions, particularly time of wetness and temperature, determine the leaching rate. In natural environments, the leaching rate is commonly low. Pollutants in the atmosphere, particularly chloride ions, also increase the rate of deterioration of the film. Chromate conversion coatings provide good corrosion resistance in a mild atmosphere, such as indoor atmospheres, and surface appearance. They also provide a good base for organic films.

Chromate conversion coatings are usually applied to zinc and its alloy coated sheets, to protect against staining during storage, and to products of zinc-die castings, aluminum and its alloys, and magnesium and its alloys.

7.12.3 Oxide Coatings

Iron or steel articles to be coated are heated in a closed retort to a temperature of 1600°F/871°C; afterwards, superheated steam is admitted. This results in the formation of red oxide (Fe_2O_3) and magnetic oxide (Fe_3O_4). Carbon monoxide is then admitted to the retort, reducing the red oxide to magnetic oxide that is resistant to corrosion. Each operation takes approximately 20 min.

Iron and steel may also be oxide coated by electrolytic means. The article to be coated is made the anode in an alkaline solution (anodic oxidation). These coatings are primarily for appearance such as for cast iron stove parts.

7.12.4 Anodized Coatings

The electrochemical treatment of a metal serving as an anode in an electrolyte is known as anodizing. Since aluminum's electrode potential is negative and its oxide film is stable in natural environments, surface treatments have been developed for the purpose of producing more stable oxide films. The anodic films formed can be either porous or nonporous depending upon which electrolyte is used.

Porous films result when electrolytes such as sulfuric acid, oxalic acid, chromic acid, and phosphoric acid are used. These films have the advantage of being able to be dyed.

Sulfuric acid is the most widely used electrolyte. A large range of operating conditions can be utilized to produce a coating to meet specific requirements. Hard protective coatings are formed that serve as a good base for dyeing. In order to obtain the maximum corrosion resistance, the porous coating must be sealed after dyeing. The anodic coating formed, using sulfuric acid as the electrolyte, is clear and transparent on pure aluminum. Aluminum alloys containing silicon or manganese and the heterogeneous aluminum–magnesium alloy yield coatings that range from gray to brown and may be patchy in some cases. The adsorptive power of these coatings make them excellent bases for dyes, especially if they are sealed in nickel or cobalt acetate solution.

It is not recommended to use sulfuric acid as the electrolyte for anodizing work containing joints that can retain the sulfuric acid after removal from the bath. The retained electrolyte will provide sites for corrosion.

When chromic acid is used as the electrolyte, the coatings produced are generally opaque, gray, and irridescent, with the quality being dependent on the concentration and purity of the electrolyte. These are unattractive as compared to those formed using sulfuric acid as the electrolyte. When 0.03% sulfate is added to the electrolyte, colorless and transparent coatings are formed. These coatings are generally thin, of low porosity, and difficult to dye. Black coatings can be obtained in concentrated solutions at elevated temperatures. Attractive opaque surfaces can be obtained by adding titanium, zirconium, and thallium compounds to the electrolyte.

The chromic acid anodizing process is the only one that can be used on structures containing blind holes, crevices, or difficult to rinse areas. Chromic acid anodizing generally increases fatigue strength, whereas, sulfuric acid anodizing may produce decreases in fatigue strength.

Boric acid electrolytes produce a film that is irridescent and oxides in the range of 2500–7500 Å. The coating is essentially nonporous.

Oxalic and other organic acids are electrolytes used to produce both protective and decorative films. Unsealed coatings are generally yellow in color. These films are harder and more abrasion resistant than the conventional sulfuric acid films. However, the specially hard coatings produced under special conditions in sulfuric acid electrolytes are superior.

The anodized coating consists of two major components: the nonporous barrier layer adjoining the metal and a porous layer extending from the barrier layer to the outer portion of the film. Sulfuric, chromic, and oxalic acid electrolytes form both barrier and porous layers, whereas, boric acid electrolytes produce only barrier films.

Anodizing of aluminum provides long-term corrosion resistance and decorative appearance. Corrosion of the anodized film is induced by SO_x gas and depositions of grime, sulfates, and chlorides. These depositions promote corrosion because they tend to absorb aggressive gases and moisture, thereby, increasing the time of wetness and decreasing the pH of the electrolyte at the interface between the depositions and the surface.

Although rain increases the time of wetness, it has the effect of cleaning the surface rather than making the surface corrosive. Some of the depositions are removed by the rain. Cleaning with water is one method that helps to protect the anodized aluminum from corrosion. In marine atmospheres, the depositions can be removed with water because the depositions are primarily soluble chlorides. However, in industrial atmospheres, detergents are needed because the deposits are greasy.

SO_x gas is the most aggressive pollutant for anodic films. The corrosive effect is dependent on the concentration, with the corrosion area increasing linearly with concentration.

7.13 Organic Coatings

Organic coatings are widely used to protect metal surfaces from corrosion. The effectiveness of such coatings is dependent not only on the properties of the coatings, that are related to the polymeric network and possible flaws in this network, but also on the character of the metal substrate, the surface pretreatment, and the application procedures. Therefore, when considering the application of a coating, it is necessary to take into account the properties of the entire system.

There are three broad classes of polymeric coatings: lacquers, varnishes, and paints. Varnishes are materials that are solutions of either a resin alone in a solvent (spirit varnishes) or an oil and resin together in a solvent (oleo-resinous varnishes). A lacquer is generally considered to be a material whose basic film former is nitrocellulose, cellulose acetate–butyrate, ethyl cellulose, acrylic resin, or another resin that dries by solvent evaporation. The term *paint* is applied to more complex formulations of a liquid mixture that dries or hardens to form a protective coating.

Organic coatings provide protection either by the formation of a barrier action from the layer or from active corrosion inhibition provided by pigments in the coating. In actual practice, the barrier properties are limited because all organic coatings are permeable to water and oxygen to some extent. The average transmission rate of water through a coating is about 10–100 times larger than the water consumption rate of a freely flowing surface, and in normal outdoor conditions, an organic coating is saturated with water at least half of its service life. For the remainder of the time, it contains a quantity of water comparable in its behavior to an atmosphere of high humidity. Table 7.13 shows the diffusion data for water through organic films.

It has also been determined that, in most cases, the diffusion of oxygen through the coating is large enough to allow unlimited corrosion. Taking these factors into account indicates that the physical barrier properties alone do not account for the protective action of coatings. Table 7.14 shows the flux of oxygen through representative free films of paint 100 μm thick.

Additional protection may be supplied by resistance inhibition that is also a part of the barrier mechanism. Retardation of the corrosion action is accomplished by inhibiting the charge transport between cathodic and anodic sites. The reaction rate may be reduced by an increase in the electrical resistance and/or the ionic resistance in the corrosion cycle. Applying an organic coating on a metal surface increases the ionic resistance.

TABLE 7.13

Diffusion Data for Water through Organic Films

Polymer	Temperature °C	$p \times 10^9$ (cm³ (STP)cm)	$D \times 109$ cm²/s
Epoxy	25	10–44	2–8
	40	—	5
Phenolic	25	166	0.2–10
Polyethylene (low density)	25	9	230
Polymethyl methacrylate	50	250	130
Polyisobutylene	30	7–22	—
Polystyrene	25	97	—
Polyvinyl acetate	40	600	150
Polyvinyl chloride	30	13	16
Vinylidene chloride/ acrylonitrile compolymer	25	1.7	0.32

Source: From H. Leidheiser Jr. 1987. Coatings, in *Corrosion Mechanisms*, F. Mansfield, Ed., New York: Marcel Dekker, pp. 165–209.

The electronic resistance may be increased by the formation of an oxide film on the metal. This is the case for aluminum substrates.

Corrosion of a substrate beneath an organic coating is an electrochemical process that follows the same principles as corrosion of an uncoated substrate. It differs from crevice corrosion because the reactants often reach the substrate through a solid. In addition, during the early stages of corrosion, small volumes of liquid are present, resulting in extreme values of pH and ion concentrations.

The total corrosion process takes place as follows:

1. Migration through the coating of water, oxygen, and ions

TABLE 7.14

Flux of Oxygen through Representative Free Films of Paint, 100 µm Thick

Paint	J (mg/cm² day)
Alkyd (15% PVC Fe₂O₃)	0.0069
Alkyd (35% PVC Fe₂O₃)	0.0081
Alkyl melamine	0.001
Chlorinated rubber (35% PVC Fe₂O₃)	0.017
Cellulose acetate	0.026 (95% RH)
Cellulose nitrate	0.115 (95% RH)
Epoxy melamine	0.008
Epoxy coal-tar	0.0041
Epoxy polyamide (35% PVC Fe₂O₃)	0.0064
Vinyl chloride/vinyl acetate copolymer	0.004 (95% RH)

Source: From H. Leidheiser Jr. 1987. Coatings, in *Corrosion Mechanisms*, F. Mansfield, Ed., New York: Marcel Dekker, pp. 165–209.

2. Development of an aqueous phase at the coating/substrate interface

3. Activation of the substrate surface for the anodic and cathodic reactions

4. Deterioration of the coating/substrate interfacial bond

7.13.1 Composition of Organic Coatings

The composition of a coating (paint) determines the degree of corrosion protection that will be supplied. A paint formulation is made up of four general classes of ingredients: vehicle (binder, resin component), pigment, filler, and additive. It is the combination of these ingredients that impart the protective properties to the coating.

7.13.1.1 *Resin Component*

The resin component is also known as the vehicle or binder, but *resin component* is more descriptive. The resin is the film-forming ingredient of the paint. It forms the matrix of the coating, the continuous polymeric phase that all other components may be incorporated in. Its density and composition largely determine the permeability, corrosion resistance, and UV resistance of the coating.

The most common resins used are vinyls, chlorinated rubber, acrylics, epoxies, urethanes, polyester, autooxidative cross-linking coatings, and water soluble resins.

7.13.1.1.1 *Vinyl Resins*

Vinyl is the general term denoting any compound containing the vinyl linkage ($-CH=CH_2$) group. However, there are many compounds containing this linkage that are not considered as vinyl coatings such as styrene and propylene.

Vinyl coatings are primarily considered to be copolymers of vinyl chloride and vinyl acetate having the chemical structure shown in Figure 7.18.

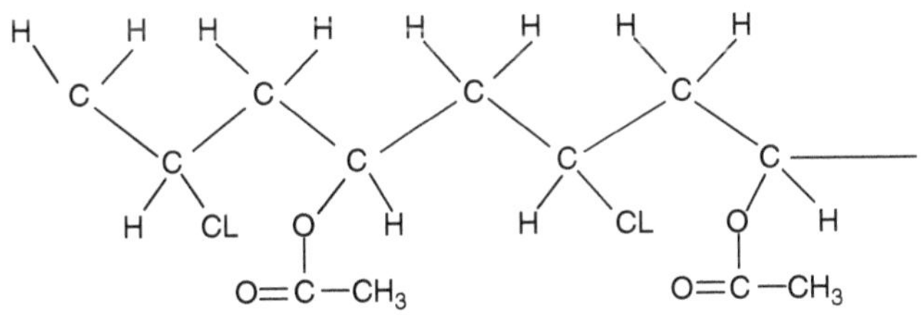

FIGURE 7.18
Chemical structure of vinyl acetate and vinyl chloride copolymer.

A relatively large amount of solvent is required to dissolve a vinyl copolymer high in vinyl chloride content. Because of this, the volume of solids in the solution is relatively low. As a result, most vinyl coatings must be applied in thin (1–1.5 mil) coats. Consequently, a vinyl coating system might require five or more coats. The system is considered highly labor intensive, even though in the proper environment excellent protection is provided.

A formulation has been developed that permits 2–2.5 mils to be applied per coat, but this advantage comes at the expense of reduced protection. The thixotropes, fillers, and additives used to permit the greater thicknesses are more susceptible to environmental and moisture penetration.

7.13.1.1.2 Chlorinated Rubber

Chlorinated rubber resins include those resins produced by the chlorination of both natural and synthetic rubbers. Chlorine is added to unsaturated double bonds until the final product contains approximately 65% chlorine with the chemical structure shown in Figure 7.19. Since the final product is a hard, brittle material with poor adhesion and elasticity, a plasticizer must be added. The volume of solids of the coating is somewhat higher than that of a vinyl; therefore, a suitably protective chlorinated rubber system often consists of only three coats.

7.13.1.1.3 Acrylics

The acrylics can be formulated as thermoplastic resins, thermosetting resins, and as a water-emulsion latex. The resins are formed from polymers of acrylate esters, primarily polymethyl methacrylate and polyethyl acrylate. Because the acrylate resins do not contain tertiary hydrogens attached directly to the polymer backbone chain, they are exceptionally stable to oxygen and UV light. The repeating units for the methacrylate and acrylate are as follows:

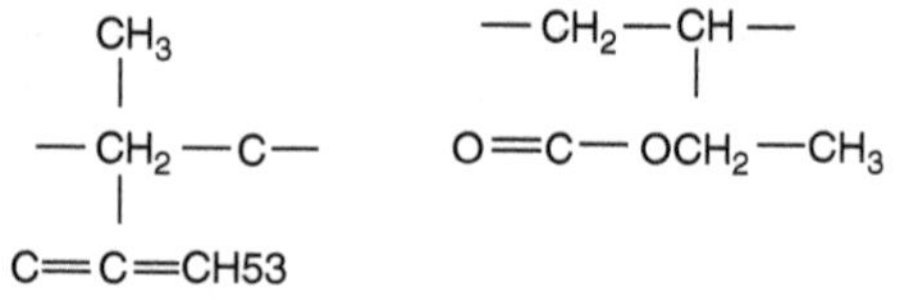

$$CH_3$$
$$|$$
$$-CH_2-C-$$
$$|$$
$$C=C=CH53$$
Polymethacrylate

$$-CH_2-CH-$$
$$|$$
$$O=C-OCH_2-CH_3$$
Polyethyl acrylate

7.13.1.1.4 Epoxies

Epoxy resins are not suitable for protective coatings because when pigmented and applied they dry to a hard, brittle film with very poor chemical resistance. When copolymerized with other resins, specifically with those of the amine or polyamide family, or with esterfied fatty acids, epoxy resins will form a durable protective coating.

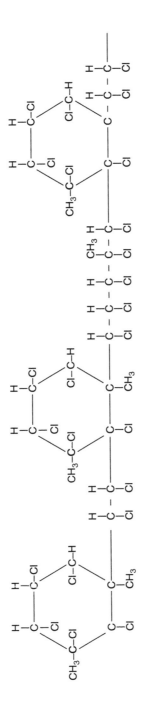

FIGURE 7.19
Chemical structure of chlorinated rubber.

7.13.1.1.5   Urethanes

Urethanes are reaction products of isocyanates with materials having hydroxyl groups and simply contain a substantial number of urethane groups (–N–C–O–) regardless of what the rest of the molecule may be. The hydroxyl containing side might consist of a variety of compounds including water (moisture cured urethanes), epoxies, polyesters, acrylics, and drying oils. Figure 7.20 shows several typical chemical structures.

Epoxy and polyester polyols, though more expensive than the acrylic polyols, are more chemically and moisture resistant. However, the acrylic polyol, when suitably reacted to form a urethane coating, is satisfactory for weathering atmospheres.

Coatings formulated with aliphatic isocyanates cost approximately twice as much as similar coatings formulated with epoxy and aromatic isocyanates. However, they have superior color and gloss retention and are resistant to deterioration by sunlight. Aliphatic urethanes will continue to maintain a glossy "wet" look years after application, whereas, other urethane coatings will chalk or yellow somewhat after exposure to the sun.

7.13.1.1.6   Polyesters

In the coating industry, polyesters are characterized by resins based on components that introduce unsaturation (–C=C–) directly into the polymer backbone. Typical is the structure of an isophthalic polyester resin as shown in Figure 7.21. The most common polyester resins are polymerized products of maleic or isophthalic anhydride or their acids. In producing paint, the polyester resin is dissolved in styrene monomer, together with pigment and small amounts of a reaction inhibitor. A free radical initiator, commonly a peroxide, and additional styrene are packaged in another container. When applied, the containers are mixed; sometimes, because of the fast initiating

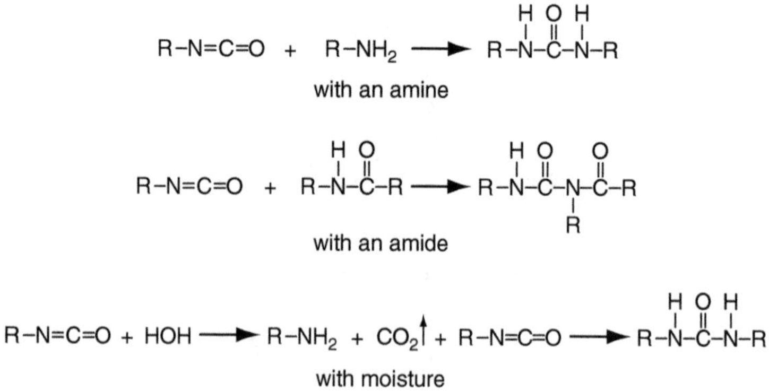

$$R{-}N{=}C{=}O \ + \ R{-}NH_2 \ \longrightarrow \ R{-}\overset{H}{\underset{}{N}}{-}\overset{O}{\underset{}{C}}{-}\overset{H}{\underset{}{N}}{-}R$$

with an amine

$$R{-}N{=}C{=}O \ + \ R{-}\overset{H}{\underset{}{N}}{-}\overset{O}{\underset{}{C}}{-}R \ \longrightarrow \ R{-}\overset{H}{\underset{}{N}}{-}\overset{O}{\underset{}{C}}{-}\overset{}{\underset{R}{N}}{-}\overset{O}{\underset{}{C}}{-}R$$

with an amide

$$R{-}N{=}C{=}O \ + \ HOH \ \longrightarrow \ R{-}NH_2 \ + \ CO_2{\uparrow} \ + \ R{-}N{=}C{=}O \ \longrightarrow \ R{-}\overset{H}{\underset{}{N}}{-}\overset{O}{\underset{}{C}}{-}\overset{H}{\underset{}{N}}{-}R$$

with moisture

FIGURE 7.20
Typical structure of urethane coating.

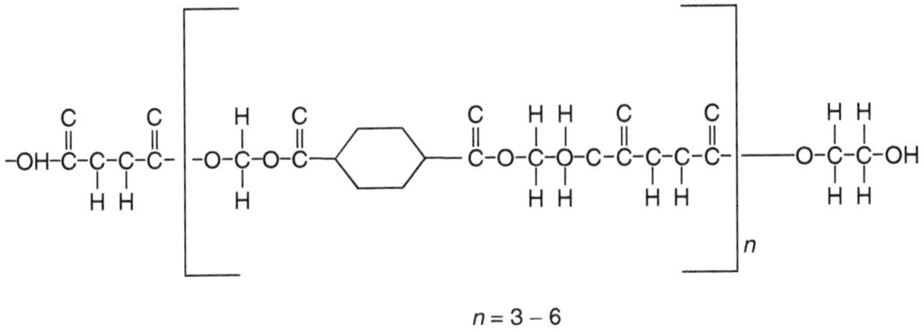

FIGURE 7.21
Chemical structure of polyester coating.

reaction (short pot life), they are mixed in an externally mixing or dual-headed spray gun. After being mixed and applied, a relatively fast reaction takes place, resulting in cross-linking and polymerization of the monomeric styrene with the polyester resin.

Another class of coatings, formulated in a similar manner, are the vinyl esters, The vinyl esters derive from a resin based on a reactive end vinyl group that can open and polymerize. The chemical structure is shown in Figure 7.22.

7.13.1.1.7 Autooxidative Cross-Linking Coatings

Autooxidative cross-linking coatings rely on a drying oil with oxygen to introduce cross-linking within the resin and attainment of final film

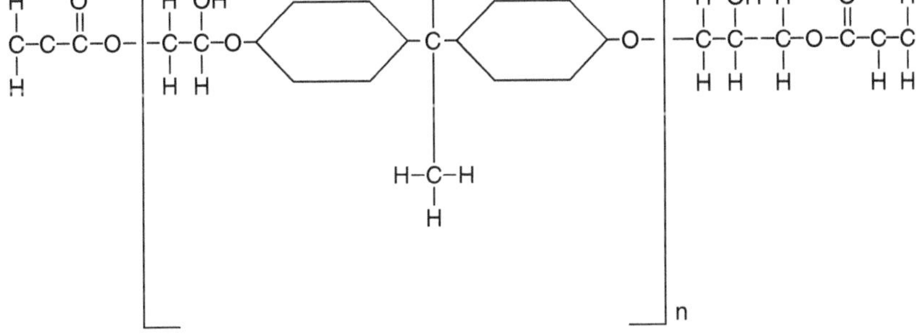

FIGURE 7.22
Chemical structure of vinyl acetate coating.

properties. The final coating is formed as a result of the drying oil reacting with the resin that is combined with pigments and solvents. The paint is packaged in a single can that can be opened, mixed, and the paint applied. Since oxygen reacts with the coating, introducing additional cross-linking, final film properties can take weeks or even months to attain.

Autooxidative cross-linking type paints include alkyds, epoxies, esters, oil-modified urethanes, and so on. These are commonly used, when properly formulated, to resist moisture and chemical fume environments. They can be applied over wood, metal, or masonry substrates.

The advantage of these coatings is their ease of application, excellent adhesion, relatively good environmental resistance (in all but immersion and high chemical fume environment), and tolerance for poorer surface preparation. Their major disadvantage, compared with other synthetic resins, is their lessened moisture and chemical resistance.

7.13.1.1.8   Water Soluble Resins and Emulsion Coatings

By introducing sufficient carboxyl groups into a polymer, any type of resin can be made water soluble. These groups are then neutralized with a volatile base such as ammonia or an amine, rendering the resin a polymeric salt, soluble in water or water/ether–alcohol mixtures. The major disadvantage to such resins is that polymers designed to be dissolved in water will remain permanently sensitive to water. Because of this, they are not widely used industrially.

However, water emulsions are widely accepted. The water-based latex emulsions are formulated with high molecular weight resins in the form of microscopically fine particles of high molecular weight copolymers of polyvinyl chloride, polyvinyl acetate, acrylic esters, stryren–butadiene, or other resins combined with pigments, plasticizers, UV stabilizers, and other ingredients. Water-based epoxy formulations are also available. The epoxy resin and a polyamide copolymer are emulsified and packaged in separate containers. After mixing, coalescence, and ultimate drying, the polyamide reacts with the epoxy to form the final film.

The major advantage of the emulsion coatings is their ease of cleanup. Their disadvantage is the permanent water sensitivity of the coating, making them unsuitable for use in continually wet environments. Water-based coatings cannot be applied to blast-cleaned steel surfaces with the same confidence as solvent-based coatings.

7.13.1.1.9   Zinc-Rich Coatings

In all of the above mentioned coatings, the final film properties of corrosion protection and environmental resistance have been attributed to organic composition of their constituent resin or binder. The pigment, although playing an important role in the final film properties, is secondary to the resistance of the organic binders.

Conversely, in a zinc-rich coating, the role of the zinc pigment predominates. The high amount of zinc dust metal in the dried film determines the coating's basic property, i.e., galvanic protection. For a coating to be considered zinc-rich, the common rule of thumb is that there must be at least 75% by weight of zinc dust in the dried film. This may change because conductive extenders (notably di-iron phosphide) have been added to improve weldability and burn-through, with supposedly equivalent protection at lower zinc loadings.

The primary advantage of zinc-rich coatings is their ability to galvanically protect. The zinc coating preferentially sacrifices itself in the electrochemical corrosion reaction to protect the underlying steel. This galvanic reaction, together with the filling and sealing effect of zinc reaction products (primarily zinc carbonate, zinc hydroxide, and complex zinc salts), provides more effective corrosion protection to steel substrates than does any other type of coating. Zinc-rich coatings can not be used outside the pH range of 6–10.5.

7.13.1.2 *Pigments*

Pigments serve several functions as part of the coating: They reinforce the film structurally, provide color and opacity, and serve special purposes for metal protection.

Typical materials used as color or hiding pigments and to provide aesthetic value, retention of gloss and color, and to help with the film structure and impermeability are iron oxides, titanium dioxide, carbon or lamp black, and others.

Pigments must be somewhat resistant to the environment and must be compatible with the resin. For example, calcium carbonate, that is attacked by acid, should not be used in an acid environment. Water-soluble salts are corrosion promoters, so special low salt containing pigments are used in primers for steel.

Primers contain one of three kinds of pigments for special protective properties as follows:

1. *Inert or chemically resistant.* These are used to form barrier coatings in severe environments such as pH less than 5 or greater than 10. They are also used as nonreactive extender, hiding, or color pigments in neutral environments.

2. *Active.* Leads, chromates, or other inhibitive pigments are used in linseed oil/alkyd primers.

3. *Galvanically sacrificial.* Zinc is employed at high concentrations to obtain electrical contact for galvanic protection in environments between pH 5 and 10.

Table 7.15 shows the type and characteristics of these pigments.

TABLE 7.15

Characteristics of Pigments for Metal Protective Paints

Pigment	Specific Gravity	Color	Opacity	Specific Contribution to Corrosion Resistance
Active pigments				
Red lead	8.8	Orange	Fair	Neutralizes film acids, insolubilizes sulfates and chlorides, renders water noncorrosive
Basic silicon lead chromate	3.9	Orange	Poor	Neutralizes film acids, insolubilizes sulfates and chlorides, renders water noncorrosive
Zinc yellow (chromate)	3.3	Yellow	Fair	Neutralizes film acids, anodic passivator, renders water noncorrosive
Zinc oxide (French process)	5.5	White	—	Neutralizes film acids, renders water noncorrosive
Zinc dust at low concentration in coatings for steel	7.1	Gray	Good	Neutralizes film acids
Galvanically protective pigments				
Zinc dust sacrificial at high concentration	7.1	Gray	Good	Makes electrical contact, galvanically sacrificial
Barrier pigments				
Quartz	2.6	Nil	Translucent	Inert, compatible with vinyl ester additives
Mica	2.8	Nil	Translucent	Impermeability and inertness
Talc	2.8	Nil	Translucent	Impermeability and inertness
Asbestine	2.8	Nil	Translucent	Impermeability and inertness
Barytes	4.1	Nil	Translucent	Impermeability and inertness
Iron oxide	4.1	Red	—	Impermeability and inertness
Iron oxide	4.1	Ochre	—	Impermeability and inertness
Iron oxide black	4.1	Black	—	Impermeability and inertness
Titanium dioxide	4.1	White	Excellent	Impermeability and inertness
Carbon black	1.8	Black	Good	Impermeability and inertness

Note: Titanium dioxide has better "hiding" than any other pigment.

Source: From K.B. Tator. 1988. Coatings, in *Corrosion and Corrosion Protection Handbook*, P.A. Schweitzer, Ed., New York: Marcel Dekker, pp. 453–489.

7.13.1.3 Fillers (Extenders)

In addition to lowering the cost, extenders also provide sag resistance to the liquid paint so that the edges remain covered. When the paint has dried, they reduce the permeability to water and oxygen and provide reinforcing structure within the film. Talc and mica are used extensively as extenders. Mica is limited to approximately 10% of the total pigment. Both talc and mica, but particularly mica, reduce the permeability through the film as platelike particles block permeation, forcing water and oxygen to seek a longer path through the binder around the particle.

7.13.1.4 Additives

Additives are used in coating formulations for specific purposes and in small, sometimes even trace amounts. Some typical additives and their purposes are as follows:

Phenylmercury, zinc, cuprous compounds: mildew inhibitors
Cobalt and manganese naphthanates: aid to surface and thorough drying
Zinc oxide: protection of the resin from heat and sun

Latex paints have a number of additives that act as stabilizers, coalescing aids, emulsion stabilizers, freeze–thaw stabilizers, and so on.

7.13.1.5 Solvents

Organic solvents are usually required only to apply the coating and after application are designed to evaporate from the wet paint film. Water is considered either as a solvent or an emulsifier. The rate that solvents evaporate at influences the application characteristics of the coating. It is imperative that the solvent completely evaporate. If the solvents are partially retained and do not evaporate completely, premature failure by blistering or pinholing is likely to occur.

Most coatings are formulated to be applied at ambient conditions of approximately 75°F/24°C and a 50% relative humidity. If ambient conditions are considerably higher or lower, it is conceivable that solvent release could pose a problem. This potential problem can be eased by changing the "solvent balance." Generally, faster evaporating solvents should be used in colder weather, and slower evaporating solvents should be used in hot weather. Classes and characteristics of some common solvents are shown in Table 7.16.

Low viscosity and two component epoxies and powder coatings are examples of paint systems referred to as solvent-free. The epoxy coatings may be mixed and applied without the use of a solvent, as the two components typically have a low viscosity.

Powder coatings are cured by sintering. Before application thermoplastic powder coatings consist of a large number of small binder particles. These particles are deposited on a metal surface using special application techniques. Subsequently, the paint is baked in an oven to form a continuous film by sintering.

7.13.2 Surface Preparation

For a paint to provide maximum protection, it is essential that the surface to which the paint is to be applied is properly prepared. The specific preparation system will be dependent on the coating system to be applied

TABLE 7.16

Characteristics of Solvent Classes

Class	Solvent Name	Polarity	Specific Gravity	Boiling Range (°F)	Flash Point of TCC	Evaporation Rate[a]
Aliphatic	VM&P naphtha	Nonpolar	0.74	246–278	52	24.5
	Mineral spirits	Nonpolar	0.76	351–395	128	9.0
Aromatic	Toluene	Intermediate	0.87	230–233	45	4.5
	Xylene	Intermediate	0.87	280–288	80	9.5
	High solvency	Intermediate	0.87	360–400	140	11.6
Ketone	Methyl ethyl ketone (MEK)	High	0.81	172–176	24	2.7
	Methyl isobutyl ketone (MIBK)	High	0.80	252–266	67	9.4
Ester	Cyclohexanone	High	0.95	313–316	112	4.1
Alcohol	Ethyl acetate	Intermediate	0.90	168–172	26	2.7
Unsaturated aromatic	Ethanol	Intermediate	0.79	167–178	50	6.8
	Styrene	Intermediate	0.90			
Glycol	Cellosolve	High	0.93	273–277	110	0.3
Ethers	Butyl cellosolve	High	0.90	336–343	137	0.06

[a] Butyl acetate = 1.

and should usually be in accord with one of the surface preparation specifications defined by the Steel Structures Painting Council that follow:

Specification	Description
SP1: Solvent cleaning	Removal of oil, grease, dirt, soil, salts and contaminants by cleaning with solvent, vapor, alkali, emulsion, or steam
SP2: Hand-tool cleaning	Removal of loose rust, mill scale, and paint, to degree specified, by hand chipping and wire brushing
SP3: Power tool cleaning	Removal of loose rust, mill scale, and paint, to degree specified, by power tool chipping, descaling, sanding, wire brushing, and grinding
SP5: White-metal blast cleaning	Removal of all visible rust, mill scale, paint, and foreign matter by blast cleaning, by wheel or nozzle, dry or wet, using sand, grit, or shot for very corrosive atmospheres where high cost of cleaning is warranted
SP6: Commercial blast cleaning	Blast cleaning until at least two-thirds of the surface area is tree of all visible residues (for severe exposure)
SP7: Brush-off blast cleaning	Blast cleaning of all except tightly adhering residues of mill scale, rust and coatings, exposing numerous evenly distributed flecks of underlying metal
SP8: Pickling	Complete removal of rust and mill scale by acid pickling, duplex pickling, or electrolyte pickling
SP10: Near white blast cleaning	Blast cleaning to near white metal cleanliness until at least 95% of the surface area is free of all visible residue (for high humidity, chemical atmosphere, marine, and other corrosive environments)

In addition to these mechanical methods, conversion coatings on metals produced by chemical or electrochemical means also increase the adhesive bonding of paint coatings. These types of coatings will be discussed later.

The mechanical processes are used to remove dirt, rust, and mill scale from the surface and are adopted to steel products such as thick plates, construction steels, and steel structures that chemical processes are not available for.

The chemical processes are applied to steel-strip products such as automotive bodies and home electrical appliances, to the internal coating of steel tubes, and to aluminum and zinc and their alloy products.

7.13.3 Application Methods

The method of application for a corrosion resistant paint will depend on:

1. Purpose that coated product is to be used for
2. Environment that the coating will be exposed to

3. The type of paint
4. The shape and size of the object to be painted
5. The period of application process
6. Cost

7.13.3.1 Brushing

This was once the main coating method, but at the present time, spray coating is more widely used. Brush coating has the following advantages:

1. Applicators are simple and inexpensive.
2. Complicated forms and shapes can be coated.
3. Thick films are obtained with one coat.
4. It provides particularly useful antirusting coating.

The disadvantage of brushing results from the nonuniformity of coating layers, especially coating layers of rapidly drying paints.

7.13.3.2 Spray

There are two types of spray coatings: air sprays and airless sprays. These coatings are subdivided into hot and cold types.

7.13.3.2.1 Air Spray

Air spray is a process for spraying fine paint particles atomized by compressed air. Drying rates are high and a uniform and decorative surface is obtained. This method is used with paints that are not suitable for brushing methods, for mass produced painted products, or for products where surface appearance is important.

7.13.3.2.2 Airless Spray

In an airless spray system, high pressure is supplied to the paint that is forced through a high pressure resistant hose and airless spray gun. High viscosity paints are warmed before spraying. This technique has the following advantages over the air spray system:

1. The sticking ratio of paint is increased to 25–40%.
2. Thicker films can be applied.
3. The running of paint on the object is reduced.

Steel structures and bridge girders can be coated in the factory by this method because the efficiency of work is several times higher than brushing. However, its paint loss is 30–40% greater than for paints applied by brushing.

7.13.3.3 Roller Coating

This method is used to coat coils and sheets by passing them through two rollers. The thickness of paint film is controlled by adjusting the rollers. One-side or both-sides coating is possible.

7.13.3.4 Powder Coating

Because of the absence of organic solvents, powder coatings have grown in popularity as antipollution coatings. Coating thicknesses of 25–250 µm can be obtained. Automotive bodies, electric components, housing materials, wires, and cables make use of this method. Polyethylene and epoxy resins are the predominent types of paints used. At the present time, the following eleven procedures are used in this coating process:

1. Pouring method (flock coating)
2. Rotational coating of pipes
3. Fluidized bed
4. Dipping in nonfluidizing powders
5. Centrifugal casting
6. Rotational molding
7. Electrostatic Powder spraying
8. Electrostatic fluidized bed
9. Pouring or flowing of fluidized powder
10. Electrogasdynamics powder spraying
11. Flame spraying of thermoplastic powders

7.13.3.5 Electrodeposition

The anodic deposition process for paint coating systems was introduced in the early 1960s and the cathodic deposition system in 1972. Electrodeposition processes are widely used since they possess the advantages of unmanned coating, automation, energy saving, and lower environmental pollution. This process is used to apply coatings to automotive bodies and parts, domestic electric components, machine parts, and architecturals such as window frames. Schematic illustrations of anodic and cathodic electrodeposition of paints are shown in Figure 7.22.

The primary paints used in the electrodeposition process are anionic type resins with a carboxyl group (RCOOH, polybutadiene resin) and cationic type resin (R–NH$_2$, epoxy resin). Hydrophilic groups and neutralizing agents are added to the water insoluble or undispersed prepolymers to convert them to soluble or dispersed materials.

The dissolution of metal substrate in the cathodic process is much less than that in the anodic process. The primary resins used in the cathodic

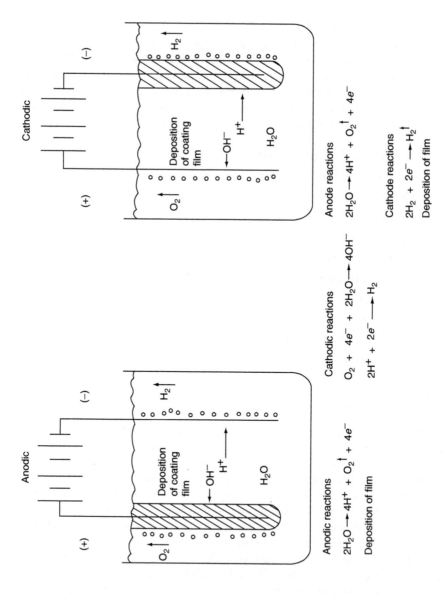

FIGURE 7.23

Electrodeposition of paints.

process are epoxy, and because epoxy resins provide good water and alkali resistance as well as adhesion, cationic paint coatings are superior in corrosion protection to anodic paint coatings.

7.13.3.6 Multilayer Coatings

The protective ability of a coating is also dependent upon its thickness. Thicker coatings provide improved protection. However, single thick paint films tend to crack because of internal stress. When long term protection is required, multilayer paint coating systems are applied.

The multilayer system to be used with a steel structure is determined by the required service life and the environmental conditions that the steel structure must exist under. General coating systems are applied to structures in mild environments, whereas, heavy-duty coating systems are applied to these structures in severe environments. Typical general coating systems and heavy duty coating systems are shown in Table 7.17 and Table 7.18, respectively.

7.13.4 Factors Affecting Life of Film

For a paint film to perform effectively, it must have mechanical resistance to external forces and resistance against environmental factors. As discussed, a paint film is a composite layer that is composed of a vehicle (resin), pigment, and additives, and tends to contain defects in its structure. The common defects existing in a resultant paint film are:

1. Nonuniformity of vehicle molecules
2. Solvent residue
3. Residual stress

TABLE 7.17

General Coating Multistage Paint Systems

Coat	System		
	A	**B**	**C**
First	Etching primer	Etching primer	Zinc rich primer
Second	Oil corrosion Preventative paint	Oil corrosion Preventative paint	Chlorinated rubber System primer
Third	Oil corrosion Preventative paint	Oil corrosion Preventative paint	Chlorinated rubber System primer
Fourth	Long oil alkyd Resin paint	Phenolic resin System MIO paint	Chlorinated rubber System paint
Fifth	Long oil alkyd Resin paint	Chlorinated rubber System paint	Chlorinated rubber System paint
Sixth		Chlorinated rubber System paint	

TABLE 7.18

Heavy-Duty Coating Multilayer Paint Systems

	System			
Coat	A	B	C	D
First	Zinc spray or zinc-rich paint	Thick type zinc-rich paint	Thick type zinc-rich paint	Zinc-rich primer
Second	Etching primer	Thick type Vinyl or chlorinated rubber system	Thick type epoxy primer	Tar epoxy resin paint
Third	Zinc chromate primer	Thick type vinyl or chlorinated rubber system	Thick type epoxy primer	Tar epoxy resin paint
Fourth	Phenolic resin system	Vinyl resin or chlorinated	Epoxy resin system paint	Tar epoxy resin paint
	MIO paint	Rubber system		
Fifth	Chlorinated rubber	Vinyl resin or chlorinated	Epoxy or poly-urethane	
	System	Rubber system	Resin system paint	
Sixth	Chlorinated rubber system paint			

4. Differences in expansion coefficients in multilayer coating systems
5. Poor adhesion of polymer to pigment

7.13.5 Strength of Paint Film

Paint films require hardness, flexibility, brittleness resistance, abrasion resistance, mar resistance, and sag resistance, Paint coatings are formulated providing a balance of these mechanical properties. The mechanical strength of a paint film is described by the words *hardness* and *plasticity* that correspond to the modulus of elasticity and to the elongation at break obtained from the stress stress–strain curve of a paint film. Typical paint films have tensile properties as follows:

Paints	Tensile Strength (g/mm)	Elongation at Break (%)
Linseed oil	14–492	2–40
Alkyd resin varnish (16% PA)	141–1266	30–50
Amino-alkyd resin varnish ($A/W = 7/3$)	2180–2602	—
NC lacquer	844–2622	2–28
Methyl-*n*-butyl-meta-acrylic resin	1758–2532	19–49

Mechanical properties of paint coatings vary depending on the type of pigment, baking temperatures, and aging times. As baking temperatures rise, the curing of paint films is promoted and elongation reduced. Tensile strength is improved by curing and the elongation at breaks is reduced with increased drying time.

TABLE 7.19

Glass Transition Temperature of Organic Films

Organic Film	Glass Transition Temperature, T_g (°C)
Phthalic acid resin	50
Acrylic lacquer	80–90
Chlorinated rubber	50
Bake type melamine resin	90–100
Anionic resin	80
Catonic resin	120
Epoxy resin	80
Tar epoxy resin	70
Polyurethane resin	40–60
Unsaturated polyester	80–90
Acrylic powder paint	100

Structural defects in a paint film causes failures that are determined by environmental conditions such as thermal reaction, oxidation, photooxidation, and photothermal reaction. An important factor in controlling the physical properties of a paint film is the glass transition temperature T_g. In the temperature range higher than T_g, the motion of the resin molecules become more active so that hardness, plasticity, and the permeability of water and oxygen vary greatly. Table 7.19 lists the glass transition temperatures of organic films.

Deterioration of paint films is promoted by photolysis, photooxidation, or photo thermal reaction as a result of exposure to natural light. Specifically, UV light (wavelength 40–400 nm) decomposes some polymer structures. Polymer films, such as vinyl chloride resins, are gradually decomposed by absorbing the energy of UV light.

The T_g of a polymer is of critical importance in the photolysis process. Radicals formed by photolysis are trapped in the matrix, but they diffuse and react at higher temperatures than T_g. The principal chains of polymers with ketone groups form radicals:

$$RCOR' \quad R + COR' \quad CO + R'$$
$$ROCOR' \quad OCOR' \quad CO_2 + R'$$

The resultant radicals accelerate the degradation of the polymer and in some cases HCl (from polyvinyl chloride) or CH_4 is produced.

7.13.6 Adhesion of Paint Film

Paint coated materials are exposed primarily to the natural atmosphere. Consequently, paint adhesion is influenced by atmospheric factors, particularly moisture and water. All organic polymers and organic coatings are permeable to water. They differ only in the degree of permeability. Table 7.13 provides typical values for the permeability and the diffusion coefficient for water in a number of different polymers.

When moisture permeates into the paint substrate interface, blisters form. There are four mechanisms whereby blisters can form:

1. Volume expansion because of swelling
2. Gas inclusion or gas formation
3. Osmotic process because of soluble impurities at the film–substrate interface
4. Electroosmotic blistering

Of the four mechanisms, osmotic blister formation is the most important. Past experience has indicated that 70% of all paint coating failures are the result of poor or inadequate surface preparation prior to application of the coating. Types of osmotically active surfaces at film-substrate interfaces are impurities such as sand particles from incomplete washing after wet sanding, water soluble salt residues from phosphating, or surface nests in a rust layer. It is also possible to entrap hydrophilic solvents to create blisters.

Regardless of the specific driving force, the blister formation process is shown in Figure 7.24 using an example of an intact coating on steel. When the coating is exposed to an aqueous solution, water vapor molecules (and to a lesser extent, oxygen), diffuse into the film toward the coating-substrate interface. Eventually, water may accumulate at the interface forming a thin water film of at least several monolayers. The accumulation takes place at sites of poor adhesion or at sites where wet adhesion problems arise. The degree that permeated water may change the adhesion properties of a coated system to is referred to as wet adhesion. No initial, poorly adhered spots need be present for chemical disbondment to take place. A corrosion reaction can be initiated by the presence of an aqueous electrolyte with an electrochemical double layer, a cathodic species (oxygen) and an anodic species (metal). When the corrosion reaction has started at the interface, the corrosion products can be responsible for the buildup of osmotic pressure, resulting in the formation of macroscopic blisters (see Figure 7.24). Depending on the specific materials and circumstances, the blisters may grow out because of the hydrodynamic pressure in combination with one of the chemical propagation mechanisms such as cathodic delamination and anodic under-coating.

Oxygen migration through a paint film coating is generally much lower than that of water. In some instances, the flow of oxygen increases with the water content of the film, probably because the water acts to swell the polymer.

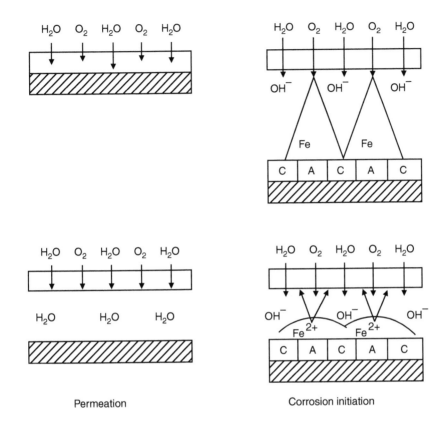

FIGURE 7.24
Introduction of a blister under an intact coating.

7.13.7 Types of Corrosion under Organic Coatings

For corrosion to take place on a metal surface under a coating, it is necessary for an electrochemical double layer to be established. In order for this to take place, it is necessary for the adhesion between the substrate and coating to be broken. This permits a separate thin water layer to be formed at the interface from water that has permeated the coating. As previously mentioned, all organic coatings are permeable to water to some extent. Water permeation occurs under the influence of various driving forces:

1. A concentration gradient, e.g., during immersion or exposure to a humid atmosphere, resulting in true diffusion through the polymer
2. Capillary forces in the coating because of poor curing, improper solvent evaporation, entrapment of air during application, or bad interaction between binder and additives
3. Osmosis because of impurities or corrosion products at the interface between the metal and coating

Water molecules will eventually reach the coating substrate interface when a coated system has been exposed to an aqueous solution or a humid atmosphere. These molecules can interfere with the bonding between the coating and substrate, eventually resulting in the loss of adhesion and corrosion initiation if a cathodic reaction can take place. A constant supply of a cathodic species, such as water or oxygen, is required for the corrosion reaction to proceed.

Water permeation may also result in the buildup of high osmotic pressure that are responsible for blistering and delamination.

7.13.7.1 Cathodic Delamination

When cathodic protection is applied to a coated metal, loss of adhesion between the substrate and paint film, adjacent to defects, often takes place. This loss of adhesion is known as cathodic delamination. Such delamination may also occur in the absence of applied potential. Separation of anodic and cathodic reaction sites under the coating results in the same local driving force as during external polarization. The propagation of a blister because of cathodic delamination under an undamaged coating on a steel substrate is schematically illustrated in Figure 7.25. Under an intact coating, corrosion may be initiated locally at sites of poor adhesion.

A similar situation develops in the case of corrosion under a defective coating. When there is a small defect in the coating, part of the substrate is directly exposed to the corrosive environment. Corrosion products are formed immediately that block the damaged site from oxygen. The defect in the coating is sealed by the corrosion products, after which corrosion propagation takes place, according to the same mechanism as for the initially undamaged coating. Refer to Figure 7.25 for the sequence of events.

7.13.7.2 Anodic Undermining

Anodic undermining results from the loss of adhesion caused by anodic dissolution of the substrate metal or its oxide. In contrast to cathodic delamination, the metal is anodic at the blister edges. Coating defects may cause anodic undermining, but in most cases it is associated with a corrosion sensitive site under the coating, such as a particle from a cleaning or blasting procedure, or a site on the metal surface with potentially increased corrosion activity (scratches). These sites become active once the corrodent has penetrated to the metal surface. The initial corrosion rate is low. However, an osmotic pressure is caused by the soluble corrosion products which stimulates blister growth. Once formed, the blister will grow because of a type of anodic crevice corrosion at the edge of the blister.

Coated aluminum is very sensitive to anodic undermining, whereas, steel is more sensitive to cathodic delamination.

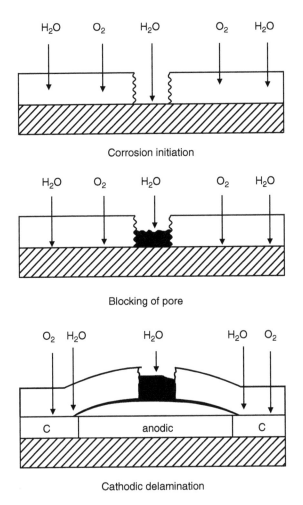

FIGURE 7.25
Blister initiation and propagation under a defective coating (cathodic delamination).

7.13.7.3 Filiform Corrosion

Metals with semipermeable coatings or films may undergo a type of corrosion, resulting in numerous meandering threadlike filaments of corrosion beneath the coatings or films. Conditions that promote this type of corrosion are:

1. High relative humidity (60–95% at room temperature).
2. The coating is permeable to water.
3. Contaminants (salts, etc.) are present on or in the coating or at the coating-substrate interface.

4. The coating has defects (mechanical damage, pores, insufficient coverage of localized areas, air bubbles).

Filiform corrosion under organic coatings is common on steel, aluminum, magnesium, and zinc (galvanized steel). It has also been observed under electroplated silver plate, gold plate, and under phosphate coatings.

This form of corrosion is more prevalent under organic coatings on aluminum than on other metallic surfaces, being a special form of anodic undermining.

A differential aeration cell is the basic driving force. The filaments have considerable length but little width and depth and consist of two parts: a head and a tail. The primary corrosion reactions and, subsequently, the delamination process of the paint film takes place in the active head, whereas, the tail is filled with the resulting corrosion products. As the head of the filiform moves, the filiform grows in length.

7.13.8 Causes of Coating Failures

Coating failures result from two basic causes:

1. Poor or inadequate surface preparation and/or application of the paint to the substrate
2. Atmospheric effects

Coating failures typical from surface preparation and application problems include:

1. Cracking, checking, and alligatoring. These type of failures develop with the aging of the paint film. Shrinkage within the film during aging causes cracking and checking. Alligatoring is a film rupture, usually caused by application of a hard brittle film over a more flexible film.
2. Peeling, flaking, and delamination. These failures are caused by poor adhesion. When peeling or flaking occurs between coats, it is called delamination.
3. Rusting. Failure of a coated surface may appear as (a) spot rusting at minute areas, (b) pinhole rusting at minute areas, (c) rust nodules breaking through the coating, and (d) underfilm rusting, that eventually causes peeling and flaking of the coating.
4. Lifting and wrinkling. When the solvent of a succeeding coat of paint too rapidly softens the previous coat, lifting results. Rapid surface drying of a coating without uniform drying throughout the rest of the film results in a phenomenon known as wrinkling.

5. Failures around weld areas. Coating adhesion can be hampered by weld flux that may also accelerate corrosion under the film. Relatively large projections of weld spatter cause possible gaps and cavities that may not be coated sufficiently to provide protection.

6. Edge failures. Edge failures usually take the form of rusting through the film at the edge where the coating is normally the thinnest. This is usually followed by eventual rust creepage under the film.

7. Pinholing, tiny holes that expose the substrate, caused by improper paint spray atomization or segregation of resin in the coating. If practical during application, brush out coating. After application and proper cure, apply additional coating.

Polymer coatings are exposed to the environment and, therefore, are subject to degradation by environmental constituents. The primary factors promoting degradation are thermal, mechanical, radiant, and chemical. Polymers may also be degraded by living organisms such as mildew. Any atmospheric environment is subject to dry and wet cycles. Since water and moisture have a decided effect on the degradation of a coating, the time of wetness of a coating is important. Moisture and water that attack organic films are from rain, fog, dew, snow, and water vapors in the atmosphere. Relative humidity is a particularly important factor. As exposure time is increased in 100% relative humidity, the bond strength of a paint coating is reduced. This is shown in Table 7.20.

Temperature fluctuations and longer time of wetness tends to produce clustered water that increases the acceleration of degradation of the organic film, particularly in a marine atmosphere. The most severe natural atmosphere for a paint film is that of a seashore environment.

The mode of degradation may involve depolymerization, generally caused by heating, splitting out of constituents in the polymer, chain scission, cross-linking, oxidation, and hydrolysis. Polymers are subject to cracking on the application of a tensile force, particularly when exposed to

TABLE 7.20

Relationship of Bond Strength to Exposure Time in 100% Relative Humidity

Exposure Time (h)	Bond Strength (psi)		
	Epoxy Ester	Polyurethane	Thermosetting Acrylic
Initial	4790	3410	5700
24	1640	1500	3650
48	1500	1430	3420
120	—	1390	2400
195	1400	1130	1850
500	1390	670	480

certain liquid environments. This phenomenon is known as environmental stress cracking or stress corrosion cracking.

Other factors in the atmosphere that cause corrosion or degradation of the coating are UV light, temperature, oxygen, ozone, pollutants, and wind. Types of failures caused by these factors are:

1. Chalking. UV light, moisture, oxygen, and chemicals degrade the coating resulting in chalk. This may be corrected by providing an additional top coat of material with proper UV inhibitor.

2. Color fading or color change. This may be caused by chalk on the surface or by breakdown of the colored pigments. Pigments can be decomposed or degraded by UV light or reaction with chemicals.

3. Blistering. Blistering may be caused by:
   a. Inadequate release of solvent during both application and drying of the coating system.
   b. Moisture vapor that passes through the film and condenses at a point of low adhesion.
   c. Poor surface preparation.
   d. Poor adhesion of the coating to the substrate or poor intercoat adhesion.
   e. A coat within the paint system that is not resistant to the environment.
   f. Application of a relatively fast drying coating over a relatively porous surface.
   g. Failures because of chemical or solvent attack. When a coating is not resistant to its chemical or solvent environment, there is apparent disintegration of the film.

4. Erosion, coating worn away. Loss of coating because of inadequate erosion protection. Provide material with greater resistance to erosion.

7.13.9 Maintenance of the Coating

Any paint job, even if properly done, does not last forever. Within the first six months or year after application, spots, inadvertent misses, or weak spots in the coating system can be detected by simple visual inspection. A thorough inspection and repair should be done at this time.

Over a period of time, the coating will break down and deteriorate as a result of the effect of the environment. Scheduled inspections should he made to determine the extent and rate of the coating breakdown. Spot touchup repair should be made at localized areas of failure before deterioration of the entire coated surface has taken place.

Extensive costs of total surface preparation (such as complete blast cleaning removal of all old coating) can be avoided if a "planned maintenance approach" of periodic spot touchup and an occasional recoat of the entire surface is followed.

Many coatings that provide good long term protection are more difficult to touch up or repair in the event of physical damage or localized failure. This is particularly true of the thermosetting and zinc-rich coatings. Application of a subsequently applied paint coat to an older aged epoxy, urethane, or other catalyzed coating often results in reduced adhesion that leads to peeling.

Thermoplastic coatings do not normally present this problem. Solvents of a freshly applied thermoplastic coating soften and allow for intermolecular mixing of the new and old coatings with good intercoat adhesion.

Heavily pigmented coatings, such as zinc-rich, require agitated pots to keep the pigment in suspension during application. Because of this, touchup and repair of large areas is not recommended using zinc-rich coatings unless it is done by spray using an agitated pot.

For the most part, oil-based coatings (alkyds, epoxy esters, and modifications thereof) have a greater tolerance for poor surface preparation and an ability to wet, penetrate, and adhere to poorly prepared surfaces or old coatings. Consequently, these coatings are often specified for these purposes, even though they do not provide as long a term of corrosion protection.

References

1. I. Suzuki. 1989. *Corrosion Resistant Coatings Technology,* New York: Marcel Dekker.
2. D. Satas. 1991. *Coatings Technology Handbook,* New York: Marcel Dekker.
3. F. Mansfield. 1987. *Corrosion Mechanisms,* New York: Marcel Dekker.
4. F.C. Porter. 1994. *Corrosion Resistance of Zinc and Zinc Alloys,* New York: Marcel Dekker.
5. P.A. Schweitzer. 1988. *Corrosion and Corrosion Protection Handbook,* 2nd ed., New York: Marcel Dekker.
6. H.R. Copson. 1995. Report of subcommittee B-3 on Atmospheric Corrosion of Nonferrous metals. ASIM Annual Meeting, Atlantic City, NJ, June 25.
7. R. Leidheiser, Jr. 1987. Coatings, in *Corrosion Mechanisms,* F. Mansfield, Ed., New York: Marcel Dekker, pp. 165–209.
8. K.B. Tator. 1988. Coatings, in *Corrosion and Corrosion Protection Handbook,* 2nd ed., P.A. Schweitzer, Ed., New York: Marcel Dekker, pp. 453–489.

8

Specific Organic Coatings

To select a suitable coating system, it is necessary to determine the environmental conditions around the structure to be painted. If the coating is to be applied to an exterior surface, are there nearby chemical plants, pulp and paper mills, or heavy industries that might provide airborne pollutants? Is the objective of the coating to provide an aesthetic appearance or to provide maximum corrosion protection that will limit the colors to the normal grays, whites, and pastels of the most corrosion resistant coatings? Will the surface be predominantly wet (salt or freshwater) or exposed to a chemical contaminant (acid or alkaline)? Is the environment predominantly a weathering environment subject to heat, cold, daily or seasonal temperature changes, precipitation, wind (flexing), exposure to sunlight, or detrimental solar rays? Will the surface be exposed to occasional spells of a corrosive or solvent? After the environmental conditions of the area have been defined, a suitable resistant coating can be selected. Characteristics, corrosion resistance, limitations, and uses of the common generic types of coatings will be discussed.

8.1 Thermoplastic Resins

Protective coatings formulated from these resins do not undergo any chemical change from the time of application until the final properties of a dried protective film have been attained. The film is formed by the evaporation of the solvent, concentrating the binder/pigment solution and causing it to precipitate as a continuous protective film.

8.1.1 Vinyls

These are polyvinyls dissolved in aromatics, ketones, or esters.

1. Resistance: Insoluble in oils, greases, aliphatic hydrocarbons, and alcohols; resistant to water and salt solutions; at room temperature, resistant to inorganic acids and alkalies; fire resistant.

2. Temperature resistance: 180°F/82°C dry; 140°F/60°C wet.

3. Limitations: Dissolved by ketones, aromatics, and ester solvents. Adhesion may be poor until all solvents have vaporized from the coating. Relatively low thickness per coat requiring several coats to provide adequate protection.

4. Features: Tasteless and odorless.

5. Applications: Used on surfaces exposed to potable water, as well as for immersion service, sanitary equipment, and widely used as industrial coatings.

When used for immersion service, it is essential that all solvents have evaporated before placing the vinyl coated object into service. The high polarity of the resin tends to retain the solvents used to dissolve the resin. Solvent evaporation is retarded, with the resultant effect of solvent voids within the coating, pinholes caused by volatilization of retained solvents upon heating or water filled blisters because of hydrogen bonding attraction of water by the retained solvents in the coating.

8.1.2 Chlorinated Rubber

Chlorinated rubber resins include those resins produced by the chlorination of both natural and synthetic rubbers. Chlorine is added to unsaturated double bonds until the final product contains approximately 65% chlorine. A plasticizer must be added because the final product is a hard, brittle material with poor adhesion and elasticity. The volume of solids of the coating dissolved in hydrocarbon solvents is somewhat higher than that of a vinyl; therefore, a suitably protective chlorinated rubber system often consists of only three coats.

1. Resistance: Chemically resistant to acids and alkalies; low permeability to water vapor; abrasion resistant; fire resistant. Nontoxic.

2. Temperature resistance: 200°F/93°C dry; 120°F/49°C wet.

3. Limitations: Degraded by ultraviolet light; attacked by hydrocarbons.

4. Applications: Excellent adherence to concrete and masonry. Used on structures exposed to water and marine atmospheres (swimming pools, etc.).

When used for water immersion service, the same care must be exercised during evaporation as was described for the vinyls.

8.1.3 Acrylics

The acrylics can be formulated as thermoplastic resins, thermosetting resins, and as a water emulsion latex.

TABLE 8.1

Glass Transition Temperature vs. Application Area

T_g (°C/°F)	Application Area
80–100/176–212	High heat resistant coatings
50–65/122–149	Floor care coatings
35–50/95–122	General industrial coatings
10–40/50–100	Decorative paints

The resins are formed from polymers of acrylate esters, primarily polymethyl methacrylate and polyethyl acrylate. Because the acrylate esters do not contain tertiary hydrogens attached directly to the polymer backbone chain, they are exceptionally stable to oxygen and UV light.

A wide range of monomers is available for use in designing a specific acrylic system. Typically, mixtures of monomers are chosen for the properties they impart to the polymer. The glass transition temperature T_g of the polymer can be varied by selection of the proper monomers. This permits a varied area of application. Table 8.1 illustrates the wide range of T_g resulting from different monomer compositions for emulsion acrylics.

Acrylics may be formulated as lacquers, enamels, and emulsions. Lacquers and baking enamels are used as automotive and appliance finishes. In both these industries, acrylics are used as top coats in multicoat finishing systems.

1. Resistance: Somewhat resistant to acids, bases, weak and moderately strong oxidizing agents, and many corrosive industrial gases and fumes. Resistant to weather and UV light.

2. Limitations: Limited penetrating power (water emulsion) because of water surface tension; may flash rust as a primer over bare steel; not suitable for immersion service or strong chemical environments. Soluble in ketones, esters, aliphatic chlorinated hydrocarbons, and aromatic hydrocarbons.

3. Application: Compatible with other resins and used to improve application, lightfastness, gloss, or color retention. Used in automotive and appliance finishes.

8.2 Thermosetting Resins

After application and solvent evaporation of a thermosetting resin, a chemical change takes place. As this chemical reaction is taking place, the coating is said to be "curing." This reaction may take place at room temperature or in baked coatings at elevated temperatures. The reaction is irreversible. Exposure to high temperature or to solvents does not cause the coating to soften or melt.

8.2.1 Epoxy

Epoxy resins on their own are not suitable for protective coatings because when pigmented and applied, they dry to a hard, brittle film with very poor chemical resistance. By copolymerizing with other resins, the epoxy resins will form a durable protective coating.

8.2.2 Polyamine Epoxy

1. Resistance: Resistant to acids, acid salts, alkalies, and organic solvents.
2. Temperature resistance: 225°F/107°C dry; 190°F/88°C wet.
3. Limitations: Not resistant to UV light, harder and less flexible than other epoxies; less tolerant of moisture during application.
4. Applications: Widest range of chemical and solvent resistance of epoxies; used for piping and vessels.

8.2.3 Polyamide Epoxy

1. Resistance: Resistance inferior to that of polyamine epoxy; only partially resistant to acids, acid salts, alkalies, and organic solvents; resistant to moisture.
2. Temperature resistance: 225°F/108°C dry; 150°F/66°C wet.
3. Limitations: Not resistant to UV light; inferior chemical resistance.
4. Applications: Used on wet surfaces under water, as in tidal zone areas of pilings, oil rigs, etc.

8.2.4 Aliphatic Polyamine

1. Resistance: Partially resistant to acids, acid salts, and organic solvents.
2. Temperature resistance: 225°F/108°C dry; 150°F/66°C wet.
3. Limitations: Film formed has greater permeability than other epoxies.
4. Application: Used for protection against mild atmospheric corrosion.

8.2.5 Esters of Epoxies and Fatty Acids

1. Resistance: Resistant to weathering and UV light.
2. Temperature resistance: 225°F/108°C dry; 150°F/66°C wet.

3. Limitations: Chemical resistance generally poor.

4. Applications: Used where the properties of a high quality oil base paint are required.

8.2.6 Coal Tar Plus Epoxy

1. Resistance: Excellent resistance to freshwater, saltwater, and inorganic acids.

2. Temperature resistance: 225°F/108°C dry; 150°F/66°C wet.

3. Limitations: Not resistant to weather and sunlight; attacked by organic solvents.

4. Applications: Used on clean blasted steel for immersion service or below grade service.

8.2.7 Urethanes

Polyurethane-resin based coatings are versatile. They are available as oil modified, moisture curing, blocked, two component, and lacquers. Properties of the various types are shown in Table 8.2.

The urethane coatings possess excellent gloss and color retention and are used as a decorative coating of tank cars and steel in highly corrosive environments. Moisture cured types require humidity during application and may yellow under UV light. They have a temperature resistance of 250°F/121°C dry; 150°F/ 66°C wet.

Catalyzed (two component) urethanes exhibit very good chemical resistance. They are not recommended for immersion or exposure to strong

TABLE 8.2

Properties of Polyurethane Coatings

	One-Component			Two-Component	Lacquer
Property	Urethane Oil	Moisture	Blocked		
Abrasion resistance	Fair–good	Excellent	Good–excellent	Excellent	Fair
Hardness	Medium	Medium–hard	Medium–hard	Soft–very hard	Soft–medium
Flexibility	Fair–good	Good–excellent	Good	Good–excellent	Excellent
Impact resistance	Good	Excellent	Good–excellent	Excellent	Excellent
Solvent resistance	Fair	Poor–fair	Good	Excellent	Poor
Chemical resistance	Fair	Fair	Good	Excellent	Fair–good
Corrosion resistance	Fair	Fair	Good	Excellent	Good–excellent
Adhesion	Good	Fair–good	Fair	Excellent	Fair–good
Toughness	Good	Excellent	Good	Excellent	Good–excellent
Elongation	Poor	Poor	Poor	Excellent	Excellent
Tensile	Fair	Good	Fair–good	Good–excellent	Excellent
Cure rate	Slow	Slow	Fast	Fast	None
Cure temperature	Room	Room	300–390°F 149–199°C	212°F/100°C	150–225°F 66–108°C

acids/alkalies. They have a temperature resistance of 225°F/108°C dry; 150°F/66°C wet.

These resins are quite expensive.

8.2.8 Polyesters

In the coating industry, polyesters are characterized by resins based on components that introduce unsaturation (–C=C–) directly into the polymer backbone. The most common polyester resins are polymerization products of maleic or isophthalic anhydride or their acids. In producing paint, the polyester resin is dissolved in styrene monomer, together with pigment and small amounts of inhibitor. A free radical initiator, commonly a peroxide, and additional styrene are packaged in another container. When applied, the containers are mixed. Sometimes, because of the fast initiating reaction (short pot life), they are mixed in an externally mixing or dual-headed spray gun. After being mixed and applied, a relatively fast reaction takes place, resulting in cross-linking and polymerization of the monomeric styrene with the polyester resin.

Polyester coatings exhibit high shrinkage after application. The effect of high shrinkage can be reduced by proper pigmentation that reinforces the coating and reduces the effect of the shrinkage.

1. Resistance: Excellent resistance to acids and aliphatic solvents; good resistance to weathering.
2. Temperature resistance: 180°F/82°C dry and wet.
3. Limitations: Not suitable for use with alkalies and most aromatic solvents because they swell and soften these coatings. They also have a short pot life and must be applied with special equipment.
4. Applications: Coatings for tanks and chemical process equipment.

8.2.9 Vinyl Esters

The vinyl esters are formulated in a manner similar to the polyesters. They are derived from a resin based on a reactive end vinyl group that can open and polymerize. The vinyl esters have better moisture and chemical resistance than the polyesters. They also have good resistance to weathering and acids.

8.3 Autooxidative Cross-Linking Paints

Paints of this type rely on the reaction of a drying oil with oxygen to introduce cross-linking within the resin and attainment of final film properties. The final coating is formed when a drying oil reacts with a resin that is then combined with pigments and solvents. These paints are packaged in a single can. The can is opened, the paint mixed and applied. As the solvents evaporate, the coating becomes hard. Since oxygen in the atmosphere reacts with the coating to produce additional cross-linking, the final film properties may be formed weeks or months after application.

Among these resin fortified "oil based paints" are included the alkyds, epoxy esters, oil modified urethanes, etc. These paints may be formulated as air-drying or baking types, and with suitable pigments, can be formulated to be resistant to a variety of moisture and chemical fume environments as well as application over wood, metal, and masonry substrates. These paints have the following advantages:

Ease of application

Excellent adhesion

Relatively good environmental resistance

Tolerance for poorer surface preparation

Their major disadvantages are:

Cannot be used for immersion or in areas of high chemical fume concentration

Reduced moisture and chemical resistance compared to other resin coatings

8.3.1 Epoxy Esters

Esters of epoxies and fatty acids modified.

1. Resistance: Resistant to weathering; limited resistance to chemicals and solvents.

2. Temperature resistance: 225°F/108°C dry; 150°F/66°C wet.

3. Limitations: Attacked by alkalies. Application time between coats is critical.

4. Applications: Used where the properties of a high quality oil based paint is required.

8.3.2 Alkyd Resin

These paints are the most widely used synthetic resins. They are produced by reactions of phthalic acid with polyhydric alcohols such as glycerol and vegetable oil (or fatty acid). Classification is by oil content into drying types and baking types. Drying types provide good weathering resistance and good adhesion to a wide variety of substrates, but relatively poor resistance to chemical attack. They have a maximum temperature resistance of 225°F/108°C dry; 150°F/66°C wet.

These paints are used on exterior wood surfaces for primers requiring penetrability and in less severe chemical environments.

8.3.3 Oil-Based Paints

These paints are mixtures of pigment and boiled linseed oil or soybean oil or other similar materials. This paint is used as a ready-mixed paint particularly on wood surfaces because of its penetrating power. It is resistant to weather, but has relatively poor chemical resistance and will be attacked by alkalies. Temperature limitations are 225°F/108°C dry; 150°F/66°C wet.

8.3.4 Water Emulsion Paints

Water-based latex emulsions consist of fine particles of high molecular weight copolymers of polyvinyl chloride or polyvinyl acetate, acrylic esters, styrene–butadiene, or other resins combined with pigments, plasticizers, UV stabilizers, and other ingredients. A variety of thickeners, coalescing aids, and other additives are present in the water phase.

The paint film is formed by water evaporation. Initial adhesion may be relatively poor as the water continues to evaporate, and coalescing aids, solvents, and surfactants evaporate or are leached from the "curing" film. Final adhesion and environmental resistant properties are reached from as little as a few days to a few months after application.

1. Resistance: Resistant to weather but poor chemical resistance.
2. Temperature resistance: 150°F/66°C dry or wet.
3. Limitations: Not suitable for immersion service.
4. Applications: Used in general decorative applications on wood.

8.3.5 Silicones

There are two types: high temperature and water repellent in water or solvent.

1. Resistance: Excellent resistance to sunlight and weathering; poor resistance to acids and alkalies; resistant to water.
2. Temperature resistance: In aluminum formulation, resistant to 1200°F/699°C and to weathering at lower temperatures.
3. Limitations: High temperature type requires baking for good cure; not chemically resistant.
4. Applications: Used on surfaces exposed to high temperatures as water repellant; water solvent formulations used on limestone,

cement, and nonsilaceous materials; solvent formulations used oil bricks and noncalcareous masonry.

8.4 Bituminous Paints

There are two paints in this category: asphalt and coal tar.

8.4.1 Asphalt Paint

Asphalt paint consists of solids from crude oil refining suspended in aliphatic solvents.

1. Resistance: Good moisture resistance; good resistance to weak acids, alkalies, and salts; better weathering properties than coal tar.
2. Temperature resistance: Softens at 100°F/38°C.
3. Limitations: Heavy, dark color hides corrosion under coating.
4. Applications: Used in above-ground weathering environments and chemical fume atmospheres.

8.4.2 Coal Tar

Coal tar is a distilled coking by-product in an aromatic solvent.

1. Resistance: Excellent resistance to moisture; good resistance to weak acids and alkalies, petroleum oils, and salts.
2. Temperature resistance: Will soften at 100°F/38°C.
3. Limitations: Will be degraded by exposure to UV light and weathering.
4. Applications: Used on submerged or buried steel and concrete.

8.5 Zinc-Rich Paints

Zinc-rich paints owe their protection to galvanic action. Whereas all of the preceding coatings owe their final film properties, corrosion resistance, and environmental resistance to the composition of their resin or binder rather than their pigment, the high amount of zinc dust metal pigment in zinc-rich paints determines the coating's fundamental property: galvanic protection. Many of the previous coatings, chlorinated rubber and epoxies in particular, are formulated as zinc-rich coatings. In doing so, the high pigment content changes the properties of the formulated coating.

Zinc-rich coatings can be classified as organic or inorganic. The organic zinc-riches have organic binders with polyamide epoxies and chlorinated rubber binders being the most common. Other types such as urethane zinc-rich are also available. These latter coatings are more easily applied than the other zinc-rich coatings.

Inorganic zinc-rich binders are based on silicate solutions that after drying or curing, crystallize and form an inorganic matrix, holding zinc dust particles together and to the steel substrate.

For a coating to be considered zinc-rich, it must contain at least 75% by weight of zinc dust in the dry film. This may change because conductive extenders (notably di-iron phosphite) have been added to improve weldability and burn-through with supposedly equivalent protection at lower zinc loadings.

The primary advantage of zinc-rich coatings is their ability to galvanically protect. The zinc pigment in the coating preferentially sacrifices itself in the electrochemical corrosion reaction to protect the underlying steel. This galvanic reaction, together with the filling and sealing effect of zinc reaction products (primarily zinc carbonate, zinc hydroxide, and complex zinc salts), provides more effective corrosion protection to steel substrates than does any other type of coating. Zinc-rich coatings cannot be used outside the pH range of 6–10.5.

Table 8.3 presents a summary of the properties of various paint formulations.

8.6 Selecting a Paint System

The first step in selecting a paint system for corrosion protection is to determine the environment around the structure or item to be painted. Is the environment predominantly a weathering environment subject to heat, cold, daily or seasonal temperature changes, precipitation, wind (flexing), exposure to sunlight, or detrimental solar rays? If the structure or item is located outdoors, are there chemical plants located nearby, or pulp and paper mills, or other industrial facilities that are apt to discharge airborne pollutants? Is color, gloss, and overall pleasing effect more important than corrosion protection, or are the normal grays, whites, and pastels of the more corrosion resistant paints satisfactory? If located in a chemical facility, what specific chemicals are used nearby; is there any chance of a chemical spill on the painted surface?

Since surface preparation is an important factor in the selection of a paint system, the suitability or availability of the surface for specific preparation techniques must be known. In some instances, certain types of surface preparation may not be permitted or practical. For example, many companies

TABLE 8.3

Properties of Paints

	Resistance To						
Coating Type	UV	Weather	Acid	Alkali	Moisture	Salt Solutions	Comments
Vinyls dissolved in esters, aromatics, or ketones			R	R	R	R	Adhesion may be poor until all solvents have vaporized from the coating
Chlorinated rubbers dissolved in hydrocarbon solvents	N		R	R	R		Excellent adhesion to metals, concrete, and masonry. Used on structures exposed to water and marine atmospheres
Epoxies, polyamine plus epoxy resin	N		PR	PR	R	PR	Harder and less flexible than other epoxies. Greatest chemical resistance of the epoxies
Polyamide plus epoxy resin (polyamide epoxy)	N		PR	PR	R	PR	Chemical resistance inferior to that of the polyamine epoxies
Aliphatic polyamine			PR	PR		PR	Flexible film
Esters of epoxies and fatty acids (epoxy esters)	R	R	PR	N		PR	On surfaces requiring the properties of a high quality oil based paint
Coal tar plus epoxy resin	N	N	R		R		Used on clean blast cleaned steel for immersion or below grade service. Attacked by organic solvents
Oil-based coatings with vehicle (alkyd epoxy, urethane)	R	R		N			Lower cost than most coatings. Used on exterior wood surfaces
Urethane, moisture cured	R	R	Weak	Weak	R		May yellow under UV light. High gloss and ease of cleaning
Urethanes catalyzed	R	R	R	R	R	R	Expensive. Used as coating on steel in highly corrosive areas

(continued)

TABLE 8.3 *Continued*

Coating Type	Resistance To						Comments
	UV	Weather	Acid	Alkali	Moisture	Salt Solutions	
Silicones, water repellent in water or solvent	R	R	N	N	R		Used on masonry surfaces
Silicones, water based aqueous emulsions of polyvinyl acetate, acrylic or styrene–butadiene latex	R	R	N	N			May flash rust as a primer on steel. Not chemically resistant
Polyesters, organic acids combined with polybasic alcohols. Styrene is a reaction diluent	PR	R	R	N			Must be applied with special equipment. Not suitable for use with most aromatic solvents
Coal tar	N	N	Weak R	Weak R	R		Used on submerged or buried steel
Asphalt. Solids from crude oil refining in aliphatic solvents	R	R	Weak R	Weak R	R	Weak R	Used in above ground weathering environments and chemical fume atmospheres
Zinc-rich metallic zinc in an organic or inorganic vehicle	Provides galvanic protection as a primer						
Acrylic-resin water emulsion base	R	R	Weak R	Weak R			Limited penetrating power. May flash rust as a primer over bare steel. Not suitable for immersion service. Soluble in ketones, esters, aliphatic chlorinated hydrocarbons, and aromatic hydrocarbons

Note: R, resistant; N, not resistant; PR, poor resistance.

do not permit open blast cleaning where there is a prevalence of electric motors or hydraulic equipment. Refineries, in general, do not permit open blast cleaning or any other method of surface preparation that might result in the possibility of a spark, static electricity buildup, or an explosion hazard.

If a new facility is being constructed, it is possible that during erection many areas may become enclosed or covered or so positioned that access is difficult or impossible. These structures must be painted prior to installation.

When all of this information has been collected, the appropriate paint systems may be selected. In most instances, it will not be practical or possible to select one single coating system for the entire plant. There will be areas requiring systems to provide protection from aggressive chemicals, whereas, other areas may require coating systems simply for esthetics. If an area is a combination of mild and aggressive conditions, a coating system should be selected that will be resistant to the most aggressive condition.

Several typical industrial environmental areas have been illustrated that coating systems may be exposed to with recommendations for paint systems to be used in these areas. The paint systems are shown in Table 8.4 through Table 8.9 with the appropriate surface preparation. The tables have been arranged based on surface preparation. Each coating system shown in a particular table requires the same surface preparation.

8.6.1   Area 1. Mild Exposure

This is an area where structural steel is embedded in concrete, encased in masonry, or protected by noncorrosive type fireproofing. In many instances, no coating will be applied to the steel. However, it is a good idea to coat the steel substrate with a protective coating to protect them during construction and in case they end up being either intentionally or unintentionally exposed.

A good practice would be to apply a general use epoxy primer 3–5 mils dry film thickness (dft) (75–125 μm). If the surface cannot be abrasive blasted, a surface tolerant epoxy mastic may be used.

Recommended systems are found in Table 8.7 systems A and C.

8.6.2   Area 2. Temporary Protection; Normally Dry Interiors

This area consists of office space or dry storage areas (warehouses) or other locations exposed to generally mild conditions or areas where oil-based paints presently last for ten or more years. If located in an industrial environment, there is the possibility of exposure to occasional fumes, splashing, or spillage of corrosive materials. Because of this, it is suggested that an industrial grade acrylic coating system or a single coat of epoxy be applied.

This recommendation is not suitable for interior surfaces that are frequently cleaned or exposed to steam cleaning. Refer to area 4. Recommended for this area are systems A and C in Table 8.7.

TABLE 8.4

Multilayer Paint Systems Requiring Commercial Blast (SSPC-SP-6) for Surface Preparation

---

System A. Inorganic Zinc/Epoxy Mastic

Paint layers
  One coat inorganic zinc: 2–3 mils dft (50–75 μm)
  One coat epoxy mastic: 4–6 mils dft (100–150 μm)

Properties
  Zinc primer provides outstanding corrosion resistance and undercutting resistance. A barrier protection for the zinc primer is provided by the finish coat of epoxy which also provides a color coat for appearance. Suitable for use on carbon steel only

Limitations
  A relatively high level of applicator competence required for the primer

System B. Inorganic Zinc/Epoxy Primer/Polyurethane Finish

Paint layers
  One coat inorganic zinc: 2–3 mils dft (50–75 μm)
  One coat epoxy primer: 4–6 mils dft (100–150 μm)
  One coat polyurethane finish: 2–4 mils dft (50–100 μm)

Properties
  Zinc primer provides outstanding corrosion resistance and undercutting resistance. The zinc primer is protected by a barrier coating of epoxy primer, while the finish coat of polyurethane provides color and gloss retention. This is a premium industrial finish for steel surfaces. Can only be used on carbon steel

Limitations
  A relatively high level of applicator competence required for the primer

System C. Inorganic Zinc/Acrylic Finish

Paint layers
  One coat inorganic zinc: 2–3 mils dft (50–75 μm)
  One coat acrylic finish: 2–3 mils dft (50–75 μm)

Properties
  Zinc primer provides outstanding corrosion resistance and undercutting resistance. Water based single package finish has excellent weathering and semigloss appearance

Limitations
  A relatively high level of applicator competence required for the primer. The finish coat has low temperature curing restrictions

System D. Aluminum Epoxy Mastic/Epoxy Finish

Paint layers
  One coat aluminum epoxy mastic: 4–6 mils dft (100–150 μm)
  One coat epoxy finish: 4–6 mils dft (100–150 μm)

Properties
  Can be used on tight rust and marginally prepared surface. The epoxy finish coat is available in a variety of colors and has good overall chemical resistance. May be used on carbon steel or concrete. Concrete must be clean, rough, and cured at least 28 days. Hand or power tool cleaning, including water blasting, may be used for surface preparation

---

(continued)

TABLE 8.4 *Continued*

System E. Aluminum Epoxy Mastic/Acrylic Finish

Paint layers
  One coat aluminum epoxy mastic: 4–6 mils dft (100–150 μm)
  One coat acrylic finish: 2–3 mils dft (50–75 μm)

Properties
  May be used on tight rust and marginally prepared surfaces. The acrylic finish coat is available in a variety of colors and has good overall chemical resistance. This is an excellent maintenance system. Normally used on carbon steel and concrete

System F. Epoxy Mastic/Epoxy Mastic

Paint layers
  One coat epoxy mastic: 4–6 mils dft (100–150 μm)
  One coat epoxy mastic: 4–6 mils dft (100–150 μm)

Properties
  May be used on tight rust and marginally prepared surfaces. The substrate is protected by the formation of a tight barrier stopping moisture from reaching the surface. Normally used on steel or concrete. Concrete must be clean, rough, and cured at least 28 days. If necessary, hand or power tools may be used for cleaning

System G. Epoxy Primer/Epoxy Finish

Paint layers
  One coat epoxy primer: 4–6 mils dft (100–150 μm)
  One coat epoxy finish: 4–6 mils dft (100–150 μm)

Properties
  An easily applied two coat high build barrier protection is provided with ease of application. Used on carbon steel or concrete

Limitations
  Since these are two component materials, they must be mixed just prior to application. They require additional equipment and more expertise to apply than a single-packed product. Most epoxy finish coats will chalk, fade, and yellow when exposed to sunlight

System H. Epoxy Primer

Paint layers
  One coat epoxy primer: 4–6 mils dft (100–150 μm)

Properties
  Normally applied to carbon steel or concrete in protected areas such as the interiors of structures, behind walls and ceilings, or for temporary protection during construction

Limitations
  This is a two component material requiring mixing just prior to application

System I. Epoxy Novalac/Epoxy Novalac

Paint layers
  One coat epoxy novalac: 6–8 mils dft (150–200 μm)
  One coat epoxy novalac: 6–8 mils dft (150–200 μm)

Properties
  An exceptional industrial coating for a wide range of chemical resistance and physical abuse resistance. Has a higher temperature resistance than standard epoxy. May be used to protect insulated piping or for secondary containment. Normally used on carbon steel and concrete surfaces

TABLE 8.5

Multilayer Paint Systems Requiring Surface to be Abrasive Blasted in Accordance with SSPC-SP-10 Near White Blast

---

System A. Aluminum–Epoxy Mastic/Aluminum–Epoxy Mastic

Paint layers
   One coat aluminum–epoxy mastic: 4–6 mils dft (100–150 μm)
   One coat aluminum-epoxy mastic: 4–6 mils dft (100–150 μm)

Properties
   Tolerates poorly prepared surfaces and provides excellent barrier protection. A third coat may be added for additional protection. Can be used on carbon steel and concrete. Concrete must be clean, rough, and cured at least 28 days. If necessary this system can be applied to surfaces that are pitted or cannot be blasted. However, the service life will be reduced.

System B. Epoxy Phenolic Primer/Epoxy Phenolic Finish/Epoxy Phenolic Finish

Paint layers
   One coat epoxy phenolic primer: 8 mils dft (200 μm)
   One coat epoxy phenolic finish: 8 mils dft (200 μm)
   One coat epoxy phenolic finish: 8 mils dft (200 μm)

Properties
   Because of this system's outstanding chemical resistance it is often used in areas subject to frequent chemical spills. The finish coats are available in a limited number of colors. Normally used on carbon steel and concrete. Concrete must be clean, rough, and cured at least 28 days.

System C. Epoxy Phenolic Primer/Epoxy Phenolic Lining/Epoxy Phenolic Lining

Paint layers
   One coat epoxy phenolic primer: 8 mils dft (200 μm)
   One coat epoxy phenolic lining: 8 mils dft (200 μm)
   One coat epoxy phenolic lining: 8 mils dft (200 μm)

Properties
   Because of the system's outstanding overall chemical resistance it is suitable for lining areas subject to flowing or constant immersion in a variety of chemicals. Normally used on carbon steel and concrete. When used on concrete the surface must be clean, rough, and cured at least 28 days.

System D. Epoxy/Epoxy

Paint layers
   One coat epoxy: 4–6 mils dft (100–150 μm)
   One coat epoxy: 4–6 mils dft (100–150 μm)

Properties
   Two coats of the same product are applied providing a high build protection. Can be used in immersion service without the addition of corrosion inhibitors. When used in potable water systems the product must meet Federal Standard 61. A third coat may be added for additional protection. Normally used on carbon steel and concrete.

System E. Coal Tar Epoxy/Coal Tar Epoxy

Paint layers
   One coat coal tar epoxy: 8 mils dft (200 μm)
   One coat coal tar epoxy: 8 mils dft (200 μm)

(*continued*)

TABLE 8.5   *Continued*

Properties
  Provides excellent barrier protection and is the most economical of the water lining systems or for water immersion. Normally used on carbon steel and concrete.

System F. Solventless Elastomeric Polyurethane

Paint layers
  One coat elastomeric polyurethane: 20–250 mils dft (500–6250 µm)

Properties
  Excellent barrier protection. Normally used on carbon steel and concrete.

Limitations
  Must be applied by a knowledgeable contractor.

8.6.3   Area 3. Normally Dry Exteriors

This includes such locations as parking lots, water storage tanks, exterior storage sheds, and lighting or power line poles that are exposed to sunlight in a relatively dry location. Under these conditions, oil based paints should last six or more years. These materials that are resistant to ultraviolet rays and are normally rated for exterior use include acrylics, alkyds, silicones, and polyurethanes.

Epoxies will lose gloss, normally chalk, and fade rapidly when exposed to UV rays. Recommended systems include A in Table 8.6, C in Table 8.7, A in Table 8.8, and A in Table 8.9.

8.6.4   Area 4. Freshwater Exposure

Under this category, the surface to be protected is frequently wetted by freshwater from condensation, splash, or spray. Included are interior or

TABLE 8.6

Multilayer Paint Systems Requiring Surface to be Clean, Dry, and Free of Loose Dirt, Oil, and Chemical Contamination

System A. Aluminum Epoxy Mastic/Polyurethane Finish

Paint layers
  One coat aluminum epoxy mastic: 4–6 mils dft (100–150 µm)
  One coat polyurethane finish: 2–4 mils dft (50–100 µm)

Properties
  Excellent over tight rust. Tolerant of minimally prepared steel. May be applied to a wide range of surfaces, but normally used on carbon steel and concrete. This is a premium system to use when cleaning must be minimal.

Limitations
  In order to cure properly, temperature must be above 50°F/10°C. For lower temperature requirements other aluminum epoxy/urethane mastics may be substituted.

TABLE 8.7

Multilayer Paint Systems for New Clean Surfaces, Free of Chemical Contamination

---

System A. Epoxy Mastic

Paint layers

One coat epoxy mastic: 3–5 mils dft (75–125 μm)

Properties

Good color selection, excellent chemical resistance, good physical characteristics, ease of maintenance. Used on carbon steel, concrete masonry units, masonry block (a filler is recommended), sheet rock (a sealer is required), wood, polyvinyl chloride, galvanized steel, and other surfaces.

Limitations

This is a two-component material that is mixed just prior to application. Additional equipment is required and more expertise to apply than a single packaged product. Epoxy solvents may be objectionable to some people.

System B. Acrylic Primer/Acrylic Intermediate/Acrylic Finish

Paint layers

One coat acrylic primer: 2–3 mils dft (50–75 μm)
One coat acrylic intermediate: 2–3 mils dft (50–75 μm)
One coat acrylic finish: 2–3 mils dft (50–75 μm)

Properties

This is a single package, water base, low odor, semigloss paint. It possesses excellent weathering and acidic acid resistance. Can be used on most surfaces including carbon steel, concrete, concrete masonry units, masonry block (a block filler is recommended), sheet rock (a sealer is required), wood, polyvinyl chloride, galvanized steel, stainless steel, copper, and fabric. May be applied over existing coatings of any type including inorganic zinc.

Limitations

Must be protected from freezing during shipping and storage. For application, temperature must be above 60°F/16°C and will remain so for 2–3 hr after application.

System C. Acrylic Primer/Acrylic Finish

Paint layers

One coat acrylic primer: 2–3 mils dft (50–75 μm)
One coat acrylic finish: 2–3 mils dft (50–75 μm)

Properties

Excellent weathering and acidic chemical resistance, with good color selection.

Limitations

For best performance metallic surfaces should be abrasive blasted. For mild conditions hand or power cleaning may be sufficient. Paint must be applied when temperature exceeds 60°F/16°C.

---

exterior areas that are frequently exposed to cleaning or washing, including steam cleaning.

The systems used for these surfaces make use of inorganic zinc as a primer. Inorganic zinc is the best coating that can be applied to steel because it provides the longest term of protection. In some situations, it may be

TABLE 8.8

Multilayer Paint Systems Requiring an Abrasive Blast to the Substrate Surface

System A. Epoxy Primer/Polyurethane Finish

Paint layers
One coat epoxy primer: 3–5 mils dft (75–125 μm)
One coat polyurethane finish: 2–3 mils dft (50–75 μm)

Properties
Two coat protection is provided with excellent high gloss finish and long-term color gloss retention. Normally applied to carbon steel and concrete.

Limitations
Since these are two component materials they must be mixed just prior to application and require additional equipment and more expertise to apply.

necessary to substitute an organic zinc (an organic binder such as epoxy or polyurethane with zinc added) for the inorganic zinc.

Recommended systems are B, C, and E of Table 8.4 and A of Table 8.6.

8.6.5 Area 5. Saltwater Exposure

This area includes interior or exterior locations on or near a seacoast or industrial environments handling brine or other salts. Under these conditions, the surfaces are frequently wet by salt water and include condensation, splash, or spray.

Conditions in this area are essentially the same as for freshwater and the comments for area 4 apply here. Because of the more severe conditions, it is recommended that two coats of the primer be applied for system E of Table 14.1 of *Fundamentals of Metallic Corrosion: Atmospheric and Media Corrosion of Metals* and system A of Table 8.6.

Recommended systems are B, C, and E of Table 8.6 and A of Table 8.6.

TABLE 8.9

Multilayer Paint Systems for Previously Painted Surfaces That Have Had Loose Paint and Rust Removed by Hand Cleaning

System A. Oleoresin

Paint layers
One coat oleoresin: 2–4 mils dft (50–100 μm)

Properties
This very slow drying material penetrates and protects existing surfaces that cannot be cleaned properly with a single coat. Provides long-term protection without peeling, cracking, and other such problems. Easy to apply by spray, brush, roller, or glove. Normally used on carbon steel and weathering galvanized steel.

Limitations
This material is designed to protect steel that will not see physical abuse. It also stays soft for an extended period of time.

8.6.6   Area 6. Freshwater Immersion

Wastewaters are also a part of this area. Included are all areas that remain underwater for periods longer than a few hours at a time. Potable and nonpotable water, sanitary sewage, and industrial waste liquids are all included.

If the systems recommended are to be used as a tank lining material, it is important that the application be done by experienced workers. In addition, if the coating to be applied is to be in contact with potable water, it is important that the material selected meets the necessary standards and is approved for use by the local health department. Two coats of epoxy (system D in Table 8.5) is frequently used in this service.

Recommended systems are F of Table 8.4 and A, D, E, and F of Table 8.5.

8.6.7   Area 7. Saltwater Immersion

Areas that remain under water in a coastal environment or industrial area or that are constantly subjected to flowing salt or brine laden water are included in this category. Because of the increased rate of corrosion, a third coat may be added to system F of Table 8.5 and systems A and E of Table 8.5 as additional protection against this more severe corrosion.

System D of Table 8.4 and systems A, E, and F of Table 8.5 are recommended for this service.

8.6.8   Area 8. Acidic Chemical Exposure (pH 2.0–5.0)

In the chemical process industries, this is one of the most severe environments to be encountered. When repainting, it is important that all surfaces are clean of any chemical residue. Inorganic zinc and zinc filled coatings are not recommended for application in this area.

The system selected will be dependent upon the quality of surface preparation, length of chemical exposure, and housekeeping procedures. Decreased cleanup and longer exposure times require a more chemical resistant coating system such as system I in Table 8.4.

Other recommendations for this area include systems D, G, and I in Table 8.4 and system B in Table 8.7.

8.6.9   Area 9. Neutral Chemical Exposure (pH 5.0–10.0)

This is an area that is not subject to direct chemical attack, but it may be subject to fumes, spillage, or splash. Under these conditions, more protection is required than will be provided by a standard painting system. This would include such locations as clean rooms, packaging areas, hallways, enclosed process areas, instrument rooms, electrical load centers, and other similar locations.

A list of potential chemicals that may contact the coating aids in the coating selection. Knowledge of cleanup procedures will also prove helpful.

It may be possible to use systems requiring less surface preparation such as system D in Table 8.4, system A in Table 8.6, system A in Table 8.7, and system A in Table 8.8.

Recommendations for area 9 are systems A and D in Table 8.4, system A in Table 8.6, systems A and C in Table 8.7, and system A in Table 8.8.

8.6.10   Area 10. Exposure to Mild Solvents

This is intended for locations subject to intermittent contact with aliphatic hydrocarbons such as mineral spirits, lower alcohols, glycols, etc. Such contact can be the result of splash, spillage, or fumes.

Cross-linked materials, such as epoxies, are best for this service because solvents will dissolve single package coatings. A single coat of inorganic zinc is an excellent choice for immersion service in solvents or for severe splashes and spills.

The gloss of a coating system is often reduced as a result of solvent splashes or spills. However, this is a surface effect that usually does not affect the overall protective properties of the coating.

Recommended systems for use in this location are A, D, and G of Table 8.4.

8.6.11   Area 11. Extreme pH Exposure

This covers locations that are exposed to strong solvents, extreme pHs, oxidizing chemicals, or combinations thereof with high temperatures. The usual choice for coating these areas are epoxy novalacs, epoxy phenolics, and high build polyurethanes. Other special coatings such as the polyesters and vinyl esters may also be considered. However, these systems require special application considerations.

Regardless of which coating system is selected, surface preparation is important. An abrasive blast, even on concrete, is required. In addition, all surface contaminants must be removed. When coating concrete, a thicker film is required. System F in Table 8.5 is recommended for optimum protection.

Recommended for this location is system I in Table 8.4 and systems B, C, and F in Table 8.5.

The foregoing have been generalizations as to what environmental conditions may be encountered, along with suggested coating systems to protect the substrate. Data presented will act as a guide in helping the reader to select the proper coating system. Keep in mind that the surface preparation is critical and should not be skimped on.

9

Cementitious Coatings

Cementitious coatings provide corrosion resistance to substrates such as steel by maintaining the pH at the metal/coating interface above 4, a pH range where steel corrodes at a low rate. The proper selection of materials and their application is necessary if the coating is to be effective. To select the proper material, it is necessary to define the problem:

1. Identify all chemicals that will be present and their concentrations. Knowing pH alone is not sufficient. All pH indicates is whether or not the environment is acid, neutral, or alkaline. It does not identify if the environment is oxidizing, organic or inorganic, alternately acid or alkaline.

2. Is the application fumes, splash, or immersion?

3. What are the minimum and maximum temperatures that the coating will be subjected to?

4. Is the installation indoors or outdoors? Thermal shock and ultraviolet exposure can be deleterious to many resins.

5. How long is the coating expected to last? This can have an effect on the cost.

Surface preparation prior to application of the coating is essential. The surface must be free of mill scale, oil, grease, and other chemical contaminants. The surface must be roughened by sandblasting and the coating applied immediately after surface preparation. An intermediate bonding coating is used when adhesion between the substrate and coating is poor or where thermal expansion characteristics are incompatible. Coatings are installed in thicknesses of 1/16 to 1/2 in. (1.5 to 13 mm). They may be applied by casting, troweling, or spraying. The spraying process, known as Gunite or Shotcrete, is particularly useful on systems with unusual geometry or with many sharp bends or corners. It has the advantage that there are no seams that are often weak points as far as corrosion resistance is concerned.

9.1   Silicates

These materials are noted for their resistance to concentrated acids, except hydrofluoric acid and similar fluorinated chemicals at elevated temperatures. They are also resistant to many aliphatic and aromatic solvents. They are not intended for use in alkaline or alternately acid and alkaline environments. This category of coatings includes:

1. Sodium silicate
2. Potassium silicate
3. Silica (silica sol)

The alkali silicates form a hard coating by a polymerization reaction, involving repeating units of the structure:

The sodium and potassium silicates are available as two-component systems: filler and binder with the setting agent in the filler. Sodium and potassium silicates are referred to as soluble silicates because of their solubility in water. This prevents their use in many dilute acid services, whereas, they are not affected by strong concentrated acids. This disadvantage becomes an advantage for formulating single component powder systems. All that is required is the addition of water at the time of use. The fillers for these materials are pure silica.

The original sodium silicate acid resisting coating uses an inorganic silicate base consisting of two components, a powder and a liquid. The powder is basically quartzite of selected gradation and a setting agent. The liquid is a special sodium silicate solution. When the coating is used, the two components are mixed together, and the hardening occurs by a chemical reaction.

This coating may be cast, poured, or applied by guniting. It has excellent acid resistance and is suitable over a pH range of 0.0–7.0.

The sodium silicates can be produced over a wide range of compositions of the liquid binder. These properties and the new hardening systems have significantly improved the water resistance of some sodium silicate coatings. These formulations are capable of resisting dilute as well as concentrated acids without compromising physical properties.

The potassium silicate materials are less versatile in terms of formulation flexibility than the sodium silicate materials. However, they are less susceptible to crystallization in high concentrations of sulfuric acid so long as metal ion contamination is minimal. Potassium silicate materials are available with halogen-free hardening systems, thereby, removing remote possibility for catalyst poisoning in certain chemical processes.

Chemical setting potassium silicate materials are supplied as two component systems that comprise the silicate solution and the filler powder and setting agent. Setting agents may be inorganic, organic, or a combination of both. The properties of the coating are determined by the setting agent and the alkali–silica ratio of the silicate used. Properties such as absorption, porosity, strength, and water resistance are affected by the choice of the setting agent. Organic setting agents will burn out at low temperatures, thereby, increasing porosity and absorption. Organic setting agents are water soluble and can be leached out if the coating is exposed to steam or moisture. Coatings that use inorganic setting agents are water and moisture resistant.

Silicate formulations will fail when exposed to mild alkaline mediums such as bicarbonate of soda. Dilute acid solutions, such as nitric acid, will have a deleterious effect on sodium silicates unless the water resistant type is used. Table 9.1 points out the differences between the various silicate coatings.

Silica, or silica sol, types of coatings are the newest of this class of material. They consist of a colloidal silica binding instead of the sodium or potassium silicates with a quartz filler. These materials are two component systems that comprise a powder composed of high quality crushed quartz and a hardening agent that are mixed with colloidal silica that solution to form the coating.

TABLE 9.1

Comparative Chemical Resistance: Silicate Coatings

	Sodium		Potassium	
Medium, RT	**Normal**	**Water Resistant**	**Normal**	**Halogen Free**
Acetic acid, glacial	G	G	R	R
Chlorine dioxide, water sol.	N	N	R	R
Hydrogen peroxide	N	R	N	N
Nitric acid, 5%	C	R	R	R
Nitric acid, 20%	C	R	R	R
Nitric acid, over 20%	R	R	R	R
Sodium bicarbonate	N	N	N	N
Sodium sulfite	R	R	N	N
Sulfates, aluminum	R	R	R	R
Sulfates, copper	G	G	R	R
Sulfates, iron	G	G	R	R
Sulfates, magnesium	G	G	R	R
Sulfates, nickel	G	G	R	R
Sulfates, zinc	G	G	R	R
Sulfuric acid, to 93%	G	G	R	R
Sulfuric acid, over 93%	G	G	R	R

Note: RT, room temperature; R, recommended; N, not recommended; G, potential failure, crystalline growth; C, conditional.

These coatings are recommended for use in the presence of hot concentrated sulfuric acid. They are also used for weak acid conditions up to a pH of 7.

9.2 Calcium Aluminate

Coatings of this type consist of calcium aluminate-based cement and various inert ingredients and are supplied in powder form to be mixed with water when used. They may be applied by casting, pouring, or guniting. Calcium aluminate-based coatings are hydraulic and consume water in their reaction mechanism to form hydrated phases. This is similar to portland cement compositions; however, their rates of hardening are very rapid. Essentially, full strength is reached within 24 h at 73°F (23°C).

They have better acid resistance than portland cement, but they are not useful in acids below pH 4.5–5. They are not recommended for alkali service above a pH of 10, nor are they recommended for halogen service. Refer to Table 9.2 for the chemical resistance of calcium aluminate and portland cement.

9.3 Portland Cement

Portland cement is made from limestone or other natural sources of calcium carbonate, clay (a source of silica), alumina, ferric oxide, and minor impurities. After grinding, the mixture is fired in a kiln at approximately 2500°F (1137°C). The final product is ground to a fineness of about 10 μm,

TABLE 9.2

Chemical Resistance of Calcium Aluminate and Portland Cement

Cement Type	Calcium Aluminate	Portland Cement
pH range	4.5–10	7–12
Water resistance	E	E
Sulfuric acid	X	X
Hydrochloric acid	X	X
Phosphoric acid	P	X
Nitric acid	X	X
Organic acids	F	F
Solvents	G	G
Ammonium hydroxide	F	G
Sodium hydroxide	F	F
Calcium hydroxide	F	G
Amines	F	G

Note: E, excellent; G, good; F, fair; P, poor; X, not recommended.

and it is mixed with gypsum to control setting. When mixed with water, the portland cement forms a hydrated phase, and it hardens. As the cement hardens, a chemical reaction takes place. The two most important reactions are the generation of calcium hydroxide and of tricalcium silicate hydrate. The calcium hydroxide generated could, theoretically, be as high as 20% of the weight of the cement, producing an alkalinity that at the solubility of lime, results in an equilibrium pH of 12.5. Steel that has been coated with cement is passivated as a result of the hardened materials. The alkalinity of the coating is provided by the presence of calcium oxide (lime). Any material that will cause the calcium oxide or hydroxide to be removed, lowering the pH, will prove detrimental and cause solution of the cement hydrates. Contact with organic or inorganic acids can cause this to happen. Organic acids can be generated when organic materials ferment.

When carbon dioxide dissolves in water that may be present on the cement, weak carbonic acid is produced. The weak carbonic acid lowers the pH of the cement solution, allowing the coated steel to corrode. This is sometimes referred to as the carbonation of cement.

Sulfates will also cause portland cement to deteriorate. In addition to being able to produce sulfuric acid that is highly corrosive to portland cement, sulfates are reactive with some additives used in the formulations. Refer to Table 9.2.

10

Coatings for Concrete

The life of concrete can be prolonged by providing a coating that will be resistant to the pollutants present. These coatings are referred to as monolithic surfacings. Before selecting an appropriate coating, consideration must be given to the condition of the concrete and the environment that the concrete will be exposed to. Proper surface preparation is essential. Surface preparations can be different for freshly placed concrete than for old concrete.

When concrete is poured, it is usually held in place by means of steel or wood forms that are removed when the concrete is still in its tender state. To facilitate their removal, release agents are applied to the forms prior to pouring. Oils, greases, and proprietary release agents are left on the surface of the concrete. These must be removed if they will interfere with the adhesion of subsequent coatings.

Quite often, curing compounds are applied to fresh concrete as soon as practical after the forms have been removed. These are liquid membranes based on waxes, resins, chlorinated rubber, or other film formers, usually in a solvent. Pretesting is necessary to determine whether or not they will interfere with the coating to be applied.

Generally, admixtures that are added to concrete mixes in order to speed up or slow down the cure, add air to the mix, or obtain special effects will not interfere with surface treatments to improve durability. The concrete supplier can furnish specific data regarding the admixtures. If in doubt, try a test patch of the coating material to be used.

It is essential that water from subslab ground sources be eliminated or minimized since migration through the concrete may create pressures at the bond line of water resistant barriers.

As discussed in Chapter 9 under portland cement coating, when water is added to portland cement, a chemical reaction takes place during the hardening. This reaction produces calcium hydroxide and tricalcium silicate hydrate. The alkalinity of concrete is provided by the presence of calcium oxide from the cement. Consequently, concrete attack can be due to chemicals that react with the portland cement binder and form conditions that physically deteriorate the material. Any material that will cause the calcium oxide or hydroxide to be removed, lowering the pH of the cement mix, will cause instability and solution of the cement hydrates.

The most common cause for the deterioration of concrete results from contact with inorganic and organic acids. Those that form soluble salts with calcium oxide or calcium hydroxide are the most aggressive. Typical compounds that can cause problems include sour milk, industrial wastes, fruit juices, sonic ultrapure waters, and organic materials that ferment and produce organic acids. Typical chemical families found in various types of chemical processing industry plants and their effect on concrete are shown in Table 10.1.

10.1 Surface Preparation

It is essential that the concrete surface be property prepared prior to application of the coating. The surfaces of cement containing materials may contain defects that require repair before the application of the coating. In general, the surface must be thoroughly cleaned and all cracks repaired.

Unlike specifications for the preparation of steel prior to coating, there are no detailed standard specifications for the preparation of concrete surfaces. In most instances, it is necessary to follow the instructions supplied by the coating manufacturer. Specifications can range from the simple surface cleaning that provides a clean surface without removing concrete from the substrate, to surface abrading that provides a clean roughened surface, to acid etching that also provides a clean roughened surface.

10.1.1 Surface Cleaning

Surface cleaning is accomplished by one of the following means:

1. Broom sweeping
2. Vacuum cleaning
3. Air blast
4. Water cleaning
5. Detergent cleaning
6. Steam cleaning

Water cleaning and detergent cleaning will not remove deep embedded soils in the pores of the concrete surfaces. In addition, these methods saturate the concrete that then requires a drying period (not always practical).

TABLE 10.1

Effects of Various Chemicals on Concrete

Chemical	Effect on Concrete
Chemical Plants	
Acid waters pH 6.5 or less	Disintegrates slowly
Ammonium nitrate	Disintegrates
Benzene	Liquid loss by penetration
Sodium hypochlorite	Disintegrates slowly
Ethylene	Disintegrates slowly
Phosphoric acid	Disintegrates slowly
Sodium hydroxide, 20% and above	Disintegrates slowly
Food and beverage plants	
Almond oil	Disintegrates slowly
Beef fat	Solid fat disintegrates slowly, malted fat more readily
Beer	May contain, as fermentation products, acetic, carbonic, lactic, or tannic acids which disintegrate slowly
Buttermilk	Disintegrates slowly
Carbonic acid (soda water)	Disintegrates slowly
Cider	Disintegrates slowly
Coconut oil	Disintegrates slowly
Corn syrup	Disintegrates slowly
Fish oil	Disintegrates slowly
Fruit juices	Disintegrates
Lard or lard oil	Lard disintegrates slowly, lard oil more quickly
Milk	No effect
Molasses	Disintegrates slowly above 120°F/49°C
Peanut oil	Disintegrates slowly
Poppyseed oil	Disintegrates slowly
Soybean oil	Disintegrates slowly
Sugar	Disintegrates slowly
Electric generating utilities	
Ammonium salts	Disintegrates
Coal	Sulfides leaching from damp coal may oxidize to sulfurous or sulfuric acid, disintegrates
Hydrogen sulfide	Dry, no effect; in moist oxidizing environments converts to sulfurous acid and disintegrates slowly
Sulfuric acid (10–80%)	Disintegrates rapidly
Sulfur dioxide	With moisture forms sulfurous acid which disintegrates rapidly
Pulp and paper mills	
Chlorine gas	Slowly disintegrates moist concrete
Sodium hypochlorite	Disintegrates slowly
Sodium hydroxide	Disintegrates concrete
Sodium sulfide	Disintegrates slowly
Sodium sulfate	Disintegrates concrete of inadequate sulfate resistance
Tanning liquor	Disintegrates if acid

10.1.2 Surface Abrading

Surface abrading can be accomplished by

1. Mechanical abrading
2. Water blasting
3. Abrasive blasting

Of the three methods to produce a roughened surface, the most technically effective is abrasive blasting by means of sandblasting or shot blasting followed by broom sweeping, vacuum cleaning, or an air blast to remove the abrasive. However, in some instances this may not be a practical approach.

10.1.3 Acid Etching

Acid etching is a popular procedure used for both new and aged concrete. It must be remembered that during this process, acid fumes will be evolved that may be objectionable. Also thorough rinsing is required that saturates the concrete. This may necessitate a long drying period, depending on the coating to be used.

10.2 Coating Selection

The coating to be selected will depend on the physical properties and conditions of the concrete as well as the environmental conditions.

Factors such as alkalinity, porosity, water adsorption and permeability, and weak tensile strength must be considered. The tendency of concrete to crack, particularly on floors, must also enter into the decision. Floor cracks may develop as a result of periodic overloading or from drying shrinkage. Drying shrinkage is not considered a working movement, whereas, periodic overloading is. In the former, a rigid flooring system could be used, whereas, in the latter, an elastomeric caulking of the moving cracks would be considered.

Selection may be influenced by the presence of substrate water during coating. If the concrete cannot be dried, one of the varieties of water based or water tolerant systems should he considered.

When aggressive environments are present, the surface profile and surface porosity of the concrete must be taken into account. If complete coverage of the substrate is required, specification of film thickness or number of coats may need to be modified. Block fillers may be required. In a nonaggressive atmosphere, an acrylic latex coating may suffice.

Specific environmental conditions will dictate the type of coating required. Not only do normal atmospheric pollutants have to be considered, but any specific local pollutants will also have to be taken into account.

Also to be considered are the local weather conditions that will result in minimum and maximum temperatures as seasons change. Also to be considered are the possibilities of spillage of chemicals on the surface. Coatings can be applied in various thicknesses depending upon the environment and contaminants.

Thin film coatings are applied at less than 20 mils dry film thickness. Commonly used are epoxies that may be formulations of polyamides, polyamines, polyesters, or phenolics. These coatings will protect against spills of hydrocarbon fuels, some weak solutions of acids and alkalies, and many agricultural chemicals. Epoxies can also be formulated to resist spills of aromatic solvents such as xylol or toluol.

Most epoxies will lose some of their gloss and develop a "chalk face" when exposed to weather. However, this does not affect their chemical resistance.

Medium film coatings are applied at approximately 20–40 mils dry film thickness. Epoxies used in this category are often flake filled to give them rigidity, impact strength, and increased chemical resistance. The flakes can be mica, glass, or other inorganic platelets.

Vinyl esters are also used in this medium film category. These coatings exhibit excellent resistance to many acids, alkalies, hypochlorites, and solvents. Vinyl esters may also be flake filled to improve their resistance.

Sonic vinyl esters require the application of a low viscosity penetrating primer to properly cleaned and profiled concrete before application, whereas, others may he applied directly.

Thick film coatings are installed by two means. The specialty epoxy types are mixed with inorganic aggregates and trowel applied. The polyesters and vinyl esters are applied with a reinforcing fiberglass mat.

Table 10.2 provides a guideline for specifying film thickness.

The most popular monolithic surfacings are formulated from the following resins:

1. Epoxy, including epoxy novolac
2. Polyester
3. Vinyl ester, including vinyl ester novolac
4. Acrylic
5. Urethane
6. Phenolic novolacs

10.3 Installation of Coatings

As with the application of any coating system, proper surface preparation and installation techniques are essential. Basic preparations for the application of a coating are as follows:

TABLE 10.2

Guidelines for Specifying Film Thickness

	Film Thickness		
Contaminant	Thin	Medium	Thick
Aliphatic hydrocarbons	X	X	X
Aromatic hydrocarbons	X	X	
Organic acids			
Weak	X		
Moderate	X	X	
Strong			X
Inorganic acids			
Weak	X		
Moderate		X	
Strong			X
Alkalies			
Weak	X		
Moderate		X	X
Strong			X
Bleach liquors	X		
Oxygenated fuels	X	X	
Fuel additives	X	X	
Deionized water			X
Methyl ethyl ketone	X	X	
Fermented beverages	X	X	
Seawater	X		
Hydraulic/brake fluids	X	X	X

1. Substrate must be free of cracks and be properly sloped to drains.

2. New, as well as existing slabs, should have a coarse surface profile, be clean and dry, and be free of contaminants.

3. In general, slab and materials to be installed should have a temperature of 65–85°F/18–29°C. If necessary, special catalysts and hardening systems are available to accommodate higher or lower temperatures.

4. Manufacturers' directions should be adhered to regarding the priming of the substrate prior to applying the monolithic surfacing.

5. Individual and combined components should be mixed thoroughly at a maximum speed of 500 rpm to minimize air entrapment.

6. Prior to curing, the surface must be protected from moisture and contamination.

Monolithic surfacings may be installed by a variety of methods, many that are the same methods used in the portland cement concrete industry.

The primary methods are:

1. Hand troweled
2. Power troweled
3. Spray
4. Pour in place/self-level
5. Broadcast

Small areas, or areas with multiple obstructions, should be hand troweled. Topcoat sealers should be applied to provide imperviousness and increased density with a smooth, easy-to-clean finish.

Large areas with a minimum of obstructions are best handled with power troweling. The minimum thickness would be 1/4 in. (6.5 mm). The density of the finish can be improved by the use of appropriate sealers.

For areas subjected to aggressive corrosion, spray application is recommended. The consistency of the material can be formulated to control slump and type of finish. This permits the material to be sprayed on vertical and overhead surfaces including structural components with thicknesses of 1/16 to 3/32 in. (1.5 to 2.4 mm) as is suitable for light duty areas.

Economical and aesthetically attractive floors can be applied by the broadcast system where resins are "squeegee" applied to the concrete stab. Filler or colored quartz aggregates of varying color and size are sprinkled or broadcast into the resin. Excess filler and quartz are vacuumed or swept from the floor after the resin has set. This results in a floor thickness of 3/32 to 1/8 in. (2 to 3 mm). This type of floor is outstanding for light industrial and interior floors.

10.4   Epoxy and Epoxy Novolac Coatings

The three most often used epoxy resins for monolithic surfacings are the bisphenol A, bisphenol F (epoxy novolac), and epoxy phenol novolac. These base components are reacted with epichlorhydrin to form resins of varying viscosity and molecular weight. The hardening system employed to effect the cure or solidification will determine the following properties of the cured system:

1. Chemical and thermal resistance
2. Physical properties
3. Moisture tolerance
4. Workability
5. Safety during use

Bisphenol A epoxy is the most popular, followed by the bisphenol F that is sometimes referred to as an epoxy novolac. The epoxy phenol novolac is a higher viscosity resin that requires various types of diluents or resin blends for formulating coatings.

The bisphenol A resin uses the following types of hardeners:

1. Aliphatic amines
2. Modified aliphatic amines
3. Aromatic amines
4. Others

Table 10.3 shows effects of the hardener on the chemical resistance of the finished coating of bisphenol A systems for typical compounds. Table 10.4 provides a comparison of the general chemical resistance of optimum chemical resistant bisphenol A, aromatic amine cured, with bisphenol F resin systems.

Amine hardening systems are the most popular for ambient temperature curing epoxy coatings. These systems are hygroscopic and can cause allergenic responses to sensitive skin. These responses can be minimized or virtually eliminated by attention to personal hygiene and the use of protective creams on exposed areas of skin, i.e., face, neck, arms, and hands. Protective garments, including gloves, are recommended when using epoxy materials.

Epoxies are economical, available in a wide range of formulations and properties, and offered from many manufacturers. Formulations are available for interior as well as exterior applications. However, when installed outside, moisture from beneath the slab can affect adhesion and cause blistering.

The typical epoxy system is installed in layers of primer, base, and finish coats. Overall, the installation can take several days. Epoxies should not be

TABLE 10.3

Types of Epoxy Hardeners and Their Effect on Chemical Resistance

	Hardeners		
Medium	Aliphatic Amines	Modified Aliphatic Amines	Aromatic Amines
Acetic acid, 5–10%	C	N	R
Benzene	N	N	R
Chromic acid, <5%	C	N	R
Sulfuric acid, 25%	R	C	R
Sulfuric acid, 50%	C	N	R
Sulfuric acid, 75%	N	N	R

R, recommended; N, not recommended; C, conditional.

Source: From A.A. Boova. 1996. Chemical resistant mortars, grouts and monolithic surfacings, in *Corrosion Engineering Handbook*, P.A. Schweitzer, Ed., New York: Marcel Dekker, pp. 459–487.

TABLE 10.4

Comparative Chemical and Thermal Resistance of Bisphenol A, Aromatic Amine-Cured vs. Bisphenol F (Epoxy Novolac)

Medium, RT	Bisphenol A	Bisphenol F
Acetone	N	N
Butyl acetate	C	E
Butyl alcohol	C	E
Chromic acid, 10%	C	E
Formaldehyde, 35%	E	G
Gasoline	E	E
Hydrochloric acid, to 36%	E	E
Nitric acid, 30%	N	C
Phosphoric acid, 50%	E	E
Sulfuric acid, to 50%	E	E
Trichloroethylene	N	G
Maximum temperature, °F/°C	160/71	160/71

RT, room temperature; C, conditional; N, not recommended; E. excellent; G, good.

Source: From A.A. Boova. 1996. Chemical resistant mortars, grouts and monolithic surfacings, in *Corrosion Engineering Handbook*, P.A. Schweitzer, Ed., New York: Marcel Dekker, pp. 459–487.

applied to new concrete before it has reached full strength (approximately twenty eight days).

Table 10.5 provides the atmospheric corrosion resistance of monolithic floor surfacings.

TABLE 10.5

Atmospheric Corrosion Resistance of Monolithic Concrete Surfacings

	Atmospheric Pollutant							
Surfacing	NO$_x$	H$_2$S	SO$_2$	CO$_2$	UV	Chloride Salt	Weather	Ozone
Epoxy–bisphenol A								
Aromatic amine hardener	R	X	X	R	R	R	R	R
Epoxy novolac	R	X	X	R	R	R	R	R
Polyesters								
Isophthalic	R	R	X	R	RS	R	R	R
Chlorendic	R	R	X	R	RS	R	R	R
Bisphenol A fumarate	R	R	X	R	RS	R	R	R
Vinyl esters	R	R	R	R	R	R	R	R
Acrylics	R		R	R	R	R	R	R
Urethanes	R		X		R	R		

R, resistant; X, not resistant; RS, resistant when stabilized.

Source: From P.A. Schweitzer. 1999. *Atmospheric Degradation and Corrosion Control*, New York: Marcel Dekker.

TABLE 10.6

Corrosion Resistance of Bisphenol A and Bisphenol F Epoxies

			Hardeners	
			Aromatic Amines	
			Bisphenol	
Corrodent, RT	Aliphatic Amines	Modified Aliphatic Amines	A	F
Acetic acid, 5–10%	C	U	R	
Acetone	U	U	U	U
Benzene	U	U	R	R
Butyl acetate	U	U	U	R
Butyl alcohol	R	R	R	R
Chromic acid, 5%	U	U	R	R
Chromic acid, 10%	U	U	U	R
Formaldehyde, 35%	R	R	R	R
Gasoline	R	R	R	R
Hydrochloric acid, to 36%	U	U	R	R
Nitric acid, 30%	U	U	U	U
Phosphoric acid, 50%	U	U	R	R
Sulfuric acid, 25%	R	U	R	R
Sulfuric acid, 50	U	U	R	R
Sulfuric acid, 75%	U	U	U	U
Trichloroethylene	U	U	U	R

RT, room temperature; R, recommended; U, unsatisfactory.

Source: From P.A. Schweitzer. 1998. *Encyclopedia of Corrosion Technology*, New York: Marcel Dekker.

The bisphenol Fs (epoxy novolacs) are essentially premium grade epoxy resins providing an increased chemical resistance. They are also available in a wide range of formulations and from many manufacturers. The primary advantages in the use of epoxy novolacs is in their improved resistance to higher concentrations of oxidizing and nonoxidizing acids, and aliphatic and aromatic solvents. Refer to Table 10.6.

The novolacs arc more expensive than the bisphenol A epoxies and can discolor from contact with sulfuric and nitric acids. For exposure to normal atmospheric pollutants, bisphenol A epoxies are satisfactory. However, if the surface is to be exposed to other more aggressive containments, then the novolacs should be considered.

Table 10.7 provides a comparison of the chemical resistance of various monolithic surfacings.

10.5 Polyester Coatings

Polyester coatings were originally developed to resist chlorine dioxide. The three types of unsaturated polyesters most commonly used are

TABLE 10.7

Comparative Chemical Resistance

Medium, RT	1			2		3	
	A	B	C	D	E	F	G
Acetic acid, to 10%	R	R	R	R	R	R	R
Acetic acid, 10–15%	C	R	C	R	R	C	R
Benzene	C	R	R	R	N	R	R
Butyl alcohol	R	C	R	R	R	N	R
Chlorine, wet, dry	C	C	C	R	R	R	R
Ethyl alcohol	R	C	R	R	R	R	R
Fatty acids	C	R	C	R	R	R	R
Formaldehyde, to 37%	R	R	R	R	R	R	R
Hydrochloric acid, to 36%	C	R	R	R	R	R	R
Kerosene	R	R	R	R	R	R	R
Methyl ethyl ketone, 100%	N	N	N	N	N	N	N
Nitric acid, to 20%	N	N	R	R	R	R	R
Nitric acid, 20–40%	N	N	R	R	N	N	C
Phosphoric acid	R	R	R	R	R	R	R
Sodium hydroxide, to 25%	R	R	R	N	R	R	R
Sodium hydroxide, 25–50%	R	C	R	N	R	C	R
Sodium hypochlorite, to 6%	C	R	R	R	R	R	R
Sulfuric acid, to 50%	R	R	R	R	R	R	R
Sulfuric acid, 50–75%	C	R	R	R	C	R	R
Xylene	N	R	R	R	R	N	R

RT, room temperature; 1-A, bisphenol A epoxy—aliphatic amine hardener; 1-B, bisphenol A epoxy—aromatic amine hardener; 1-C, bisphenol F epoxy (epoxy novolac); 2-D, polyester resin—chlorendic acid type; 2-E, polyester resin—bisphenol A fumarate type; 3-F, vinyl ester resin; 3-G, vinyl ester novolac resin; R, recommended; N, not recommended; C, conditional.

Source: From A.A. Boova. 1996. Chemical resistant mortars, grouts and monolithic surfacings, in *Corrosion Engineering Handbook*, P.A. Schweitzer, Ed., New York: Marcel Dekker, pp. 459–487.

1. Isophthalic
2. Chlorendic acid
3. Bisplienol A fumarate

The chlorendic and bisphenol A resins offer improved chemical resistance, higher thermal capabilities, and improved ductility with less shrinkage. The bisphenol A resins provide improved resistance to alkalies and essentially equivalent resistance to oxidizing mediums. Refer to Table 10.5 for the resistance of the polyester resins to atmospheric containments and to Table 10.8 for the comparative chemical resistance of the various polyester resins.

Polyester resins are easily pigmented for aesthetic considerations. The essentially neutral curing systems provide compatibility for application to many substrates including concrete. Properly formulated polyester resins

TABLE 10.8

Comparative Chemical Resistance of Various Polyester Resins

Medium, RT	Isophthalic	Chlorendic	Bisphenol A Fumarate
Acids, oxidizing	R	R	R
Acids, nonoxidizing	R	R	R
Alkalies	N	N	R
Salts	R	R	R
Bleaches	R	R	R
Maximum temperature, °F/°C	225/107	260/127	250/121

RT, room temperature; R, recommended; N, not recommended.

Source: From A.A. Boova. 1996. Chemical resistant mortars, grouts and monolithic surfacings, in *Corrosion Engineering Handbook*, P.A. Schweitzer, Ed., New York: Marcel Dekker, pp. 459–487.

provide installation flexibility to a wide range of temperatures, humidities, and containments; encountered on most construction sites.

Polyester formulations have limitations such as

1. Strong aromatic odor that can be offensive for certain indoor and confined space applications

2. Shelf life limitations that can be controlled by low temperature storage (below 60°F/15°C) of the resin component.

Table 10.9 provides the chemical resistance of chlorendic and bisphenol A fumarate rosins in the presence of selected corrodents.

10.5.1 Vinyl Ester/Vinyl Ester Novolac Coatings

Vinyl ester resins are addition reactions of methacrylic acid and epoxy resin. These resins have many of the same properties as the epoxy, acrylic, and bisphenol A fumarate resins.

The vinyl ester resins are the most corrosion resistant of any of the monolithic surfacing systems, and they are also the most expensive and difficult to install. They are used when extremely corrosive conditions are present. The finished flooring is vulnerable to hydrostatic pressure and vapor moisture transmission. Refer to Table 10.5 for their resistance to atmospheric corrosion and Table 10.10 for their resistance to selected corrodents.

The major advantage of these resins are their resistance to most oxidizing mediums and high concentrations of sulfuric acid, sodium hydroxide, and many solvents.

TABLE 10.9

Chemical Resistance of Chlorendic and Bisphenol A Fumarate Resins

	Polyester	
Corrodent, RT	Chlorendic	Bisphenol A Fumarate
Acetic acid, glacial	U	U
Benzene	U	U
Chlorine dioxide	R	R
Ethyl alcohol	R	R
Hydrochloric acid, 36%	R	R
Hydrogen peroxide	R	U
Methanol	R	R
Methyl ethyl ketone	U	U
Motor oil and gasoline	R	R
Nitric acid, 40%	R	U
Phenol, 5%	R	R
Sodium hydroxide, 50%	U	R
Sulfuric acid, 75%	R	U
Toluene	U	U
Triethanolamine	U	R
Vinyl toluene	U	U

RT, room temperature; R, recommended; U, unsatisfactory.

Source: From P.A. Schweitzer. 1998. *Encyclopedia of Corrosion Technology*, New York: Marcel Dekker.

Vinyl ester resins also have the disadvantages of having

1. Strong aromatic odor for indoor or confined space applications
2. Shelf life limitation of the resins require refrigerated storage below 60°F/15°C to extend its useful life.

10.6 Acrylic Coating

Acrylic monolithic coatings are suitable for interior or exterior applications with relatively benign atmospheric exposures. Refer to Table 18.5 of *Fundamentals of Metallic Corrosion: Atmospheric and Media Corrosion of Metals*. They excel at water and weather resistance, and are best at "breathing" in the presence of a moisture transmissive problem in the stab. They are intended for protection against moderate corrosion environments. The advantages for their use are as follows:

1. They are the easiest of the resin systems to mix and apply by using pour-in-place and self-leveling techniques.
2. Because of their outstanding weather resistance, they are equally appropriate for indoor or outdoor applications.

TABLE 10.10

Resistance of Vinyl Ester and Vinyl Ester Novolac to Selected
Corrodents

	Vinyl Ester	
Corrodent	Vinyl Ester	Novolac
Acetic acid, glacial	U	R
Benzene	R	R
Chlorine dioxide	R	R
Ethyl alcohol	R	R
Hydrochloric acid, 36%	R	R
Hydrogen peroxide	R	R
Methanol	U	R
Methyl ethyl ketone	U	U
Motor oil and gasoline	R	R
Nitric acid, 40%	U	R
Phenol, 5%	R	R
Sodium hydroxide, 50%	R	R
Sulfuric acid, 75%	R	R
Toluene	U	R
Triethanolamine	R	R
Vinyl toluene	U	R
Maximum temperature, °F/°C	220/104	230/110

R, recommended; U, unsatisfactory.

Source: From P.A. Schweitzer. 1998. *Encyclopedia of Corrosion Technology*, New
York: Marcel Dekker.

3. They are the only system that can be installed at below freezing
   temperatures (25°F/−4°C) without having to use special hardening or catalyst systems.

4. They are the fastest set and cure of all the resins.

5. They are the easiest to pigment and, with the addition of various
   types of aggregate, can be aesthetically attractive.

6. They are equally appropriate for maintenance and new construction, and they bond well to concrete.

The disadvantage of the acrylic coating system is the aromatic odor in indoor
or confined spaces.

10.7 Urethane Coatings

Monolithic urethane flooring systems offer the following advantages:

1. They are easy to mix and apply using the pour-in-place, self-
   level technique.

2. Systems are available for indoor or outdoor installations.
3. The elastomeric quality of the systems provides underfoot comfort.
4. They have excellent sound-deadening properties.
5. They have excellent resistance to impact and abrasion.
6. They are excellent waterproof flooring systems for above-grade light and heavy duty floors.
7. They are capable of bridging Cracks in concrete 1/16 in. (1.5 mm) wide.

As with the acrylics, the urethanes are intended for protection against moderate to light corrosion environments. Standard systems are effective at temperatures of 10 to 140°F($-$24 to 60°C). High temperature systems are available with a range of 10 to 180°F($-$24 to 82°C).

Refer to Table 10.11 for the comparative chemical resistance of urethane and acrylic systems.

The acrylic and urethane systems have substantially different physical properties. The acrylic flooring systems are extremely hard and are too brittle for applications subjected to excessive physical abuse such as impact, whereas, the inherent flexibility and impact resistance of the urethanes offer potential for this type of application. Table 10.12 provides

TABLE 10.11

Comparative Chemical Resistance: Urethane vs. Acrylic Systems

		Urethane	
Medium, RT	Acrylic	Standard	High Temperature
Acetic acid, 10%	G	G	C
Animal oils	G	G	N
Boric acid	E	E	E
Butter	G	F	N
Chromic acid, 5–10%	C	C	C
Ethyl alcohol	N	N	N
Fatty acids	F	F	N
Gasoline	E	N	N
Hydrochloric acid, 20–36%	F	C	C
Lactic acid, above, 10%	F	C	C
Methyl ethyl ketone, 100%	N	N	N
Nitric acid, 5–10%	G	C	F
Sulfuric acid, 20–50%	G	C	C
Water, fresh	E	E	E
Wine	G	G	F

RT, room temperature; E, excellent; G, good; F, fair; C, conditional; N, not recommended.

Source: From A.A. Boova. 1996. Chemical resistant mortars, grouts and monolithic surfacings, in *Corrosion Engineering Handbook*, P.A. Schweitzer, Ed., New York: Marcel Dekker, pp. 459–487.

TABLE 10.12

Minimum Physical and Thermal Properties of Acrylic Monolithic Surfacing and Urethane Monolithic Surfacings

	Acrylics	Urethanes	
Property	Monolithic	Standard	High Temperature
Tensile, psi (MPa) ASTM test method C-307	1000 (7)	650 (5)	550 (5)
Flexural, psi (MPa) ASTM test method C-580	2500 (17)	1100 (8)	860 (6)
Compressive, psi (MPa) ASTM test method C-579	8000 (55)	2500 (17)	1500 (10)
Bond to concrete	Concrete fails	Concrete fails	Concrete fails
Maximum temperature, °F/°C	150/66	140/60	180/82

Source: From A.A. Boova. 1996. Chemical resistant mortars, grouts and monolithic surfacings, in *Corrosion Engineering Handbook*, P.A. Schweitzer, Ed., New York: Marcel Dekker, pp. 459–487.

physical and thermal properties for the acrylic and urethane flooring systems.

10.8   Phenolic/Epoxy Novolac Coatings

In order to satisfy the need to provide a coating system that has ability to bridge cracks and provide improved corrosion resistance, medium build coating systems have been developed. The phenolic/epoxy novolac system is capable of bridging cracks and providing outstanding corrosion resistance.

One such system uses a low viscosity penetrating epoxy primer. Low viscosity allow the primer to be used for areas that need a fast turnaround by quickly "wetting out" the substrate. If surface deterioration or preparation presents an unacceptably rough surface, the surface can be smoothed further by using a pigmented, high solid/high build epoxy polyamide filler/sealer.

For deep pits, the crack filler used is a two component epoxy paste developed specifically for sealing and smoothing out applications on concrete.

The crack filler can be used to fill and smooth hairline cracks, bug holes, gouges, or divots when minimal movement of the substrate is expected.

TABLE 10.13

Compatibility of Phenolic with Selected Corrodents

Chemical	Maximum Temperature	
	°F	°C
Acetic acid, 10%	212	100
Acetic acid, glacial	70	21
Acetic anhydride	70	21
Acetone	X	X
Aluminum sulfate	300	149
Ammonium carbonate	90	32
Ammonium chloride, to sat.	80	27
Ammonium hydroxide, 25%	X	X
Ammonium nitrate	160	71
Ammonium sulfate	300	149
Aniline	X	X
Benzene	160	71
Butyl acetate	X	X
Calcium chloride	300	149
Calcium hypochlorite	X	X
Carbonic acid	200	93
Chromic acid	X	X
Citric acid, conc.	160	71
Copper sulfate	300	149
Hydrobromic acid, to 50%	200	93
Hydrochloric acid, to 38%	300	149
Hydrofluoric acid	X	X
Lactic acid, 25%	160	71
Methyl isobutyl ketone	160	71
Muriatic acid	300	149
Nitric acid	X	X
Phenol	X	X
Phosphoric acid, 50–80%	212	100
Sodium chloride	300	149
Sodium hydroxide	X	X
Sodium hypochlorite	X	X
Sulfuric acid, 10%	250	121
Sulfuric acid, 50%	250	121
Sulfuric acid, 70%	200	93
Sulfuric acid, 90%	70	21
Sulfuric acid, 98%	X	X
Sulfurous acid	80	27

The chemicals listed are in the pure state or in a saturated solution unless otherwise indicated. Compatibility is shown to the maximum allowable temperature that data is available for. Incompatibility is shown by an X.

To provide maximum corrosion resistance a two component phenolic/epoxy novolac coating can be used. The phenolic coating provides resistance to high concentrations of acids, particularly to sulfuric acid at elevated temperatures, Refer to Table 10.13 for the compatibility of phenolic with selected corrodents.

References

1. P.A. Schweitzer. 1999. *Atmospheric Degradation and Corrosion Control*, New York: Marcel Dekker.
2. A.A. Boova. 1996. Chemical resistant mortars, grouts and monolithic surfacings, in *Corrosion Engineering Handbook*, P.A. Schweitzer, Ed., New York: Marcel Dekker, pp. 459–487.
3. P.A. Schweitzer. 1998. *Encyclopedia of Corrosion Technology*, New York: Marcel Dekker.

11

Corrosion Monitoring

It is essential for operators of industrial process plants to have a program for controlling corrosion. For such a program to be effective, it should facilitate safe operation of the process unit over its intended life and maximize the economic return to the owner. To accomplish these objectives, a corrosion control program should consider: characteristics (toxicity, flammability, etc.) of the process; selection of construction materials; control of operating conditions; addition of neutralizing or inhibition chemicals; and monitoring to ensure that anticipated results are obtained.

Selection of construction materials almost always involves a compromise between expensive and often hard-to-obtain alloys highly resistant to corrosion by the process under any conceivable operating conditions and less expensive, more available materials that are more susceptible to corrosion. Often, a material can have acceptable resistance to a process at design conditions but corrode at very high rates during process upsets. Particularly in these cases, on-line monitoring and control of process operating parameters and direct on-line monitoring of corrosion rates are essential to an effective program. Several types of materials are often used in conjunction in the construction of process facilities. For instance, corrosion-resistant alloys are used in particularly susceptible equipment (heat exchanger tubes, for example) with less expensive materials used elsewhere in the process. It is, of course, essential that the behavior of various metals in the anticipated operating environment be known in advance of selection. whereas there is a vast array of available literature on the subject, laboratory and field trials are often performed to verify theoretical predictions or tryout new materials before a commitment is made to use them in the process unit.

Sometimes a less expensive material, that would otherwise be unacceptable because of low corrosion resistance can still be the best choice when used in conjunction with a chemical corrosion inhibitor injection system or an anodic or cathodic protection system.

There are many operating and environmental parameters that can affect the corrosion of a metal—changes in any that can greatly accelerate the corrosion rate. Without early detection and accurate measurement of the impact of these changes, costly and potentially hazardous damage can occur before corrections are made. Accurate and timely corrosion measurement is, therefore, an essential part of almost all corrosion control programs. The technology of corrosion detection and measurement has rapidly advanced in the recent past, so today's corrosion engineer has many proven techniques available for both laboratory analysis and on-line monitoring to devise and manage an effective corrosion control program for his (her) particular plant process and environmental conditions.

11.1 Measuring and Monitoring Corrosion

As used herein, measurement of corrosion refers to any technique that can be used to determine the effects of corrosion:

1. While the facilities are in operation
2. During shutdowns
3. While the laboratory analyses are performed outside the process equipment

Monitoring refers to a special group of measurement techniques that are suitable for use while the facility is in operation.

A comprehensive corrosion control program should include several techniques because no single technique is capable of providing all the information necessary in a timely manner. An overview of measurement techniques is presented below with emphasis on those useful in practical on-line monitoring.

11.1.1 Radiography

Radiography (x-ray) permits two-dimensional views of the piping or equipment walls and is suitable for detecting major flaws or a severe corrosive attack when results are compared to an earlier (or baseline) study. Radiography is not suitable for detecting small changes in residual wall thickness because of accuracy limitations. Radiography is a measurement technique that requires specialized equipment and trained operators, usually provided by contractor personnel, and is not currently used in continuous on-line monitoring.

11.1.2 Ultrasonic Measurement

Ultrasonic measurement techniques are similar to radiography in that specialized equipment is required, and they are almost exclusively used for measurement rather than monitoring. There are several types of ultrasonic equipment. A scan provides a simple depth measurement from the exterior surface of a pipe or vessel to the next interface that reflects sound waves. Generally, this measures wall thickness, but A scan can be occasionally fooled, by midwall pipe flaws. B scan instruments are much more powerful because they produce cross-sectional images similar to x-rays. The C-scan systems produce a three-dimensional view of a surface using complex and expensive equipment. C-scan systems can be very useful for large, critical surfaces such as aircraft skins, but they are much less commonly used in process plants at this time because of cost, speed of coverage, and very large quantity of data produced.

11.1.3 Visual Inspection

Visual inspection is rarely practical in a working process plant. However, whenever an opportunity presents itself, full advantage should be taken of it. Only through direct visual inspection can one be absolutely certain as to the condition of a process unit. Consequently, during any turnaround activities, internal visual inspection should be carried out to verify the results of on-line monitoring programs, choose new locations for monitoring, etc.

11.1.4 Destructive Analysis

Destructive analysis is less practical than visual inspection but can be important and should, therefore, be performed as often as possible. The most common opportunity arises when piping is replaced for one reason or another. Sections of the replaced piping can then be examined for evidence of stress corrosion cracking (SCC), hydrogen-induced corrosion (HIC), pitting, and any other unusual or unexpected corrosive activity.

11.1.5 Chemical Analysis

Chemical analysis is the monitoring of the chemical composition or other chemical characteristics of the process fluids. Because corrosion is an electrochemical phenomenon, chemical analysis can be a useful indicator of corrosivity of the process. In water systems, for instance, pH and conductivity are two parameters frequently monitored to give an indication of corrosivity. Increases in concentrations of corrodents such as H_2S or CO_2 can also be useful in determining corrosivity, but such measurements require more complex equipment and are usually infrequently made.

Another technique that can be useful involves analyzing process streams for the presence of corrosion products, e.g., the quantity of metal and metallic compounds. This type of analysis can yield a good estimate of the current metal loss from corrosion. When compared with previous data, it can be used to establish trends and approximate corrosion rates. Such data must be used with caution, however. For instance, "iron counts" in fluid produced by oil wells can be a good indicator of the amount of corrosion taking place on the casing, tubing, and completion equipment. If iron is present naturally in the producing formation, however, information gained from such chemical analysis will most likely be meaningless.

11.1.6 Coupons

Coupons are the oldest and simplest device used in the monitoring of corrosion. Coupons are small pieces of metal, usually of a rectangular shape that are inserted in the process stream and removed after a period of time for study. The most common and basic use of coupons is to determine average corrosion rate over the period of exposure. This is accomplished by weighing the coupon before and after exposure (coupons must first be cleaned following exposure to remove corrosion products and any other deposits) and determining the weight loss. The average corrosion rate can easily be calculated from the weight loss, the initial surface area of the coupon, and the time exposed. It is advisable to leave a coupon exposed for at least 30 days to obtain valid corrosion rate information. There are two reasons for this recommended practice. First, a clean coupon generally corrodes much faster than one that has reached equilibrium with its environment. This will cause a higher corrosion rate to be reported in a short test than is actually being experienced on the pipe or vessel. Second, there is an unavoidable potential for error as a result of the cleaning operation. Coupon cleaning procedures are designed to remove all of the deposits without disturbing the remaining uncorroded metal of the coupon. A small amount of the underlying metal is often removed with the deposits, however, and if the actual metal loss from corrosion is small (as would be the case in a short test), the effect of metal removed during cleaning would create a significant error. Care must be taken to correct for this effect. It should be recognized that a coupon can only provide corrosion rate data based on the total weight loss divided by the total time of exposure. A major shortcoming of coupon monitoring is that high corrosion rates for short periods of time may be undetectable and cannot be correlated to process upset conditions. If more frequent information on weight loss is desired, the use of *electrical resistance* monitoring systems is recommended. In many cases, it is also recommended that coupons and electrical resistance (ER) probes be used in conjunction.

One of the most important roles of coupons is to provide information about the type of corrosion present. Unlike ER probes that only detect the amount of metal removed, coupons can be examined for evidence of pitting and other localized forms of attack. It is also important to remember that coupons or monitoring probes indicate the attack of the environment only at the point of exposure. It is important, therefore, that the coupon or probe be installed at *representative locations* as close as possible to critical points where corrosion measurements are desired (e.g., vessel or pipe walls, tube sheets, trays, lateral lines upstream to major tie-ins). Conditions of flow, temperature, concentration, etc., may change considerably only a few inches away from any given location with resulting differences in corrosion rates. Coupons suspended in the center of a pipeline or vessel may corrode at different rates than coupons suspended near the wall of the vessel or pipe, for example. Since corrosive conditions can significantly change from one location to another, coupon data is best used for relative comparisons (changes in the rate or characteristics of corrosion over time at a single point) and to obtain an approximate corrosion rate at a particular point in the system rather than to precisely calculate the corrosion rate.

There are several other types of coupons available for specialized analysis. These include disk coupons that can be mounted flush to the pipe wall; ladder coupon holders for mounting coupons at several depths at the same point in the piping; prestressed coupons that are for the investigation of possible stress cracking mechanisms; and welded coupons that are used to detect preferential corrosive action on weldments.

11.1.7 Hydrogen Probes

Hydrogen probes are used to detect the penetration of elemental hydrogen into metal, pipe, and vessel. This can occur in cathodic reactions in acid solutions, particularly in the presence of hydrogen sulfide. There are three basic types of hydrogen probes. The simplest and most common consists of a thin-walled carbon steel tube inserted into the flow stream with a solid rod inside the tube forming a small annular space. Hydrogen atoms small enough to permeate the carbon steel collect in the annular space and combine to form molecular hydrogen gas that is too large to pass back into the process. As hydrogen gas collects, pressure builds in the annular space and registers on a pressure gauge located outside the piping. "Patch probes" operate identically except that the "patch" is sealed to the outside of the pipe or vessel and collects hydrogen atoms that penetrate the pipe wall. The third type of hydrogen probe is the palladium foil type that produces an electrical output proportional to hydrogen evolution rate.

All three types of hydrogen probes provide useful information about changes in the corrosive environment in processes where HIC is present or could occur. If pressure is accumulating at a certain rate per day and then increases by a factor of 10, this indicates a significant change that should be addressed. Hydrogen probe readings, however, have not been found to be suitable for direct determination of corrosion rate, only of changes in the process, and it is recommend that they be applied carefully as part of a complete corrosion control program in applications where HIC is a concern.

11.1.8 Polarization Studies

Polarization studies have been primarily laboratory electrochemical techniques to study corrosion phenomena, especially pitting, by disturbing the natural corrosion potential of a system, frequently by substantial amounts (a few volts compared with a few millivolts in linear polarization measurements) and measuring the external current flowing. With advances in computer technology, some of these systems are now being used in the field. In the laboratory, the metal/fluid environment can be carefully controlled. Such control is more difficult in the field, and the technique must be used more judiciously there. A variety of methods and equipment are used for such studies that are generally classified under the following types:

Potentiostatic   Potential held constant

Galvanostatic   Current held constant

Potentiodynamic (a) Potential changed continuously at a specified rate

                              (b) Potential changed in steps and held constant at each step

Galvanodynamic (a) Current changed continuously at a specified rate

                              (b) Current changed in steps and held constant at each step (galvano staircase)

The potentials or currents applied with these techniques may make irreversible changes to the metal-fluid interface because they may substantially polarize the metal away from its natural state in the fluid environment, generally preventing reuse of the test electrodes after one scan. In common with all electrochemical methods, they can only be made in sufficiently conductive media where the wetted electrode area is known. Although corrosion rates can be estimated with this technique, it tends to be less precise than linear polarization resistance (LPR) because of the high polarization potentials required.

11.1.9 Electrical Impedance Spectroscopy

Electrochemical impedance spectroscopy (AC impedance) is a now well-established laboratory technique used to determine the electrical impedance of the metal–electrolyte interface at various AC excitation frequencies. Impedance measurements combine the effects of DC resistance with capacitance and inductance. In order to make impedance measurements, it is necessary to have a corrosion cell of known geometry, a reference electrode, and instrumentation capable of measuring and recording the electrical response of the test corrosion cell over a wide rate of AC excitation frequencies. Again, with the evolution of more rugged computers, some investigation of this method is now being made in the field. AC impedance is capable of characterizing the corrosion interface more comprehensively and with good quality equipment specifications of achieving measurements in lower conductivity solution or high-resistivity coatings. AC impedance measurements can be used to predict corrosion rates and characterize systems under study, and they are commonly used for performance studies of chemical inhibitors and protective coatings to evaluate the resistance of alloys to specific environments, etc.

11.1.10 Electrochemical Noise

Electrochemical noise is a monitoring technique that directly measures naturally occurring electrochemical potential and current disturbances because of ongoing corrosion activity. It has the same media conductivity limitations and requirement for a known electrode area as the other electrochemical techniques. It is generally less quantitative than LPR for corrosion rate calculations, and it is more useful in detection of transient effects in marginally conductive situations. Laboratory and field interpretation is still rather developmental. Its proponents claim this technique can provide a large quantity of information that can be useful in determining what is actually happening in real time with corrosion activity in the piping or equipment being monitored. Detractors contend that there is, in fact, too much information and especially analyze when there is a contact but acceptable level of corrosion activity in the system or another source of potential electrical disturbance in the system.

This technique may have merit as part of an on-site corrosion survey conducted by specialists trained to interpret the data, but by no means is it recommended as a permanent monitor or a substitute for other techniques. The CORRATER® instruments operating in the "pitting index" or "imbalance" mode provide a simple electrochemical current noise sample that is useful as an indicator of pitting tendency or surface instability.

11.1.11 Electrical Resistance

Electrical resistance (ER) probes and instruments are basically "automatic coupons" and share many characteristics with coupons, as discussed above, when it comes to advantages and limitations. ER systems work by measuring the ER of a thin metal probe. As corrosion causes metal to be removed from the probe, its resistance increases. The major advantage of the ER method compared to coupons is that measurements can be obtained on a far more frequent basis and require much less effort to make. With automated systems, continuous readings are, in fact, made, and sophisticated data analysis techniques are now available that permit the detection of significant changes in corrosion rate in as little as 2 h without generating nuisance alarms.

11.1.12 CORROSOMETER® Systems

CORROSOMETER (ER) monitoring systems can be applied to all processes. CORROSOMETER (ER) probe elements are available in a variety of styles. A selection of the available styles is shown in Figure 11.1. Wire loop, tube loop, and strip loop styles all have a loop of metal exposed to the process. The loop protrudes from the end of the probe body through either a hermetic glass seal or a Teflon/ceramic, Teflon/epoxy, or epoxy seal/packing system.

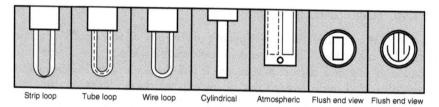

Strip loop   Tube loop   Wire loop   Cylindrical   Atmospheric   Flush end view   Flush end view

FIGURE 11.1

Selection of available CORROSOMETER probe element styles.

Choice of materials is dependent on stream composition, process conditions, and performance requirements. Cylindrical elements utilize specially made, thin wall tubing as the measurement element. Cylindrical probes are generally "all metal," i.e., there is no other material exposed to the process. There are, however, some cylindrical probes available that join the probe body at a hermetic glass seal. A variety of flush-mounted probes are also available, so called because the measuring element is mounted parallel to the flow stream, flush with the inside pipe wall.

Some types of CORROSOMETER (ER) probes are better suited to the requirements of particular applications than others. Where pitting or substantial iron sulfide (Fe_xS_y) deposition are expected to be problems, cylindrical probes should be chosen over loop style probes if possible. It is possible that wire loop and tube loop elements may be severed by a pitting attack, necessitating probe replacement. Wire loop and tube loop elements also have a tendency to be electrically shorted by a bridge of iron sulfide corrosion product. This is especially prevalent in low-velocity streams over an extended period. The effect of such bridging is to reduce the measured metal loss of the probe, creating a misleadingly low corrosion rate.

Cylindrical elements, on the other hand, are affected to a much lesser degree by pitting because of the much larger circumference of the measuring element. Cylindrical probes also demonstrate more resistance to iron-sulfide bridging because of their construction and lower inherent resistance per unit length, thus, minimizing the effect of the shunt resistance.

Most cylindrical probes are of all-welded construction in order to eliminate the need for sealing metal elements to nonmetallic glass, epoxy, or ceramic. This all-welded construction gives the probe superior resistance to leaking. Probes with higher temperature ratings can also be constructed in the all-welded style. A drawback to the all-welded style is that the element is electrically connected to the pipe wall that can, under certain conditions, interfere with the corrosion reaction on the probe. Also, because cylindrical probes are welded, under some conditions, preferential corrosion can occur in the heat-affected zones near the weld or end grain attack can occur.

Flush probe elements are thin, flat metal sections embedded in epoxy or a hermetic glass seal inside a metal probe body. Flush probes also experience certain characteristic problems, most notably lack of adhesion of the metal element to the epoxy; cracking of glass seals because of differential expansion in changing temperature environments; and erosion of the epoxy or glass because of high velocities, abrasive materials in the flowstream, or both. Flush ER probes mounted on the bottom of the line have been shown to provide good results in a sour gas–gathering system.

Because the measurement element is part of the primary pressure seal and because it is designed to corrode, ER probes have a reduced resistance to leaking after prolonged exposure. Once the measurement element has corroded through, the internals of the probe body are exposed to the process fluid. Although materials are chosen in part for their strength and lack of permeability, experience shows that process fluids will permeate throughout the probe packing materials. For this reason, quality probes are constructed of corrosion-resistant body materials and include an outboard pressure seal, often consisting of a hermetic glass-sealed connector. Other sealing materials are utilized in special cases, especially where process fluids will attack glass (e.g., hydrofluoric acid service), and secondary seals are also used in some probes, particularly when highly dangerous or toxic fluids are to be contained.

The reference/check elements are protected from the process that the measurement element is directly exposed to.

Because of the very low resistances involved, these changes can significantly affect the metal loss. Temperature changes in the process will, therefore, affect the measure element before the reference and check element readings. ER probes incorporate special design features to minimize the thermal resistance of the materials, insulating the reference and check elements from the process. It should also be noted that cylindrical probes are inherently better able to react to temperature changes because of location of the reference and check elements concentrically inside the measure element.

11.1.13 Linear Polarization Resistance

LPR is an electrochemical technique that measures the DC current (i_{meas}) through the metal–fluid interface that results from polarization of one or two electrodes of the material under study by application of a small electrical potential. Because i_{meas} is related to i_{corr} by a factor based on the anodic and cathodic Tafel slopes that are relatively constant for a given metal–fluid system, and because i_{corr} is directly proportional to corrosion rate, LPR techniques result in instantaneous corrosion rate readings. This is a significant advantage over ER or coupon monitoring in that a series of readings, over a period of time, is required to determine corrosion rate. LPR measurements cannot, however, be made in nonconductive fluids or fluids that contain compounds that coat the electrodes (e.g., crude oil). LPR techniques are, therefore, used most commonly in industrial water systems.

11.1.14 CORRATER Systems

CORRATER systems measure the instantaneous corrosion rate of a metal in a conductive fluid using the LPR measurement technique. As described earlier, corrosion is an electrochemical process where electrons are transferred between anodic and cathodic areas on the corroding metal resulting in oxidation (corrosion) of the metal at the anode and reduction of cations in the fluid at the cathode.

Sterns and Geary originally demonstrated that the application of a small polarizing potential difference (ΔE) from the corrosion potential (E_{corr}) of a corroding electrode resulted in a measured current density (i_{meas}) that is related to the corrosion current density (i_{corr}) by Equation 11.1:

$$\frac{\Delta E}{i_{meas}} = \frac{b_a b_c}{(2,303\, i_{corr})(b_a + b_c)},$$ (11.1)

where

b_a = anodic Tafel slope
b_c = cathodic Tafel slope.

Because the Tafel coefficients are more or less constant for a given metal–fluid combination, i_{meas} is proportional to i_{corr} that is proportional to the corrosion rate. Equation 11.1 and the entire LPR technique are only valid when the polarizing potential difference is very low (typically up to 20 mV). In this region, the curves are linear, hence the term LPR. Inspection of Equation 11.1 shows that the result is a resistance, the polarization resistance, R_p. Strictly speaking, there are both anodic and cathodic R_p values that can differ, they are usually assumed to be equal. The resistance to current flow between anode and cathode on the LPR probe is the sum of both polarization resistance values and the resistance of the solution between the electrodes (R_s) as shown in Equation 11.2:

$$\Delta E = i_{meas}(2R_p + R_s)$$ (11.2)

From Equation 11.1 and Equation 11.2, obtaining results from the LPR technique would seem to require only instantaneous readings of resistance. In practice, however, the determination of polarization resistance is complicated by a capacitance effect at the metal–fluid interface (double-layer capacitance). Figure 11.2 is an equivalent electrical circuit of the corrosion cell formed by the measuring electrodes and the fluid, showing the importance of R_s and double-layer capacitance effects. The effect of the double-layer capacitance is to require the direct current flow to initially charge up the capacitors, resulting in a decaying exponential current flow curve vs. time, after application of the polarizing potential difference. A typical LPR current vs. time curve is shown in Figure 11.3. Each metal–fluid interface has its own characteristic capacitance that in turn, determines the amount of time required to obtain valid measurements of i_{corr} and

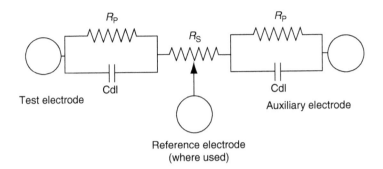

FIGURE 11.2
Equivalent circuit of LPR probe.

corrosion rate. The actual time required can vary from a few seconds up to 20 min, depending on the metal–process combination being measured. Choosing too short a polarization time can result in current readings much higher than the true i_{corr}, thus, causing measured corrosion rate to be lower than actual, sometimes by a significant amount.

Solution resistance can also have a significant effect on accuracy if it is relatively high compared to the polarization resistance. In most industrial water applications, conductivity of the solution is high and solution resistance is low compared to the polarization resistance, so i_{meas} is an accurate measure of polarization resistance and, therefore, corrosion rate.

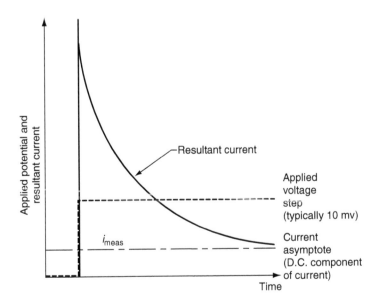

FIGURE 11.3
Type LPR current vs. time decay curve.

A serious problem develops, however, when the solution resistance increases or the polarization resistance decreases enough to make the solution resistance a significant portion of the total resistance to current flow between the electrodes. In these cases, the accuracy of the LPR measurement is affected.

This situation tends to occur at high corrosion rates (low polarization resistance) and in solutions with low conductivity (high solution resistance) and is manifested by the indicated (measured) corrosion rate being lower than the actual corrosion rate. The graph in Figure 11.4 shows the effect of this limitation on the recommended operating range of LPR instruments.

Several techniques have been used over the years to minimize the impact of solution resistance on LPR measurements. The most common technique involved the use of a three-electrode probe. The effectiveness of the reference electrode in reducing the effect of solution resistance has been shown to be dependent on the proximity of the reference electrode to the measurement electrode.

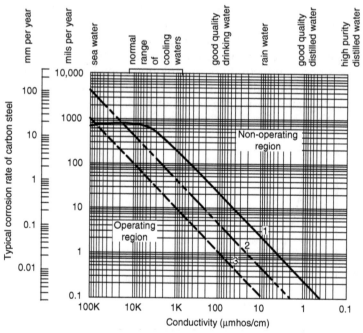

1 ——————— 2 electrode probes with Model AQUAMATE™, 9030, AQUACORR®, CORRDATA®, RCS8, and 9134 instruments
2 — — — — 2 electrode probes on Model 9000 and SCA-1 with conductivity correction
3 — — — — 2 electrode probes on Model 9000 and SCA-1 with conductivity correction

NOTES: 1. Operating range limits based on nominal 20% error in measurement of R_p.
          2. For copper alloys, typical corrosion rate of carbon steel is increased
             by alloy multiplier.

FIGURE 11.4
Operating range of LPR instruments: corrosion rate vs. solution conductions.

A better way to deal with this problem, however, is to directly measure and compensate for the solution resistance. In this method, a high-frequency AC voltage signal is applied between the electrodes short-circuiting R_p through the double-layer capacitance, thereby, directly measuring the solution resistance. The state-of-the-art, patented SRC technology also eliminates the need for a third electrode, even in low-conductivity solutions. Consequently, the two-electrode probes have become the standard with the three-electrode probe available on special order only. The above points are clearly indicated in ASTM Standard Guide G96 that states:

3.2.8 Two-electrode probes and three-electrode probes with the reference electrode equidistant from the test and auxiliary electrode do not correct for effects of solution resistance without special electronic solution resistance compensation. With high to moderate conductivity environments, this effect of solution resistance is not normally significant.

3.2.9 Three-electrode probes compensate for the solution resistance R_s by varying degrees depending on the position and proximity of the reference electrode to the test electrode. With a close-spaced reference electrode, the effects of R_s can be reduced up to approximately ten-fold. This extends the operating range over that adequate determination of the polarization resistance can be made.

3.2.10 A two-electrode probe with electrochemical impedance measurement technique at high frequency short circuits the double-layer capacitance. Cdl, so that a measurement of solution resistance R_s can be made for application as a correction. This also extends the operating range over which adequate determination of polarization resistance can be made.

11.2 Comparison of CORROSOMETER (ER) and CORRATER (LPR) Measurement Techniques

The CORROSOMETER (ER) and CORRATER (LPR) techniques are both extremely effective corrosion monitoring methods. The choice of which is better is dependent on the application. The principal attributes and limitations of each are briefly summarized below:

1. CORRATER systems require a sufficiently conductive liquid; CORROSOMETER systems are suitable for all environments.

2. CORROSOMETER systems measure metal loss directly, and their accuracy is comparable to the coupon method. CORRATER systems determine corrosion rate based on electrochemical theory and depend on average calibration constants to determine corrosion rate.

3. CORRATER systems measure corrosion rate virtually instantaneously. CORROSOMETER systems measure metal loss so that corrosion rate must be determined over a period of time, usually a few hours to a few days.

4. CORRATER systems supply pitting tendency information; CORROSOMETER systems do not at this time.

11.3 Applications Suited to CORRATER (LPR) and CORROSOMETER (ER) Techniques

Because of their intrinsic properties, CORRATER and CORROSOMETER systems are generally used in different applications. Occasionally, both are used to obtain complementary information; however, application requirements will usually dictate the choice of one or the other.

The following are the principal uses of and applications for CORRATER products:

1. Control of inhibitor addition in water systems
   a. For control of injection to minimize corrosion rises
   b. Evaluate different inhibitors and optimize their effectiveness over time
2. Evaluation and prescreening of inhibitors in the laboratory before field trials
3. Cooling towers—to optimize inhibition control, detect process leakage or oxygen activity and microbiologically-induced corrosion effects on the alloys of construction
4. Oil field waterfloods—to protect the injection equipment, particularly from oxygen ingress
5. Oil field drilling mud—to control inhibitor additions
6. Geothermal systems—to monitor the condition of produced brine and changes occurring before reinjection
7. Desalination systems—to monitor water quality
8. Process waters, scrubbers—to detect particularly corrosive contaminants
9. Evaluation of various alloys to solve material selection options, particularly in pilot or full-scale plant environments
10. Potable water systems—to test effectiveness of inhibitors and corrosiveness of biocides

The CORROSOMETER has much broader application and can be used in any medium, specifically

1. In oil and gas production
2. In gas sweetening, storage, and transportation systems
3. In refineries and petrochemical plants
4. In inhibitor evaluation/optimization programs
5. In power plants—for cooling water, feedwater, and scrubber systems. In bag houses and stacks for special function such as dew point alarming in flue gas systems
6. In chemical processes
7. In cathodically protected systems
8. In air-cleaning systems in control rooms
9. In paper mills or other plants with a volatile inherently corrosive process stream

Index